Development of Food Chemistry, Natural Products, and Nutrition Research

Development of Food Chemistry, Natural Products, and Nutrition Research

Editors

Antonello Santini
Nicola Cicero

MDPI • Basel • Beijing • Wuhan • Barcelona • Belgrade • Manchester • Tokyo • Cluj • Tianjin

Editors
Antonello Santini
University of Napoli Federico II
Italy

Nicola Cicero
University of Messina
Italy

Editorial Office
MDPI
St. Alban-Anlage 66
4052 Basel, Switzerland

This is a reprint of articles from the Special Issue published online in the open access journal *Foods* (ISSN 2304-8158) (available at: https://www.mdpi.com/journal/foods/special_issues/food_chemistry_natural_products_nutrition).

For citation purposes, cite each article independently as indicated on the article page online and as indicated below:

LastName, A.A.; LastName, B.B.; LastName, C.C. Article Title. *Journal Name* **Year**, *Article Number*, Page Range.

ISBN 978-3-03936-461-9 (Pbk)
ISBN 978-3-03936-462-6 (PDF)

© 2020 by the authors. Articles in this book are Open Access and distributed under the Creative Commons Attribution (CC BY) license, which allows users to download, copy and build upon published articles, as long as the author and publisher are properly credited, which ensures maximum dissemination and a wider impact of our publications.

The book as a whole is distributed by MDPI under the terms and conditions of the Creative Commons license CC BY-NC-ND.

Contents

About the Editors vii

Antonello Santini and Nicola Cicero
Development of Food Chemistry, Natural Products, and Nutrition Research: Targeting New Frontiers
Reprinted from: Foods 2020, 9, 482, doi:10.3390/foods9040482 1

Francesco Giuseppe Galluzzo, Gaetano Cammilleri, Alessandro Ulrici, Rosalba Calvini, Andrea Pulvirenti, Giovanni Lo Cascio, Andrea Macaluso, Antonio Vella, Nicola Cicero, Antonella Amato and Vincenzo Ferrantelli
Land Snails as a Valuable Source of Fatty Acids: A Multivariate Statistical Approach
Reprinted from: Foods 2019, 8, 676, doi:10.3390/foods8120676 7

Helena Maria Pinheiro-Sant'Ana, Pamella Cristine Anunciação, Clarice Silva e Souza, Galdino Xavier de Paula Filho, Andrea Salvo, Giacomo Dugo and Daniele Giuffrida
Quali-Quantitative Profile of Native Carotenoids in Kumquat from Brazil by HPLC-DAD-APCI/MS
Reprinted from: Foods 2019, 8, 166, doi:10.3390/foods8050166 17

Gyung-Rim Yong, Yoseph Asmelash Gebru, Dae-Woon Kim, Da-Ham Kim, Hyun-Ah Han, Young-Hoi Kim and Myung-Kon Kim
Chemical Composition and Antioxidant Activity of Steam-Distilled Essential Oil and Glycosidically Bound Volatiles from *Maclura Tricuspidata* Fruit
Reprinted from: Foods 2019, 8, 659, doi:10.3390/foods8120659 27

Archimede Rotondo, Giovanna Loredana La Torre, Giacomo Dugo, Nicola Cicero, Antonello Santini and Andrea Salvo
Oleic Acid Is not the Only Relevant Mono-Unsaturated Fatty Ester in Olive Oil
Reprinted from: Foods 2020, 9, 384, doi:10.3390/foods9040384 45

Ping-Chen Tu, Chih-Ju Chan, Yi-Chen Liu, Yueh-Hsiung Kuo, Ming-Kuem Lin and Meng-Shiou Lee
Bioactivity-Guided Fractionation and NMR-Based Identification of the Immunomodulatory Isoflavone from the Roots of *Uraria crinita* (L.) Desv. ex DC
Reprinted from: Foods 2019, 8, 543, doi:10.3390/foods8110543 57

Inga Matulyte, Aiste Jekabsone, Lina Jankauskaite, Paulina Zavistanaviciute, Vytaute Sakiene, Elena Bartkiene, Modestas Ruzauskas, Dalia M. Kopustinskiene, Antonello Santini and Jurga Bernatoniene
The Essential Oil and Hydrolats from *Myristica fragrans* Seeds with Magnesium Aluminometasilicate as Excipient: Antioxidant, Antibacterial, and Anti-inflammatory Activity
Reprinted from: Foods 2020, 9, 37, doi:10.3390/foods9010037 71

Joanna Zielińska-Wasielica, Anna Olejnik, Katarzyna Kowalska, Mariola Olkowicz and Radosław Dembczyński
Elderberry (*Sambucus nigra* L.) Fruit Extract Alleviates Oxidative Stress, Insulin Resistance, and Inflammation in Hypertrophied 3T3-L1 Adipocytes and Activated RAW 264.7 Macrophages
Reprinted from: Foods 2019, 8, 326, doi:10.3390/foods8080326 83

Jin-Woo Jeong, Seon Yeong Ji, Hyesook Lee, Su Hyun Hong, Gi-Young Kim, Cheol Park, Bae-Jin Lee, Eui Kyun Park, Jin Won Hyun, You-Jin Jeon and Yung Hyun Choi
Fermented Sea Tangle (*Laminaria japonica* Aresch) Suppresses RANKL-Induced Osteoclastogenesis by Scavenging ROS in RAW 264.7 Cells
Reprinted from: *Foods* 2019, *8*, 290, doi:10.3390/foods8080290 103

Agata Campisi, Rosaria Acquaviva, Giuseppina Raciti, Anna Duro, Milena Rizzo and Natale Alfredo Santagati
Antioxidant Activities of *Solanum nigrum* L. Leaf Extracts Determined in In Vitro Cellular Models
Reprinted from: *Foods* 2019, *8*, 63, doi:10.3390/foods8020063 117

Gabriel López-García, Antonio Cilla, Reyes Barberá, Amparo Alegría and María C. Recio
Effect of a Milk-Based Fruit Beverage Enriched with Plant Sterols and/or Galactooligosaccharides in a Murine Chronic Colitis Model
Reprinted from: *Foods* 2019, *8*, 114, doi:10.3390/foods8040114 129

Wan-Sup Sim, Sun-Il Choi, Bong-Yeon Cho, Seung-Hyun Choi, Xionggao Han, Hyun-Duk Cho, Seung-Hyung Kim, Boo-Yong Lee, Il-Jun Kang, Ju-Hyun Cho and Ok-Hwan Lee
Anti-Obesity Effect of Extract from *Nelumbo Nucifera* L., *Morus Alba* L., and *Raphanus Sativus* Mixture in 3T3-L1 Adipocytes and C57BL/6J Obese Mice
Reprinted from: *Foods* 2019, *8*, 170, doi:10.3390/foods8050170 145

Alexandra Galetović, Francisca Seura, Valeska Gallardo, Rocío Graves, Juan Cortés, Carolina Valdivia, Javier Núñez, Claudia Tapia, Iván Neira, Sigrid Sanzana and Benito Gómez-Silva
Use of Phycobiliproteins from Atacama Cyanobacteria as Food Colorants in a Dairy Beverage Prototype
Reprinted from: *Foods* 2020, *9*, 244, doi:10.3390/foods9020244 163

Mailing Rivera, Alexandra Galetović, Romina Licuime and Benito Gómez-Silva
A Microethnographic and Ethnobotanical Approach to Llayta Consumption among Andes Feeding Practices
Reprinted from: *Foods* 2018, *7*, 202, doi:10.3390/foods7120202 177

About the Editors

Antonello Santini, Ph.D., is a Professor of Food Chemistry and Food Chemistry and Analysis of Food and Nutraceuticals with the Department of Pharmacy and the Department of Agriculture of the University of Napoli Federico II, Napoli, Italy. He is also a Visiting Professor at the Albanian University of Tirana, Albania. He holds a Ph.D. in Chemical Sciences. His research areas of interest are supported by many international collaborations, mainly in the fields of food; food chemistry, nutraceuticals, and functional food; supplements; recovery of natural compounds bioactive using eco-sustainable and environmentally friendly techniques from agro-food by products; nanocompounds; nanonutraceuticals; food risk assessment, safety, and contaminants; mycotoxins and secondary metabolites; food analysis; and chemistry and food education. He is responsible for funded research projects and for general cultural agreements established between the University of Napoli Federico II and many universities worldwide. Through his research activities, he has published more than 200 papers in reputed and peer-reviewed international journals. He is a member of the European Food Safety Authority EFSA, ERWG, Parma, Italy; the Italian Authority for Food Safety (CNSA), Italian Ministry of Health, Rome Italy; the Managing Board, Italian Chemistry Society (SCI) Division of Teaching (DD-SCI), Rome, Italy; and expert member for Chemistry, EurSchool, European Commission, Bruxelles, Belgium.

Nicola Cicero is Senior Researcher in Food Chemistry with the Department of Biomedical Sciences, Dental and Morphological and Functional images, section S.A.S.T.A.S., at the University of Messina, Italy. He holds two Ph.D. degrees: Enogastronomical Sciences and Tourism, Territory, and Environment. He is teaching Oil and Wine in Mediterranean Food Habits and Fermentation Biotechnology programs at the University of Messina, Italy. He is an expert in the quality and safety of agro-food products. He has published about 100 publications, all in relevant peer-reviewed internationally reputed journals. His research interests mainly focus on food, their quality assessment and safety including the food and agro-food chain products, contaminants of organic and inorganic origin, and analysis of food matrices. He has a relevant background of participation in international congresses in the food area as chair or organizing and scientific committee member. He is an external evaluator for the assessment of the scientific research activity of the Italian research system and has a wide structured network of international collaboration in the food and environment area of interest. He is a funder member of the scientific and technical committee of the spin-off company named Science4Life srl and is an active consultant in the food industry.

Editorial

Development of Food Chemistry, Natural Products, and Nutrition Research: Targeting New Frontiers

Antonello Santini [1],* and Nicola Cicero [2]

1. Department of Pharmacy, University of Napoli Federico II, Via D. Montesano 49, 80131 Napoli, Italy
2. Department of Biomedical and Dental Sciences and Morphofunctional Imaging, University of Messina, Polo Universitario Annunziata, 98125 Messina, Italy; ncicero@unime.it
* Correspondence: asantini@unina.it; Tel.: +39-81-253-9317

Received: 3 April 2020; Accepted: 4 April 2020; Published: 12 April 2020

Abstract: The Special Issue entitled: "Development of Food Chemistry, Natural Products, and Nutrition Research" is focused on the recent development of food chemistry research, including natural products' sources and nutrition research, with the objectives of triggering interest towards new perspectives related to foods and opening a novel horizon for research in the food area. The published papers collected in this Special Issue are studies that refer to different aspects of food, ranging from food chemistry and analytical aspects, to composition, natural products, and nutrition, all examined from different perspectives and points of view. Overall, this Special Issue gives a current picture of the main topics of interest in the research and proposes studies and analyses that may prompt and address the efforts of research in the food area to find novel foods and novel applications and stimulate an environmentally-friendly approach for the re-use of the by-products of the agro-food area. This notwithstanding, the main challenge is currently addressed to achieve a full comprehension of the mechanisms of action of food components, the nutrients, outlining their high potential impact as preventive and/or therapeutic tools, not only as a source of macro- and/or micro-nutrients, which are necessary for all the metabolic and body functions.

Keywords: foods; food analysis; nutrition; natural products; food supplements; nanocompounds; nutraceuticals

---

The complete understanding of food matrices encompasses the analytical aspects and the composition analysis, both of which are of paramount importance considering that emerging new technologies and techniques in food analysis, chemometric techniques, and methods for food authentication can allow obtaining a great amount of accurate and precise data. This potentially affects the changes in consumer preferences and expectations, as well as the analysis of food innovations and their impact on the global market [1,2].

Nonetheless, the frontier of the food chemistry has impacts with new challenges, which range from: (i) novel foods; (ii) how adequate food safety may be determined; (iii) how nutritional intakes evolve over time and are influenced by global dynamics; (iv) the novel delivery systems of the food containing health beneficial compounds, which can have a great impact on health conditions, besides their nutritional value and importance; (iv) natural sources and their waste or by-products' recovery and re-use [3,4].

The challenge to understand the mechanism of action of food micronutrients and of the secondary metabolites involved in the chemistry of the food, especially when it is ingested, is currently triggering the interest of researchers worldwide. The complete understanding of the metabolic pathway of foodstuffs, which are complex matrices formed by many different substances, as well as the complete comprehension and assessment of the effects that food has on the body's metabolism are still open challenges.

The analytical details and knowledge of all the minor food components and/or contaminants of different origins at a very high resolution give important information, especially on food safety and quality parameters. Nonetheless, the actual perspectives of the research in food chemistry reveal an emerging interest in a new interdisciplinary approach that involves the contribution from different disciplines both in the food and natural products areas. Natural products are and have been a primary source in many cases, not only of nutrients, but also of remedies for millennia. An example is the growing interest towards the re-use of by-products from industrial processing of food and foodstuff to recover biologically-active substances to obtain derived products originating from food matrices and that may be useful to support/supplement the diet. Nutraceuticals are an outstanding example of this emerging trend in the food chemistry area. The use or re-use of food industry by-products, as well as the recovery of biologically-active compounds are receiving growing attention in view of the great interest towards the green economy and the optimization of the available resources. In this perspective, foodstuff and agro-food industry by-products' re-use can play a major role. A new challenging opportunity is to explore and substantiate with detailed chemical composition data and clinical data the mechanisms and modes of action of the active substances contained in food.

These aspects are relevant for maintaining well-being and preventing, by their use, the onset of diseases due to poor diet/food habits [5–16]. Safety is also a major challenge, as well as obtaining the complete or increased bioavailability of substances derived from food. From this perspective, the interest towards nanomaterials is emerging. Due to their remarkable properties, these are currently considered novel emerging tools to be used in the food area [17]. An interesting work has been reported recently regarding nanomaterials' application to foodstuff and outlining possible beneficial effects on health [18].

Nanopharmaceuticals can be considered as an illuminating example, which have led to a great change in the pharmaceutical industry and have had a great impact also on nutraceuticals. There is increasing growth in the study of nanocompounds including nutraceuticals derived from food matrices as phytocomplexes to obtain improved delivery, bioavailability, and effects. As a consequence, many recent research works are addressed towards the use of nanotechnologies applied to food-derived nutraceuticals, building up the innovative area of new emerging products: nanonutraceuticals [19,20].

Nanotechnology could be used for the proficient delivery of bioactive substances contained in food with the aim to improve their bioavailability, thereby increasing the possible health benefits. The advantages of nanotechnology applied to nutraceuticals are efficient encapsulation, smart delivery to the target, and release from a nanoformulation. For instance, research on the encapsulation of nutraceuticals into biodegradable, environmentally-friendly nanocarriers is ongoing to increase their absorption and therapeutic potential [21].

Nanonutraceuticals are a promising tool and a new frontier for the future research in the food area, widening the horizon of foods to a new perspective, focusing on the active substances contained in food matrices and on complete understanding of their mechanisms of action in the body. These food-derived novel compounds should be assessed in order to maintain their properties at the nano level to target better bioavailability and efficacy, naturally, notwithstanding the due attention to guarantee both safety and efficacy. Follow-up studies, as well as clinical and nutritional studies to evaluate possible unwanted effects would be necessary, and there is a long way to go in targeting the above-mentioned points [22–27].

The papers that make up the Special Issue cover a wide range of topics. The application of an innovative analytical technique based on Nuclear Magnetic Resonance (NMR) experiments called Multi-Assignment Recovered Analysis (MARA)-NMR to extra-virgin olive oil allowed the quantitative assessment of the oil's chemical composition, opening a wide range of applications [28].

The study of the Fatty Acid (FA) profile of wild *Theba pisana*, *Cornu aspersum*, and *Eobania vermiculata* land snail samples, examined by Gas Chromatography with a Flame Ionization Detector (GC-FID), put into evidence a high content of Polyunsaturated Fatty Acids (PUFAs), indicating their potential as functional food constituents [29].

The study of the native carotenoid composition in kumquat (*Fortunella margarita*) from Brazil determined for the first time by a HPLC-DAD-APCI/MS (High Performance Liquid Chromatography-Diode Array Detector-Atmospheric Pressure Chemical Ionization/Mass Spectrometry) allowed identifying and quantifying eleven carotenoids, some present in the free form and some in their esterified form [30].

The Special Issue includes studies addressing natural compounds and essential oils. In particular, nutmeg (*Myristica fragrans*) has been studied with the aim of comparing the antioxidant, antimicrobial, and anti-inflammatory activity of the hydrolats and essential oil obtained by hydrodistillation in the presence and absence of magnesium aluminometasilicate as an excipient [31].

The essential oil obtained from *Maclura tricuspidata* fruit revealed the relevant antioxidant activities of the steam-distilled essential oil and the glycosidically-bound aglycone fraction when studied with the Gas Chromatography–Mass Spectrometry (GC–MS) technique [32].

Functional food ingredients were exploited in the study on *Uraria crinita* by screening its metabolites using immunomodulatory fractions from the root methanolic extract in combination with bioactivity-guided fractionation and NMR-based identification [33].

Other manuscripts published in the present Special Issue evaluated the capacity of Elderberry fruit (EDB) extract to decrease the elevated production of reactive oxygen species in hypertrophied 3T3-L1 adipocytes, evidencing a crucial role in the development of obesity and accompanying metabolic dysfunctions [34].

A study on the sea tangle (*Laminaria japonica* Aresch), a brown alga, used as a functional food ingredient in the Asia-Pacific region, allowed assessing how fermented sea tangle extract was effective on the receptor activator of the nuclear factor-κB (NF-κB) ligand using RAW 264.7 mouse macrophage cells [35].

In addition, another interesting study contained in the Special Issue evaluated the antioxidant and anti-adipogenic activities of another vegetal matrix, namely a mixture of *Nelumbo nucifera* L., *Morus alba* L., and *Raphanus sativus*, with a complete updated in vitro and in vivo study [36].

The anti-inflammatory potential effect of plant sterols from enriched milk-based fruit beverages (with or without galactooligosaccharides in an experimental mouse model of chronic ulcerative colitis) was proposed, evidencing a great beneficial effect in mice against colitis [37].

Along the same lines, another interesting paper addressed foods in traditional medicine with antioxidant potential, in particular the assessment of the antioxidant effect of leaf extracts of *Solanum nigrum* L. [38].

The topics of the Special Issue expand the horizon of food research, also examining other applications of vegetal matrices. An example is the paper dedicated to the study of Llayta, a biomass of the colonies of *Nostoc cyanobacterium* grown in the wetlands of the Andean highlands, harvested, sun-dried, and used as an ingredient for human consumption, which revealed great potential as a functional food ingredient due to its relevant content of essential amino acids and polyunsaturated fatty acids [39].

The collection of papers is completed with one study addressing also industrial food applications, especially for industries interested in replacing artificial dyes with natural pigments; cyanobacterial phycobiliproteins as water-soluble colored proteins to be used as natural eco-sustainable pigments were shown to have great potential in this area of interest [40].

Author Contributions: All the authors contributed equally in the conceptualization, assessment, visualization, and writing of the text.

Conflicts of Interest: No conflicts of interest, financial or otherwise, are declared by the authors.

References

1. Ferreira, S.L.C.; Silva, M.M.; Felix, C.S.A.; da Silva, D.L.F.; Santos, A.S.; Neto, J.A.; de Souza, C.T.; Cruz, R.A., Jr.; Souza, A.S. Multivariate optimization techniques in food analysis—A review. *Food Chem.* **2019**, *273*, 3–8. [CrossRef] [PubMed]

2. Zederkop Ballin, N.; Laursen, K.H. To target or not to target? Definitions and nomenclature for targeted versus non-targeted analytical food authentication. *Trends Food Sci. Technol.* **2019**, *86*, 537–543. [CrossRef]
3. Santeramo, F.G.; Carlucci, D.; De Devitiis, B.; Seccia, A.; Stasi, A.; Viscecchia, R.; Nardone, G. Emerging trends in European food, diets and food industry. *Food Res. Int.* **2018**, *104*, 39–47. [CrossRef] [PubMed]
4. Santini, A.; Novellino, E. Nutraceuticals: Beyond the diet before the drugs. *Curr. Bioact. Compd.* **2014**, *10*, 1–12. [CrossRef]
5. Santini, A.; Novellino, E. To Nutraceuticals and back: Rethinking a concept. *Foods* **2017**, *6*, 74. [CrossRef]
6. Santini, A.; Tenore, G.C.; Novellino, E. Nutraceuticals: A paradigm of proactive medicine. *Eur. J. Pharm. Sci.* **2017**, *96*, 53–61. [CrossRef]
7. Abenavoli, L.; Izzo, A.A.; Milić, N.; Cicala, C.; Santini, A.; Capasso, R. Milk thistle (*Silybum marianum*): A concise overview on its chemistry, pharmacological, and nutraceutical uses in liver diseases. *Phytother. Res.* **2018**, *32*, 2202–2213. [CrossRef]
8. Daliu, P.; Santini, A.; Novellino, E. A decade of nutraceutical patents: Where are we now in 2018? *Expert Opin. Ther. Pat.* **2018**, *28*, 875–882. [CrossRef]
9. Durazzo, A.; D'Addezio, L.; Camilli, E.; Piccinelli, R.; Turrini, A.; Marletta, L.; Marconi, S.; Lucarini, M.; Lisciani, S.; Gabrielli, P.; et al. From plant compounds to botanicals and back: A current snapshot. *Molecules* **2018**, *23*, 1844. [CrossRef]
10. Durazzo, A. Extractable and Non-extractable polyphenols: An overview. In *Non-Extractable Polyphenols and Carotenoids: Importance in Human Nutrition and Health*; Saura-Calixto, F., Pérez-Jiménez, J., Eds.; Royal Society of Chemistry: London, UK, 2018; pp. 1–37.
11. Durazzo, A.; Lucarini, M. A current shot and re-thinking of antioxidant research strategy. *Braz. J. Anal. Chem.* **2018**, *5*, 9–11. [CrossRef]
12. Santini, A.; Novellino, E. Nutraceuticals-shedding light on the grey area between pharmaceuticals and food. *Expert Rev. Clin. Pharmacol.* **2018**, *11*, 545–547. [CrossRef]
13. Santini, A.; Cammarata, S.M.; Capone, G.; Ianaro, A.; Tenore, G.C.; Pani, L.; Novellino, E. Nutraceuticals: Opening the debate for a regulatory framework. *Br. J. Clin. Pharmacol.* **2018**, *84*, 659–672. [CrossRef] [PubMed]
14. Daliu, P.; Santini, A.; Novellino, E. From pharmaceuticals to nutraceuticals: Bridging disease prevention and management. *Expert Rev. Clin. Pharmacol.* **2019**, *12*, 1–7. [CrossRef] [PubMed]
15. Durazzo, A.; Lucarini, M. Extractable and Non-extractable antioxidants. *Molecules* **2019**, *24*, 1933. [CrossRef]
16. Durazzo, A.; Lucarini, M.; Souto, E.B.; Cicala, C.; Caiazzo, E.; Izzo, A.A.; Novellino, E.; Santini, A. Polyphenols: A concise overview on the chemistry, occurrence and human health. *Phytother. Res.* **2019**, *33*, 2221–2243. [CrossRef] [PubMed]
17. Jeevanandam, J.; Barhoum, A.; Chan, Y.S.; Dufresne, A.; Danquah, M.K. Review on nanoparticles and nanostructured materials: History, sources, toxicity and regulations. *Beilstein J. Nanotechnol.* **2018**, *9*, 1050–1074. [CrossRef] [PubMed]
18. Farokhzad, O.C.; Langer, R. Nanomedicine: Developing smarter therapeutic and diagnostic modalities. *Adv. Drug Deliv. Rev.* **2006**, *58*, 1456–1459. [CrossRef] [PubMed]
19. Watkins, R.; Wu, L.; Zhang, C.; Davis, R.M.; Xu, B. Natural product-based nanomedicine: Recent advances and issues. *Int. J. Nanomed.* **2015**, *10*, 6055–6074.
20. Pimentel-Moral, S.; Teixeira, M.C.; Fernandes, A.R.; Arráez-Román, D.; Martínez-Férez, A.; Segura-Carretero, A.; Souto, E.B. Lipid nanocarriers for the loading of polyphenols—A comprehensive review. *Adv. Colloid Interf. Sci.* **2018**, *260*, 85–94. [CrossRef]
21. Assadpour, E.; Jafari, S.M. A systematic review on nanoencapsulation of food bioactive ingredients and nutraceuticals by various nanocarrier. *Crit. Rev. Food Sci. Nutr.* **2019**, *59*, 3129–3151. [CrossRef]
22. Helal, N.A.; Eassa, H.A.; Amer, A.M.; Eltokhy, M.A.; Edafiogho, I.; Nounou, M.I. Nutraceuticals' Novel Formulations: The Good, the Bad, the Unknown and Patents Involved. *Recent Pat. Drug Deliv. Formul.* **2019**, *13*, 105–156. [CrossRef] [PubMed]
23. Jones, D.; Caballero, S.; Davidov-Pardo, G. Bioavailability of nanotechnology-based bioactives and nutraceuticals. *Adv. Food Nutr. Res.* **2019**, *88*, 235–273. [PubMed]
24. He, X.; Deng, H.; Hwang, H.M. The current application of nanotechnology in food and agriculture. *J. Food Drug Anal.* **2019**, *27*, 1–21. [CrossRef] [PubMed]

25. Das, G.; Patra, J.K.; Paramithiotis, S.; Shin, H.S. The sustainability challenge of food and environmental nanotechnology: Current status and imminent perceptions. *Int. J. Environ. Res. Public Health* **2019**, *16*, 4848. [CrossRef]
26. Peters, R.J.B.; Bouwmeester, H.; Gottardo, S.; Amenta, V.; Arena, M.; Brandho, P.; Marvin, H.J.P.; Mech, A.; Moniz, F.B.; Pesudo, L.Q.; et al. Nanomaterials for products and application in agriculture, feed and food. *Trends Food Sci. Technol.* **2016**, *54*, 155–164. [CrossRef]
27. McClements, D.J.; Xiao, H. Is nano safe in foods? Establishing the factors impacting the gastrointestinal fate and toxicity of organic and inorganic food-grade nanoparticles. *NPJ Sci. Food* **2017**, *1*, 6. [CrossRef]
28. Rotondo, A.; La Torre, G.; Dugo, G.; Cicero, N.; Santini, A.; Salvo, A. Oleic Acid is Not the Only Relevant Mono-Unsaturated Fatty Ester in Olive Oil. *Foods* **2020**, *9*, 384. [CrossRef]
29. Galluzzo, F.; Cammilleri, G.; Ulrici, A.; Calvini, R.; Pulvirenti, A.; Lo Cascio, G.; Macaluso, A.; Vella, A.; Cicero, N.; Amato, A.; et al. Land Snails as a Valuable Source of Fatty Acids: A Multivariate Statistical Approach. *Foods* **2019**, *8*, 676. [CrossRef]
30. Pinheiro-Sant'Ana, H.; Anunciação, P.; Souza, C.; de Paula Filho, G.; Salvo, A.; Dugo, G.; Giuffrida, D. Quali-Quantitative Profile of Native Carotenoids in Kumquat from Brazil by HPLC-DAD-APCI/MS. *Foods* **2019**, *8*, 166. [CrossRef]
31. Matulyte, I.; Jekabsone, A.; Jankauskaite, L.; Zavistanaviciute, P.; Sakiene, V.; Bartkiene, E.; Ruzauskas, M.; Kopustinskiene, D.; Santini, A.; Bernatoniene, J. The Essential Oil and Hydrolats from Myristica fragrans Seeds with Magnesium Aluminometasilicate as Excipient: Antioxidant, Antibacterial, and Anti-inflammatory Activity. *Foods* **2020**, *9*, 37. [CrossRef]
32. Yong, G.R.; Gebru, Y.A.; Kim, D.W.; Kim, D.-H.; Han, H.A.; Kim, Y.H.; Kim, M.K. Chemical Composition and Antioxidant Activity of Steam-Distilled Essential Oil and Glycosidically Bound Volatiles from Maclura Tricuspidata Fruit. *Foods* **2019**, *8*, 659. [CrossRef] [PubMed]
33. Tu, P.; Chan, C.; Liu, Y.; Kuo, Y.; Lin, M.; Lee, M. Bioactivity-Guided Fractionation and NMR-Based Identification of the Immunomodulatory Isoflavone from the Roots of *Uraria crinita* (L.) Desv. ex DC. *Foods* **2019**, *8*, 543. [CrossRef] [PubMed]
34. Zielińska-Wasielica, J.; Olejnik, A.; Kowalska, K.; Olkowicz, M.; Dembczyński, R. Elderberry (*Sambucus nigra* L.) Fruit Extract Alleviates Oxidative Stress, Insulin Resistance, and Inflammation in Hypertrophied 3T3-L1 Adipocytes and Activated RAW 264.7 Macrophages. *Foods* **2019**, *8*, 326. [CrossRef]
35. Jeong, J.; Ji, S.; Lee, H.; Hong, S.; Kim, G.; Park, C.; Lee, B.; Park, E.; Hyun, J.; Jeon, Y.; et al. Fermented Sea Tangle (*Laminaria japonica* Aresch) Suppresses RANKL-Induced Osteoclastogenesis by Scavenging ROS in RAW 264.7 Cells. *Foods* **2019**, *8*, 290. [CrossRef] [PubMed]
36. Sim, W.; Choi, S.; Cho, B.; Choi, S.; Han, X.; Cho, H.; Kim, S.; Lee, B.; Kang, I.; Cho, J.; et al. Anti-Obesity Effect of Extract from Nelumbo Nucifera L., Morus Alba L., and Raphanus Sativus Mixture in 3T3-L1 Adipocytes and C57BL/6J Obese Mice. *Foods* **2019**, *8*, 170. [CrossRef]
37. López-García, G.; Cilla, A.; Barberá, R.; Alegría, A.; Recio, M. Effect of a Milk-Based Fruit Beverage Enriched with Plant Sterols and/or Galactooligosaccharides in a Murine Chronic Colitis Model. *Foods* **2019**, *8*, 114. [CrossRef]
38. Campisi, A.; Acquaviva, R.; Raciti, G.; Duro, A.; Rizzo, M.; Santagati, N. Antioxidant Activities of Solanum nigrum L. Leaf Extracts Determined in In Vitro Cellular Models. *Foods* **2019**, *8*, 63. [CrossRef]
39. Rivera, M.; Galetović, A.; Licuime, R.; Gómez-Silva, B. A Microethnographic and Ethnobotanical Approach to Llayta Consumption among Andes Feeding Practices. *Foods* **2018**, *7*, 202. [CrossRef]
40. Galetović, A.; Seura, F.; Gallardo, V.; Graves, R.; Cortés, J.; Valdivia, C.; Núñez, J.; Tapia, C.; Neira, I.; Sanzana, S.; et al. Use of Phycobiliproteins from Atacama Cyanobacteria as Food Colorants in a Dairy Beverage Prototype. *Foods* **2020**, *9*, 244. [CrossRef]

© 2020 by the authors. Licensee MDPI, Basel, Switzerland. This article is an open access article distributed under the terms and conditions of the Creative Commons Attribution (CC BY) license (http://creativecommons.org/licenses/by/4.0/).

Article

Land Snails as a Valuable Source of Fatty Acids: A Multivariate Statistical Approach

Francesco Giuseppe Galluzzo [1], Gaetano Cammilleri [1,2,*], Alessandro Ulrici [2], Rosalba Calvini [2], Andrea Pulvirenti [2], Giovanni Lo Cascio [1], Andrea Macaluso [1,2], Antonio Vella [1], Nicola Cicero [3], Antonella Amato [4] and Vincenzo Ferrantelli [1,2]

1. Istituto Zooprofilattico Sperimentale della Sicilia, via Gino Marinuzzi 3, 90129 Palermo, Italy; francescogiuseppe92@gmail.com (F.G.G.); giovanni.locascio71@gmail.com (G.L.C.); andrea.macaluso@izssicilia.it (A.M.); laboratorio.residui@gmail.com (A.V.); vincenzo.ferrantelli@izssicilia.it (V.F.)
2. Dipartimento di Scienze della Vita, Università degli studi di Modena e Reggio Emilia, Via Università 4, 41121 Modena, Italy; alessandro.ulrici@unimore.it (A.U.); rosalba.calvini@unimore.it (R.C.); andrea.pulvirenti@unimore.it (A.P.)
3. Dipartimento SASTAS, Università degli studi di Messina, Polo Universitario dell'Annunziata, 98168 Messina, Italy; ncicero@unime.it
4. Dipartimento di Scienze e Tecnologie Biologiche Chimiche e Farmaceutiche, Università degli Studi di Palermo, Viale delle Scienze, 90128 Palermo, Italy; antonella.amato@unipa.it
* Correspondence: gaetano.cammilleri86@gmail.com; Tel.: +39-328-8048262

Received: 4 November 2019; Accepted: 5 December 2019; Published: 12 December 2019

Abstract: The fatty acid (FA) profile of wild *Theba pisana*, *Cornu aspersum*, and *Eobania vermiculata* land snail samples, collected in Sicily (Southern Italy), before and after heat treatment at +100 °C were examined by gas chromatography with a flame ionization detector (GC-FID). The results show a higher content of polyunsaturated fatty acids (PUFAs) in all of the examined raw snails samples, representing up to 48.10% of the total fatty acids contents, followed by monounsaturated fatty acids (MUFAs). The thermal processing of the snail samples examined determined an overall reduction of PUFA levels (8.13%, 7.75%, and 4.62% for *T. pisana*, *C. aspersum* and *E. vermiculata* samples, respectively) and a species-specific variation of saturated fatty acid (SFA) contents. Oleic acid remained the most abundant FA of all of the snails species examined, accounting for up to 29.95% of the total FA content. A relevant decrease of ω3/ω6 ratio was found only for *T. pisana* samples. The principal component analysis (PCA) showed a separation of the snail samples in terms of species and heat treatment. The results of this work suggest land snails as a valuable source of MUFA and PUFA contents and boiling as appropriate treatment, according to the maintenance of healthy properties.

Keywords: fatty acids; land snails; GC-FID; heat processing; principal component analysis

1. Introduction

Terrestrial gastropods, commonly named land snails, constitute a niche food product traditionally appreciated by many European countries, especially France and Italy. The use of land snails as food is still steadily growing, and 26,000 tons of snails were imported from Africa and countries in the Middle East [1]. *Cornu aspersum*, *Eobania vermiculata*, and *Theba pisana* are the land snail species most consumed in Italy [2]. Land snails are consumed in different ways all over the world, but the principal cooking procedures recognized are roasting and boiling, according to the traditions of the countries. According to Milinsk et al. [3], there is a correlation between land snails' diet and their nutritional values. Recently, increasing attention was paid to the fatty acid composition, due to nutritional and health-related aspects [4–8]. However, few studies are available about the fatty acid (FA) profile in land snails [3,9,10] and, as far as we know, no data have been reported regarding the fatty acid profile of *T. pisana*. Snails are commonly consumed in different ways after boiling due to the risk posed by the possible presence of potentially pathogenic microorganisms [2]. The cooking temperature can influence the nutritional aspect of mollusks [11]. At present, there are too few studies about the influence of heat processing (such as boiling) on the nutritional composition of land snails.

In this context, the present work aimed at evaluating the fatty acids content of wild *C. apsersum*, *T. pisana*, and *E. vermiculata* samples collected in Sicily (Southern Italy). Furthermore, the effect of boiling on the fatty acid composition was evaluated to have a comprehensive nutritional evaluation of this product after processing.

2. Materials and Methods

2.1. Reagents and Standards

All chemicals, solvents, and reagents employed were of analytical grade (≥99.9%). Acetone, hexane, diatomaceous earth, sodium sulfide nonahydrate, methanol, and hydrochloric acid were purchased from Sigma-Aldrich (Amsterdam, Holland). All of the gas used for gas chromatography (GC) analysis was pure (≥99.9995%). Water used for the separation of fatty acid methyl esters (FAMEs) phase was bidistilled in Milli-Q® Integral 5 (Merck KGaA, Darmstadt, Germany). FA standards were purchased from Sigma-Aldrich (Amsterdam, Holland). The 10,000 mg/L standards were prepared by diluting 100 mg of a pure standard solution with 10 ml of n-hexane. A mixture of FA standards was used for the identification of each peak.

2.2. Sample Collection and Preparation

A total of 128 samples of *C. aspersum*, 400 samples of *T. pisana*, and 162 samples of *E. vermiculata*, were collected from Palermo provinces (Sicily, Southern Italy) in 2018 during July for *C. aspersum*, August for *T. pisana*, and September for *E. vermiculata* to have the maximum assimilation efficiency according to the literature [12–14]. The shell of the snail samples was removed and only the meat was considered for the chemical analysis. The meat of the snail samples was grouped into three pools according to the species, then homogenized by a vertical mixer B-400 (Büchi, Flawil, Switzerland) and stored at −10 °C for 24 h to prevent a decrease in fatty acid content during the storage period [15]. The FA content of each sample pool was determined both raw and after cooking at 100 °C with boiled water for 30 s. The entire procedure of analysis is shown in Figure 1.

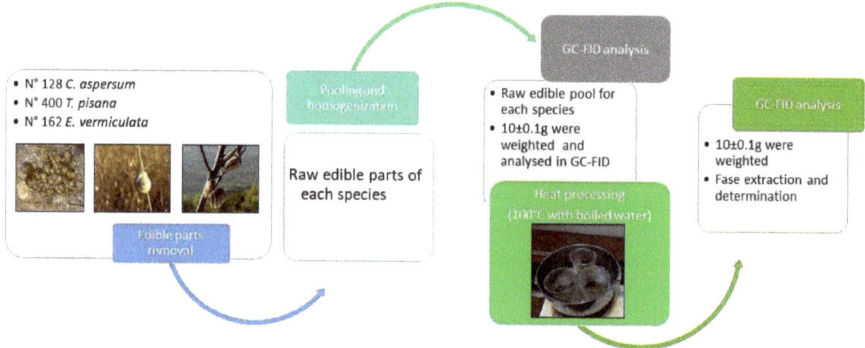

Figure 1. Scheme of the cooking process of the land snails samples collected.

2.3. Extraction of Fatty Acids and Gas Chromatography with a Flame Ionization Detector (GC-FID) Analysis

An amount of 10 ± 0.1 g of each pool of samples was placed in a glass of polypropylene and mixed with diatomaceous earth (Sigma-Aldrich, Amsterdam, Holland). The mixture was transferred in an accelerated solvent extraction (ASE) ASE 200 cell (Thermo Fisher, Waltham, Massachusetts, USA). The ASE operating conditions were set up as follows: 20 mL of hexane/acetone, 70:30; extraction temperature 120 °C for 6 min with a pressure of 120 pound per square (PSI).

The extract was filtered (size 240 nm) and dehydrated in rotavapor (Büchi, Flawil, Switzerland) at +40 °C. For the preparation of FAME, 100 mg of the oil extracted was trans-esterified in a pyrex tube by using 2 mL of HCl/MeOH (2:98 v/v) to obtain the fatty acid methyl esters (FAMEs). The solution was mixed in a vortex for 1 min and put in the oven at 120 °C for 1 h. After cooling, 2 mL of bidistilled water and 1 mL of hexane were added, and the mixture was centrifuged at 300 rpm for 1 min. Approximately 1 mL of the upper n-hexane phase was transferred in a vial and injected in gas chromatography (GC) with a flame ionization detector (FID).

Each pool of samples was examined in triplicate by GC-FID analysis. The analysis was carried out by a Trace GC/ULTRA HP 5890 GC + 7673 A/S (Thermo Fisher, Waltham, Massachusetts, USA); a Famexax column (30m × 0.25 mm i.d. × 0.25 µm df) was used for the separation. A flame ionization detector (FID) and ChromQuest 4.2.1 software (Thermo Fisher Scientific, Waltham, Massachusetts, USA) were used for the qualification and quantification of the analytes. The injector port and the detector temperatures were 220 °C and 230 °C, respectively. The split ratio was 1:20. The flow rates of compressed air and hydrogen were 350 mL min^{-1} and 35 mL min^{-1}, respectively. The carrier gas was helium (1.5 mL min^{-1}). The oven temperature was programmed at a rate of 6.0 °C min^{-1} from 130 to 225 °C, held for 15 s.

Individual FAME was identified by comparison with the chromatographic behavior of authentic standards by the formula:

$$TR = TR_{st} \pm 0.5 \quad (1)$$

where TR is the determined retention time (min), and TR_{st} is the retention time for each FA standard. The relative percentages of the fatty acids were also determined. Quantitation of individual FAs is thus based on the comparison of their peak areas (Ai), and the peak area of a suitable standard. The relative percentages of fatty acids (C) were determined by the formula:

$$C = \frac{A}{\Sigma A} 100 \quad (2)$$

2.4. Validation of the GC-FID Method

The repeatability mean and standard deviation of the analytical procedure were all calculated according to Taverniers et al. [16]. Separated FA standards were used to calculate the mean retention times (RTs) in the FID detector. The precision of the quantitative method was checked by the repeatability test, based on ten series of experiments [17]. The area of each peak was measured and corrected manually. The relative percentage of the fatty acids was also determined by comparing their peak areas.

2.5. Data Collection and Statistical Analysis

The data were expressed as g/100g FA in fat extracted and grouped according to species and treatment (raw *T. pisana*, *C. aspersum*, and *E. vermiculata* and cooked *T. pisana*, *C. aspersum*, and *E. vermiculata*). The variation of fatty acids after heat treatment was calculated as follow:

$$Fa_\% = 100 - \left(\frac{Fa_b}{Fa_{raw}}\right) 100 \quad (3)$$

where $Fa_\%$ is the variation (expressed as a percentage), Fa_b and Fa_{raw} are the fatty acid content in boiled and raw samples, respectively (expressed as mg/100 g).

Before calculating the principal component analysis (PCA) model, the erucic acid variable was removed from statistical analysis because its presence was found only in raw *T. pisana* samples. All of the variables were pre-treated by Pareto scaling [18,19], in order to have a compromise between highlighting the contribution of the most abundant analytes and keeping at the same time the information brought by the less abundant ones. The PCA model was calculated using the software PLS-Toolbox ver. 8.6 (Eigenvector Research Inc., Wenatchee, WA, USA), running in the MATLAB environment (ver. 9.3, The Mathworks Inc., Natick, MA, USA).

3. Results

3.1. Fatty Acid Profiles

The fatty acid contents of the land snail samples examined are shown in Table 1. Seventeen FAs were found in all of the species examined: Six saturated fatty acids ($C_{14:0}$, $C_{16:0}$, $C_{17:0}$, $C_{18:0}$, $C_{20:0}$, $C_{22:0}$), six monounsaturated ($C_{14:1}$, $C_{16:1}$, $C_{17:1}$, $C_{20:1}$, $C_{18:1\omega:9}$), and six polyunsaturated ($C_{18:2\omega:6}$, $C_{18:3\omega:3}$, $C_{20:2}$, $C_{20:4}$, $C_{20:5}$, $C_{22:6}$). Erucic acid ($C_{22:1}$) was found only in raw *T. pisana* samples.

Table 1. Fatty acids contents (mean ± SD; g/100 g FA) in fat extracted from the land snails species examined (n = 3 replicates of each pool of samples). SFA = Saturated fatty acids, MUFA = monounsatured fatty acids, PUFA = polyunsaturated fatty acids.

Fatty Acid	*T. pisana* Raw	*T. pisana* Boiled	*C. aspersum* Raw	*C. aspersum* Boiled	*E. vermiculata* Raw	*E. vermiculata* Boiled
Myristic ($C_{14:0}$)	0.73 ± 0.01	1.06 ± 0.01	0.76 ± 0.08	0.59 ± 0.03	0.81 ± 0.01	0.63 ± 0.00
Palmitic ($C_{16:0}$)	12.63 ± 0.04	15.75 ± 0.16	16.02 ± 0.27	13.24 ± 0.36	14.63 ± 0.13	13.31 ± 0.00
Margaric ($C_{17:0}$)	1.02 ± 0.01	1.22 ± 0.07	1.13 ± 0.05	1.35 ± 0.07	1.36 ± 0.07	1.04 ± 0.03
Stearic ($C_{18:0}$)	5.41 ± 0.08	6.31 ± 0.96	7.72 ± 0.2	7.24 ± 0.14	7.66 ± 0.03	7.59 ± 0.01
Arachidic ($C_{20:0}$)	0.63 ± 0.00	0.69 ± 0.04	0.71 ± 0.03	0.39 ± 0.02	0.81 ± 0.02	0.37 ± 0.00
Behenic ($C_{22:0}$)	0.31 ± 0.00	0.29 ± 0.01	0.64 ± 0.04	0.19 ± 0.00	0.31 ± 0.02	0.39 ± 0.00
ΣSFA	20.72	25.32	26.97	23.01	25.58	23.35
Myristoleic ($C_{14:1}$)	0.53 ± 0.00	0.59 ± 0.02	0.52 ± 0.03	0.65 ± 0.03	0.23 ± 0.01	0.58 ± 0.02
Palmitoleic ($C_{16:1}$)	0.50 ± 0.02	0.27 ± 0.05	0.32 ± 0.08	1.35 ± 0.06	0.40 ± 0.04	0.37 ± 0.01
Eptadecenoic ($C_{17:1}$)	0.52 ± 0.01	0.54 ± 0.04	0.81 ± 0.04	1.08 ± 0.04	1.01 ± 0.03	1.10 ± 0.00
Eicosenoic ($C_{20:1}$)	0.37 ± 0.00	0.26 ± 0.03	0.37 ± 0.04	0.50 ± 0.02	0.17 ± 0.05	0.48 ± 0.00
Erucic ($C_{22:1}$)	0.52 ± 0.00	-	-	-	-	-
Oleic ($C_{18:1\omega:9}$)	28.83 ± 0.08	28.83 ± 0.46	23.79 ± 1.72	29.95 ± 0.46	26.03 ± 0.63	29.71 ± 0.07
ΣMUFA	31.27	30.49	25.81	33.53	27.86	32.24
Linoleic ($C_{18:2\,\omega6}$)	18.78 ± 0.02	21.35 ± 0.10	22.15 ± 0.13	19.07 ± 0.16	21.94 ± 0.13	19.20 ± 0.02
Linolenic ($C_{18:3\,\omega3}$)	7.64 ± 0.02	8.87 ± 0.16	15.78 ± 0.09	15.14 ± 0.20	12.40 ± 0.41	15.53 ± 0.00
Eicosadienoic ($C_{20:2}$)	5.29 ± 0.02	5.44 ± 0.04	3.98 ± 0.02	4.64 ± 0.02	4.21 ± 0.40	4.69 ± 0.00
Arachidonic ($C_{20:4}$)	6.28 ± 0.08	7.17 ± 0.19	4.26 ± 0.20	4.18 ± 0.01	6.40 ± 0.52	4.27 ± 0.03
Eicosapentaenoic ($C_{20:5}$)	9.85 ± 0.01	1.15 ± 0.51	0.62 ± 0.07	0.40 ± 0.05	1.30 ± 0.06	0.43 ± 0.01
Docosahexaenoic ($C_{22:6}$)	0.18 ± 0.01	0.21 ± 0.00	0.43 ± 0.01	0.13 ± 0.00	0.33 ± 0.03	0.29 ± 0.01
ΣPUFA	48.10	44.19	47.22	43.56	46.56	44.41
ω3/ω6	0.58	0.30	0.55	0.56	0.43	0.58
PUFA/SFA	2.32	1.75	1.75	1.89	1.82	1.90

3.2. Fatty Acids of Raw Samples

The polyunsaturated fatty acid (PUFA) content of raw *T. pisana*, *C. aspersum*, and *E. vermiculata* was 48.10 g/100 g, 47.22 g/100 g, and 46.56 g/100 g, respectively, representing the most abundant class of fatty acids, followed by monounsaturated fatty acid (MUFA) in *T. pisana* and *E. vermiculata* (31.27 g/100 g and 27.86 g/100 g, respectively) and saturated fatty acid (SFA) in *C. aspersum* (26.97 g/100 g).

The main PUFA components were C18:2ω6 (18.78–22.15 g/100 g), C18:3ω3 (7.64–15.78 g/100 g), and C20:2 (3.98–5.29 g/100 g). Linoleic acid (C18:2 ω6) represents the most abundant PUFA showing a range between 7.64 and 15.78 g/100 g. A high level of eicosapentaenoic acid (C20:5) was determined in *T. pisana* samples (9.85 g/100 g).

The MUFA profiles obtained for all the species examined consisted of C14:1 (0.23–0.53 g/100 g), C16:1 (0.32–0.5 g/100 g), C17:1 (0.52–1.01 g/100 g), C18:1ω:9 (23.79–28.83 g/100 g), and C20:1 (0.17–0.37 g/100 g). Oleic acid was the main component of all the samples examined, representing 25% of the total fatty acid content. Erucic acid was found only in raw *T. pisana* samples at low concentrations (0.52 g/100 g).

Among the SFA, palmitic acid (C16:0) was the most abundant in all of the samples examined (from 12.63 to 16.02 g/100 g), followed by stearic acid (5.41–7.66 g/100 g). The SFA profiles in all of the species consisted of C14:0 (ranging from 0.73 to 0.81 g/100 g), C16:0 (12.63–16.02 g/100 g), C17:0 (1.02–1.36 g/100 g), C18:0 (5.41–7.72 g/100 g), C20:0 (0.63–0.81 g/100 g), and C22:0 (0.31–0.64 g/100 g).

The raw *T. pisana* samples showed the highest ω3/ω6 ratio (0.58), followed by *E. vermiculata* (0.55) and *C. aspersum* (0.43).

3.3. Fatty Acids after Heat Treatment

After boiling at +100 °C, the FA profile of all of the species examined verified a decrease of PUFA content up to 8.2%. Differently from PUFA, a species-specific modification of MUFA and SFA contents was found.

In particular, *T. pisana* samples showed an increase of SFA content by 22.20%; only C22:0 verified a decrease after boiling (from 0.31 to 0.29 g/100 g).

The MUFA contents decreased by 2.49%. No erucic acid was found after boiling. The PUFA content decreased from 48.10 to 44.19 g/100 g (8.13%), with a significant reduction of C20:5 content (from 9.85 to 1.15 g/100 g).

Regarding *C. aspersum*, the SFA content decreased by 14.68%, showing a reduction of C20:0 from 0.71 to 0.39 g/100 g; only C17:0 showed an increase from 1.13 to 1.35 g/100 g.

The total MUFA content increased by 29%. Palmitoleic acid (C16:1) showed a significant increase from 0.32 to 1.35 g/100 g, followed by C14:1, C17:1, and C30:1.

The *E. vermiculata* samples verified a reduction of SFA content after boiling (8.72%), especially for arachidic acid (C20:0) (from 0.81 to 0.37 g/100 g), followed by C17:0 (1.36–1.04 g/100 g) and C14:0 (0.81–0.63 g/100 g). The behenic acid (C22:0) content increased by 25.81%. The MUFA content verified an increase from 27.86 to 32.24 g/100 g. Palmitoleic acid (C16:1) was the only MUFA that decreased, from 0.40 to 0.37 g/100 g. The heat treatment of the *E. vermiculata* samples determined a decrease of PUFA components of 4.61%. Eicosapentaenoic acid (C22:6) decreased significantly from 1.30 to 0.34 g/100 g. The MUFA fatty acids more compromised after boiling were palmitoleic acid (C16:1) for *E. vermiculata* and *T. pisana* samples and myristoleic acid (C14:1) for *C. aspersum*. Oleic acid (C18:1ω9) remained the main component of all the land snails species examined even after boiling. Moreover, no oleic acid content variation was found after boiling for *T. pisana* samples.

Among the PUFA group, eicosapentaenoic acid (C20:5) showed the highest decrease in *T. pisana* and *E. vermiculata* samples after the heat treatment, whereas docosahexaenoic acid (C22:6) decreased up to 70% in *C. aspersum*.

A decrease of the ω3/ω6 ratio was found only for the *T. pisana* samples, whereas the *E. vermiculata* and *C. aspersum* samples increased the ω3/ω6 ratio up to 35%. Finally, a reduction of the PUFA/SFA ratio was found only for *T. pisana* samples.

3.4. Multivariate Analysis

Given the high number of fatty acids examined as variables, principal component analysis (PCA) was used to explore the dataset structure and to obtain more information on the variables that mainly influence sample similarities and differences after heat treatment. The PCA model calculated after Pareto scaling showed that the data group variation is visible in the first two principal components, accounting for 97.82% of total data variance. Figure 2, which reports the PC1 vs. PC2 score plot (Figure 2a), together with the corresponding loading plot, highlights that PC1 alone explains about 94% of data variance. This extremely high value, together with the fact that all the variables have positive loading values along PC1 (Figure 2b), reflect the high positive correlation among the most significant part of the considered variables, so PC1 describes the prevailing trend of the analyzed FAs for the samples examined.

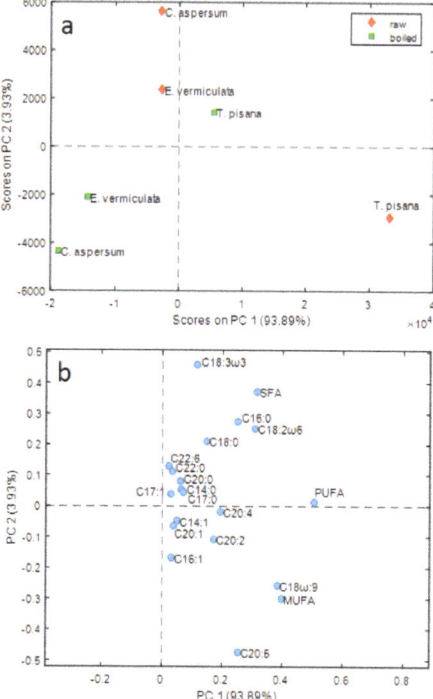

Figure 2. PC1 vs. PC2 score (**a**) and loading (**b**) plots of FA contents of the land snail samples examined, according to the species and treatment (raw vs. boiled).

The score plot shows differences related both to species and to heat treatment. PC1 describes a decrease in the number of fatty acids for each species after boiling. The variation is more marked for the *T. pisana* samples, which in general show the highest amount of unsaturated fatty acids (UFAs). In fact, both raw and boiled *T. pisana* samples lie at positive values of PC1, whereas all *E. vermiculata* and *C. aspersum* samples lie at negative values of PC1; this is due to the fact that, notwithstanding the significant decrease of FA content, boiled *T. pisana* still has a UFA content higher than raw *E. vermiculata* and raw *C. aspersum*. These last two samples have essentially the same overall amount of FAs and boiling leads to a more marked decrease for *C. aspersum* than for *E. vermiculata*.

T. pisana shows an opposite trend compared to the two other species considering the position along PC2 of the samples before and after boiling. Recalling that the percentage of variance explained by PC2 (3.93%) is much lower than the percentage of variance explained by PC1 (93.89%), PC2 accounts for the differences between the different species in the variations of the FA compositions after boiling, beyond the overall decrease, explained by PC1.

These variations of the FA profile can be examined more in-depth by means of the corresponding loading plot reported in Figure 2b. In this plot, the variables that are far from the origin contributed to the systematic variability explained by the PCA model, whereas variables close to the origin (like, e.g., C17:1, C14:0, C20:2, C20:1, C22:6, etc.) did not show a systematic trend.

The significant variations after boiling for all of the three species examined are related to PUFA, MUFA, oleic acid, SFA, and linoleic acid, which show higher PC1 values. PC2 showed a different variation of the fatty acid contents between the *T. pisana* samples and the other two species. A more considerable variation of eicosapentaenoic (C20:5) acid, MUFA, and oleic acid was found for *T. pisana* samples, whereas *E. vermiculata* and *C. aspersum* showed a more significant variation of SFA, linoleic acid, palmitic acid, and linolenic acid.

4. Discussion

Fatty acids are ubiquitous molecules in biological systems. They play several roles in metabolism, as structural components in membrane lipids, and as precursors of some molecules like prostaglandins and eicosanoids [20]. The dietary intake in favor of PUFA and MUFA instead of SFA is correlated to a significant minor risk of cardiovascular disease (CDV) and can lead to health benefits [21]. Fatty acids are a minor nutritional parameter of snail meat [4,9,10]. However, all of the raw snail samples examined in this work showed a low SFA content, in accordance to what was reported by Szkucik et al. [10] in farmed *C. aspersum* samples from Poland, but in contrast to what was found in wild *Helix pomatia* samples from Southern Turkey [9].

According to the literature, the factors of critical importance for the snail meat FA profile are the snail genus and its collection site. Interspecies differences in fatty acid composition were also confirmed in this work for the MUFA contents. In particular, the *T. pisana* raw meat samples showed C20:5 contents up to 16 times higher than *C. aspersum* and *E. vermiculata* samples; these differences could be due to the different ecological aspects of the species examined. It is well known that *T. pisana* is an agricultural pest in many parts of the world [21,22], feeding on a wide range of agricultural plants, including cereals with high UFA contents [14,23]. Differently from *T. pisana*, *C. aspersum* and *E. vermiculata* appear to be selective polyphagous organisms, preferring plants of the Poaceae family [12,24,25]. Other studies have confirmed how the feeding regimen would affect the fatty acid composition [4], verifying significant variations of MUFA contents related to the increase of soybean oil as feed supply in reared *C. aspersum* samples. Feeds that include corn, sunflower, or soybean rich in ω6 acids were shown to increase the content of these FAs in meat.

Nevertheless, the wild snail samples examined in this work showed high contents of UFA, constituting up to 79% of the total fatty acids. The level of PUFA in edible snails was found to be higher than SFA and MUFA, according to what was found in *Helix lucorum* and *Limax flavus* [26]. Our results are contrary to what was reported by Ekin et al. [27] in free-living *Melanopsis praemorsa* snails of Anatolia (Turkey) that showed a lower level of UFA and higher SFA contents. Essential fatty acids such as linoleic acid, α linolenic acid, docosahexaenoic acid (DHA), and eicosapentaenoic acid (EPA) were determined. These fatty acids show protection effects against cardiovascular disease [28,29]; however, the current intakes of EPA and DHA in European populations appear to be below the recommended daily allowance (RDA) [29].

The thermal processing of the snail samples analyzed determined an overall reduction of PUFA levels and a species-specific variation of MUFA and SFA contents, in contrast to what was found by Szkucik et al. [10] in farmed *C. aspersum* samples from Poland, verifying a significant increase of the SFA levels. The PUFA amounts of the samples analyzed decreased up to 7.98%, with a significant decrease of the C 20:5 contents in *T. pisana* samples. Nevertheless, PUFA remained the principal component, accounting for 44% of the total fatty acid contents in all of the species examined.

Among the MUFA, oleic acid (C18:1) remained the most abundant fatty acid of all of the snails species examined, even after heat processing.

Regarding the SFA contents, our results appear to comply with what was reported by Purwaningsih et al. [11] in mollusk muscles, showing that the SFA composition depends primarily on the snail species, rather than the way of cooking. The heat treatment of *T. pisana* samples determined a decrease of ω3/ω6 ratio from 0.58 to 0.3, reaching a value lower than the minimal ratio recommended by the WHO [30]. However, the heat treatment allowed obtaining a total degradation of toxic fatty acid as the erucic acid. Animal tests showed that the ingestion of oils containing erucic acid could lead to a heart disease called myocardial lipidosis. Other potential effects observed in animals (changes in liver, kidney, and skeletal muscle weight) occur at slightly higher doses.

Contrary to what was found by Szkucik et al. [10], the *C. aspersum* samples examined in this work showed a considerable decrease of the relative amounts of SFA after heat treatment, favoring an increase of the relative amounts of MUFA. The *E. vermiculata* samples showed a similar behavior of *C. aspersum* samples after heat treatment, showing a decrease of 8.72% for SFA and 4.66% for PUFA, and an increase of the amounts of MUFA. A reduced PUFA content could be caused by the autoxidation mechanisms initiated by temperature rise in meat during it is cooking [10,31]. Furthermore, these modifications appear to be related to the process temperature, the cooking time, and the internal temperature reached by the meat [31–34].

Principal component analysis allowed to depict the significant sources of variability of the dataset analyzed using two Principal Components (PCs), which showed a clear separation of the land snail samples according to species and heat treatment. The high percentage of variance explained by PC1 (93.89%) reflects the fact that the investigated variables were highly correlated, showing a general decrease due to heat treatment; this variation was much more pronounced for *T. pisana* than for the species *C. aspersum* and *E. vermiculata*. The highest variation after heat treatment common to all the three species was related to PUFA, MUFA, oleic acid, and linoleic acid.

5. Conclusions

To the best of our knowledge, this work reports, for the first time, the fatty acid composition of *T. pisana* samples and their variation as a result of heat processing, and the first time of fatty acid composition in *E. vermiculata*, *T. pisana*, and *C. aspersum* collected in Sicily (Southern Italy). The results showed a species-specific variation of FA contents in the land snails samples examined after boiling, showing the highest UFA decrease in *T. pisana* samples.

The results demonstrate that the land snail species examined could be a good source of MUFA and PUFA and their contents are species-specific. Boiling could be an adequate cooking procedure for land snail consumption according to retained nutritional and healthy criteria (PUFA contents and ω-6/ω-3 ratio). Furthermore, the boiling process can safeguard consumers against potentially pathogenic microorganisms. Given that boiling losses seem to be related to cooking time and temperature, further studies are needed to find the best cooking condition in order to preserve the best nutritional composition criteria of land snails.

Author Contributions: Conceptualization, F.G.G. and G.C.; methodology, F.G.G., A.M. and G.C.; validation, A.M., A.A., N.C. and G.L.C.; formal analysis F.G.G. and G.C.; investigation, A.P., A.U. and N.C.; data curation, A.U., R.C., G.C. and F.G.G.; writing—original draft preparation, F.G.G. and G.C.; writing—review and editing N.C., A.P., A.A., V.F. and A.V.; visualization, V.F. and A.A.; supervision, V.F.; project administration, V.F.; funding acquisition, V.F.

Funding: This research was funded by MINISTERO DELLA SALUTE, grant number RC IZS SI 16/15.

Acknowledgments: The authors thank Barbara Randisi for the technical support.

Conflicts of Interest: The authors declare no conflicts of interest. The funders had no role in the design of the study; in the collection, analyses, or interpretation of data; in the writing of the manuscript; or in the decision to publish the results.

References

1. Pellati, R. Lumache e Consumi. Available online: http://www.fosan.it/notiziario/31_lumache_e_consumi.html (accessed on 4 November 2019).
2. Cicero, A.; Giangrosso, G.; Cammilleri, G.; Macaluso, A.; Currò, V.; Galuppo, L.; Vargetto, D.; Vicari, D.; Ferrantelli, V. Microbiological and Chemical Analysis of Land Snails Commercialised in Sicily. *Ital. J. Food Saf.* **2015**, *4*, 2. [CrossRef] [PubMed]
3. Milinsk, M.C.; das Graças Padre, R.; Hayashi, C.; de Oliveira, C.C.; Visentainer, J.V.; de Souza, N.E.; Matsushita, M. Effects of feed protein and lipid contents on fatty acid profile of snail (Helix aspersa maxima) meat. *J. Food Compos. Anal.* **2006**, *19*, 212–216. [CrossRef]
4. Alasalvar, C.; Pelvan, E.; Topal, B. Effects of roasting on oil and fatty acid composition of Turkish hazelnut varieties (*Corylus avellana* L.). *Int. J. Food Sci. Nutr.* **2010**, *61*, 630–642. [CrossRef] [PubMed]
5. Ayas, D.; Ozogul, Y.; Ozogul, İ.; Uçar, Y. The effects of season and sex on fat, fatty acids and protein contents of Sepia officinalis in the northeastern Mediterranean Sea. *Int. J. Food Sci. Nutr.* **2012**, *63*, 440–445. [CrossRef]
6. Özogul, Y.; Özogul, F.; Çiçek, E.; Polat, A.; Kuley, E. Fat content and fatty acid compositions of 34 marine water fish species from the Mediterranean Sea. *Int. J. Food Sci. Nutr.* **2009**, *60*, 464–475. [CrossRef]

7. Tangolar, S.G.; Özoğul, Y.; Tangolar, S.; Torun, A. Evaluation of fatty acid profiles and mineral content of grape seed oil of some grape genotypes. *Int. J. Food Sci. Nutr.* **2009**, *60*, 32–39. [CrossRef]
8. Tokuşoğlu, Ö. The quality properties and saturated and unsaturated fatty acid profiles of quail egg: The alterations of fatty acids with process effects. *Int. J. Food Sci. Nutr.* **2006**, *57*, 537–545. [CrossRef]
9. Özogul, Y.; Özogul, F.; Olgunoglu, A.I. Fatty acid profile and mineral content of the wild snail (Helix pomatia) from the region of the south of the Turkey. *Eur. Food Res. Technol.* **2005**, *221*, 547–549. [CrossRef]
10. Szkucik, K.; Ziomek, M.; Paszkiewicz, W.; Drozd, Ł.; Gondek, M.; Knysz, P. Fatty acid profile in fat obtained from edible part of land snails harvested in Poland. *J. Vet. Res.* **2018**, *62*, 519–526. [CrossRef]
11. Purwaningsih, S.; Suseno, S.H.; Salamah, E.; Mulyaningtyas, J.R.; Dewi, Y.P. Effect of boiling and steaming on the profile fatty acids and cholesterol in muscle tissue of molluscs. *Int. Food Res. J.* **2015**, *22*, 1087–1094.
12. Lazaridou-Dimitriadou, M.; Kattoulas, M.E. Energy flux in a natural population of the land snail *Eobania vermiculata (Müller) (Gastropoda: Pulmonata: Stylommatophora)* in Greece. *Can. J. Zool.* **1991**, *69*, 881–891. [CrossRef]
13. Nicolai, A. The Impact of Diet Treatment on Reproduction and Thermophysiological Processes in the Land Snails Cornu Aspersum and Helix Pomatia. Ph.D. Thesis, Universität Bremen, Bremen, Germany, 2010.
14. Odendaal, L.J.; Haupt, T.M.; Griffiths, C.L. The alien invasive land snail Theba pisana in the West Coast National Park: Is there cause for concern? *Koedoe* **2008**, *50*, 93–98. [CrossRef]
15. Bertino, E.; Giribaldi, M.; Baro, C.; Giancotti, V.; Pazzi, M.; Peila, C.; Tonetto, P.; Arslanoglu, S.; Moro, G.E.; Cavallarin, L.; et al. Effect of prolonged refrigeration on the lipid profile, lipase activity, and oxidative status of human milk. *J. Pediatr. Gastroenterol. Nutr.* **2013**, *56*, 390–396. [CrossRef] [PubMed]
16. Taverniers, I.; De Loose, M.; Van Bockstaele, E. Trends in quality in the analytical laboratory. I. Traceability and measurement uncertainty of analytical results. *TrAC Trends Anal. Chem.* **2004**, *23*, 480–490. [CrossRef]
17. Pantano, L.; Cascio, G.L.; Alongi, A.; Cammilleri, G.; Vella, A.; Macaluso, A.; Cicero, N.; Migliazzo, A.; Ferrantelli, V. Fatty acids determination in Bronte pistachios by gas chromatographic method. *Nat. Prod. Res.* **2016**, *30*, 2378–2382. [CrossRef]
18. Raimondi, S.; Luciani, R.; Sirangelo, T.M.; Amaretti, A.; Leonardi, A.; Ulrici, A.; Foca, G.; D'Auria, G.; Moya, A.; Zuliani, V.; et al. Microbiota of sliced cooked ham packaged in modified atmosphere throughout the shelf life: Microbiota of sliced cooked ham in MAP. *Int. J. Food Microbiol.* **2019**, *289*, 200–208. [CrossRef]
19. van den Berg, R.A.; Hoefsloot, H.C.; Westerhuis, J.A.; Smilde, A.K.; van der Werf, M.J. Centering, scaling, and transformations: Improving the biological information content of metabolomics data. *BMC Genom.* **2006**, *7*, 142. [CrossRef]
20. Ekin, İ. A comparative study on fatty acid content of main organs and lipid classes of land snails *Assyriella escheriana* and *Assyriella guttata* distributed in southeastern Anatolia. *Ital. J. Food Sci.* **2015**, *27*, 75–81.
21. Baker, G.H. *The Biology and Control of White Snails (Mollusca: Helicidae), Introduced Pests in Australia*; Commonwealth Scientific and Industrial Research Organization: Melbourne, Australia, 1986.
22. Baker, G.H. *Damage, Population Dynamics, Movement and Control of Pest Helicid Snails in Southern Australia*; Commonwealth Scientific and Industrial Research Organization: Melbourne, Australia, 1989; pp. 175–185.
23. Cowie, R.H. The Life-Cycle and Productivity of the Land Snail Theba pisana (*Mollusca: Helicidae*). *J. Anim. Ecol.* **1984**, *53*, 311–325. [CrossRef]
24. Chevalier, L.; Coz, M.; Charrier, M. Influence of inorganic compounds on food selection by the brown garden snail *Cornu aspersum (Müller) (Gastropoda: Pulmonata)*. *Malacologia* **2003**, *45*, 125–132.
25. Chevalier, L.; Desbuquois, C.; Le Lannic, J.; Charrier, M. Poaceae in the natural diet of the snail Helix aspersa Müller (*Gastropoda, Pulmonata*). *Comptes Rendus de l'Académie des Sci.-Series III-Sci. de la Vie* **2001**, *324*, 979–987. [CrossRef]
26. Ekin, I.; Şeşen, R. Investigation of the Fatty Acid Contents of Edible Snails Helix lucorum, Eobania vermiculata and Non-Edible Slug *Limax flavus*. *Rec. Nat. Prod.* **2017**, *11*, 562–567. [CrossRef]
27. Ekin, İ.; Başhan, M.; Şeşen, R. Possible seasonal variation of the fatty acid composition from Melanopsis praemorsa (L., 1758) (*Gastropoda: Prosobranchia*), from southeast Anatolia, Turkey. *Turk. J. Biol.* **2011**, *35*, 203–213.
28. Harper, P.C. Breeding biology of the fairy prion (*Pachyptila turtur*) at the Poor Knights Islands, New Zealand. *N. Z. J. Zool.* **1976**, *3*, 351–371. [CrossRef]

29. Givens, D.I.; Gibbs, R.A. Current intakes of EPA and DHA in European populations and the potential of animal-derived foods to increase them: Symposium on 'How can the n-3 content of the diet be improved?'. *Proc. Nutr. Soc.* **2008**, *67*, 273–280. [CrossRef] [PubMed]
30. WHO. *Diet, Nutrition, and the Prevention of Chronic Diseases: Report of A WHO-FAO Expert Consultation; [Joint WHO-FAO Expert Consultation on Diet, Nutrition, and the Prevention of Chronic Diseases, 2002, Geneva, Switzerland]*; WHO technical report series; Expert Consultation on Diet, Nutrition, and the Prevention of Chronic Diseases, Weltgesundheitsorganisation, FAO, Eds.; World Health Organization: Geneva, Switzerland, 2003; ISBN 978-92-4-120916-8.
31. Weber, J.; Bochi, V.C.; Ribeiro, C.P.; Victório, A.D.M.; Emanuelli, T. Effect of different cooking methods on the oxidation, proximate and fatty acid composition of silver catfish (*Rhamdia quelen*) fillets. *Food Chem.* **2008**, *106*, 140–146. [CrossRef]
32. Alfaia, C.M.M.; Alves, S.P.; Lopes, A.F.; Fernandes, M.J.E.; Costa, A.S.H.; Fontes, C.M.G.A.; Castro, M.L.F.; Bessa, R.J.B.; Prates, J.A.M. Effect of cooking methods on fatty acids, conjugated isomers of linoleic acid and nutritional quality of beef intramuscular fat. *Meat Sci.* **2010**, *84*, 769–777. [CrossRef]
33. Domínguez, R.; Gómez, M.; Fonseca, S.; Lorenzo, J.M. Influence of thermal treatment on formation of volatile compounds, cooking loss and lipid oxidation in foal meat. *LWT-Food Sci. Technol.* **2014**, *58*, 439–445. [CrossRef]
34. Rasinska, E.; Rutkowska, J.; Czarniecka-Skubina, E.; Tambor, K. Effects of cooking methods on changes in fatty acids contents, lipid oxidation and volatile compounds of rabbit meat. *LWT* **2019**, *110*, 64–70. [CrossRef]

© 2019 by the authors. Licensee MDPI, Basel, Switzerland. This article is an open access article distributed under the terms and conditions of the Creative Commons Attribution (CC BY) license (http://creativecommons.org/licenses/by/4.0/).

Article

Quali-Quantitative Profile of Native Carotenoids in Kumquat from Brazil by HPLC-DAD-APCI/MS

Helena Maria Pinheiro-Sant'Ana [1], Pamella Cristine Anunciação [1], Clarice Silva e Souza [1], Galdino Xavier de Paula Filho [2], Andrea Salvo [3,*], Giacomo Dugo [3] and Daniele Giuffrida [3]

[1] Departamento de Nutrição e Saúde, Universidade Federal de Viçosa, Avenida P.H. Rolfs, s/n, Viçosa 36571-000, Brazil; helena.santana@ufv.br (H.M.P.-S.); nutripamella@gmail.com (P.C.A.); cla_souzabio@yahoo.com.br (C.S.e.S.)
[2] Departamento de Educação, Universidade Federal do Amapá, Rodovia Juscelino Kubitschek, Km 02, Jardim Marco Zero, Macapá 68903-419, Brazil; galdinoxpf@gmail.com
[3] Department, of Biomedical and Dental Sciences and Morphofunctional Imaging, University of Messina (Italy), V.le Annunziata, 98168 Messina, Italy; dugog@unime.it (G.D.); dgiuffrida@unime.it (D.G.)
* Correspondence: asalvo@unime.it; Tel.: +39-090-676-6880

Received: 3 April 2019; Accepted: 14 May 2019; Published: 16 May 2019

Abstract: In this study the native carotenoids composition in kumquat (Fortunella margarita) (peel + pulp) from Brazil was determined for the first time by a HPLC-DAD-APCI/MS (high performance liquid chromatography-diode array detector-atmospheric pressure chemical ionization/mass spectrometry), methodology. Eleven carotenoids were successfully identified and quantified in kumquat: four carotenoids in the free form and seven carotenoids in the esterified form. β-citraurin-laurate was the carotenoid found in the highest content (607.33 µg/100 g fresh matter), followed by β-cryptoxanthin-laurate (552.59 µg/100 g). The different esterified forms of β-citraurin and β-cryptoxanthin represented 84.34% of the carotenoids found, which demonstrates the importance of esterification in natural fruits. β-carotene and free xanthophylls (β-cryptoxanthin, lutein and zeaxanthin) represented 5.50% and 14.96%, respectively, of total carotenoids in kumquat. The total carotenoid content of kumquat from Brazil was very high (2185.16 µg/100 g), suggesting that this fruit could contribute significantly to the intake of important bioactive compounds by the population.

Keywords: *Fortunella margarita*; citrus; carotenes; xanthophylls; β-citraurin-laurate; β-cryptoxanthin-laurate

1. Introduction

The citrus family is one of the first crops in the world, it is estimated that half of the marketed production comes from the Americas and 12% comes from the Mediterranean basin. Citrus cultivation is thought to date back at least 4000 years and is mainly from the Asiatic south-east territories. The estimated global citrus traffic for 2017–2018 was around 6 million tons. The most representative cultures were *Citrus sinensis* (61%), *Citrus reticulata* (22%), *Citrus limon* (11%) and *Citrus paradisi* (6%). In the Americas, the primacy of citrus production lies with Brazil followed by the United States. Sweet oranges are grown in Brazil, mainly in the state of São Paulo, over an area of about 584000 hectares, but also in the Amazonas area of northern Brazil, over about 2.7 hectares. [1].

The *Citrus japonica*, known by the common names of kincan (from Japanese *kinkan*) or cunquate (from Chinese *kumquat*), is a small citrus fruit of the Rutaceae family [2]. It has four major cultivated types, including *Fortunella japonica*, *Fortunella margarita*, *Fortunella crassifolia*, and *Fortunella hindsii* [3].

In eastern countries, this fruit is a part of the regular food habits of the population [4], but in Brazil it is considered exotic, in addition to being little known and commercialized. Among the Brazilian states, São Paulo has the largest production and commercialization of this fruit [5].

Kumquats are native to Central China. They are oval or round fruits with peel and orange smooth. Its flavor varies from acid to sweet. The fruit is rich in vitamins, carotene, pectin, calcium, phosphorus, iron and flavonoids [6]. Kumquats are consumed preferably in natura, whole and in shell. It is also used to make jellies, mousse, jams, marmalades, liqueurs and cachaça [7], preparation of syrups, sauces and also, accompaniment in fruit salads and for landscaping purposes and ornamentation [8,9].

Citrus japonica has been used as a traditional folk medicine in Asian countries to reduce alcohol intoxication and as antidepressants, so they are used either as medicines or as edible fruit [10]. Many studies on antioxidant, antimicrobial and antitumor effects have been carried out on kumquats, however identification of the bioactive compounds in the fruit has received little attention [11].

The most elucidated chemical components in kumquat described in the literature are phenolics compounds and flavonoids. Different phenolic compounds and flavonoids are described in *Fortunella* sp. by HPLC-MS [12,13]. These studies have shown a higher concentration of phenolics in fruit peels, with luteolin and kaempferol being the main flavonoids found in *Fortunella* sp. [8,14–17].

Studies on carotenoids in kumquat are extremely limited [9,18–20]. Agócs et al. [18] studied the qualitative and quantitative composition of carotenoids of kumquat and other citrus species. However, the sample preparation of the carotenoids involved a saponification step, a procedure that does not allow us to evaluate the native composition of the carotenoids.

Studies on the composition of carotenoids in foods are very important because they participate in various biological processes in plants, such as photosynthesis, photomorphogenesis, photoprotection, and development [21]. In animals, provitamin A carotenoids play an essential role in the synthesis of retinol (vitamin A) [22], whereas the xanthophylls lutein and zeaxanthin have been associated in humans with the prevention of age-ralated eye degenerations [23,24].

Carotenoids are molecules made up of a long chain of usually forty carbon atoms; they can by divided into two classes: (a) non oxygenated one named carotenes [25] and (b) oxygenated one named xanthophylls [26]. Moreover, the xanthophylls are usually esterified with fatty acids in nature.

The studies on the content of carotenoids in kumquats concern plants grown in Asia but data are not available for fruits harvested in Brazil [18,19]. Therefore, this work aimed to determine the complete qualitative and quantitative profile of the kumquat carotenoid native composition for fruits collected in the rural area of Viçosa, Minas Gerais, Brazil, through liquid chromatography coupled to the mass detector (HPLC-DAD-APCI-MS).

2. Materials and Methods

2.1. Chemicals

The standards of β-carotene, β-cryptoxanthin, lutein, zeaxanthin and physalein, Standard purity was above 98% were purchased from Extrasynthese (Genay, France), and the solvents MeOH (Methanol), MTBE (Methyl-t-butyl ether) and H_2O (Water) from Sigma-Aldrich (Milan, Italy).

2.2. Collection and Preparation of the Samples

The fruits of kumquat (*Fortunella margarita*) (Figure 1) were collected in the morning, in May 2017, in the rural area of Viçosa (latitude 20° 44' 05" S and longitude 42° 51' 27" W), Minas Gerais, Brazil. Samples were collected in four repetitions of approximately 1 kg each. The fruit maturation was determined according to Donadio et al. [27] and defined by the red-orange peel color and the characteristic smell. In addition, ripe fruits were considered as those obtained after their natural fall of the trees or fall after being lightly touched by the hands.

The species was identified with the help of taxonomists from the Universidade Federal de Viçosa Herbarium through the Angiosperm Phylogeny Group IV [28], where it has already been cataloged and registered in the Virtual Herbarium network with the following records: EAC 48987, HUCO 5197, HPL 8977 and SP 42766.

The samples were transported from the harvest site to the laboratory protected in styrofoam boxes with blocks of ice, within two hours after collection. In the laboratory, the samples were selected for appearance, excluding those with any epidermis injury or mechanical damage due to transport. The fruits were removed from the seeds (peel + pulp) were homogenized in a food processor (RI 7625, Philips, São Paulo Brazil), lyophilized (Liotop-LP510, Liobras, São Carlos, Brazil) and stored in plastic containers with screw caps, covered with aluminum foil stored at $-18 \pm 1\ °C$ until further analyses.

Figure 1. Kumquat (whole fruit and cross-section) from Viçosa, Minas Gerais, Brazil.

2.3. Moisture Analysis

Moisture was determined in triplicate in the oven (SP 200, SP Labor®, São Paulo, Brazil), at $65 \pm 1\ °C$, for approximately 72 h [29].

2.4. Extraction of Carotenoids

The carotenoid pigments were extracted from the lyophilized material (peel + pulp), according to the recommended procedures by Rodrigues-Amaya et al. [30]. Three grams of the edible portion of the samples (peel + pulp) were crushed by the use of a mortar and pestle, and a few drops of distilled water were added and extracted to color exhaustion with 20 mL of acetone (7 times) in an ultrasonic bath (Labsonic LBS 1-H22.5, Treviglio, Bergamo, Italy) for 10 min each time. Then, the extracts were individually centrifuged (Awel MF20-R, Multifunction Refrigerated Centrifuge, Blain, France) at 4000 rpm, 5 °C, for 10 min in order to withdraw clear solution on the top. The acetonic extracts were pooled together was concentrated to about 25 mL, in a rotary evaporator (Buchi-heating bath B-491, Buchi, Milan, Italy) at temperature below 35 °C. The dry product was diluted with equal volumes (25 mL) of a mixture of ethyl ether and hexane (1:1) and distilled water (50 mL) and worked up with a separating funnel. The lipofilic phase, cleared by the hydrophilic impurities, was evaporated to dryness using a rotary evaporator (Buchi-heating bath B-491 at 35 °C, and the residue was dissolved in 2 mL of MeOH/MTBE (1:1) and filtered in filter units (PTFE, 0.45µm, 13mm, Sigma Aldrich, Milan, Italy) prior to HPLC analysis. Samples were stored at $-20\ °C$ until they were analyzed.

2.5. Analysis of Carotenoids by HPLC-DAD-APCI-MS

The analysis was performed on an HPLC system (Shimadzu, Kyoto, Japan) equipped with a CBM-20A controller, two LC-20AD pumps, a DGU-20A3R deaerator, a SIL-20AC autosampler, a CTO

20AC column oven and an SPD-M20A photo diode array detector. The data were processed with the Labsolution software. For MS analysis a mass spectrometer detector (LCMS-8040) was used, equipped with an APCI (atmospheric pressure chemical ionization) interface, both in positive and negative ionization mode.

The column used was YMC C30 (250 mm × 4.6 mm × 5 µm); the mobile phases: MeOH/MTBE/H2O (81:15:4, solvent A), MeOH/MTBE/H2O (6:90:4, solvent B); the linear gradient used was: 0–100% B from 0 to 140 min. The column temperature was maintained at 30 °C. The flow was 0.8 mL/min and the injection volume was 20 µL.

The UV-Vis spectra were acquired in the range 220–700 nm, while the chromatograms were extracted at 450 nm (sampling frequency: 4.16 Hz, time constant: 0.64 s). The MS was set up as follows: Scan, both APCI positive (+) and negative (-); atomized gas flow (N2): 2.0 L/min; drying gas flow: 5 L/min; Time of the event: 0.06 s; range m/z: 300–1200; interface temperature: 350 °C; Desolvation line (DL) temperature: 300 °C; thermal block: 300 °C. The samples were analyzed in triplicate.

2.6. Identification and Quantification of Carotenoids

Carotenoids were identified by their UV-Vis spectra, MS spectra, elution order, comparison with the available standard and literature data.

The kumquat carotenoids quantification was performed from the analytical curves. External standards quantitative determination of each compound was performed using all reference materials listed in Section 2.1, in the concentration range from 5 to 50 µg/mL at six concentration levels. The results were obtained from the average of three determinations and the CV% was below 8% in all the LC measurements. The R coefficient for the calibration curves was always above 0.9962, with LOD and LOQ values of 0.07 and 0.22 ppm for β-carotene, 0.1 and 0.33 ppm for β-cryptoxanthin, 0.06 and 0.18 ppm for lutein, 0.08 and 0.3 ppm for zeaxanthin, and 0.12 and 0.24 ppm for physalein, respectively.

3. Results and Discussion

3.1. Carotenoids Qualitative Profile of Brazilian Kumquat

Figure 2 shows the chromatographic profile of the carotenoid composition in not saponified kumquat fruits extracts. The identified compounds are shown in Table 1, together with the UV–Vis and MS spectra information. In Figure 3 are reported the UV-Vis (PDA) and mass spectrum of β-citraurin-laurate and β-citraurin-myristate, detected in the kumquat carotenoid extracts. It can be appreciated that the esterification does not affect the PDA spectra of β-citraurin.

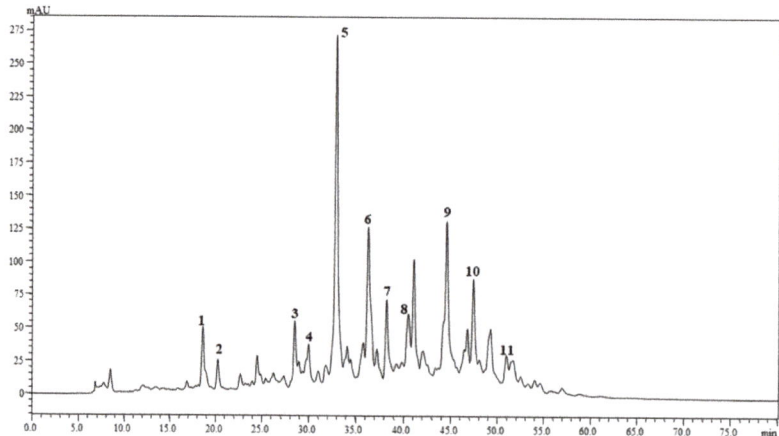

Figure 2. Chromatographic profile of native carotenoids of kumquat from Brazil: peel + pulp. Identification of the compounds are in Table 1.

Table 1. Compounds identification for Figure 2 (kumquat from Brazil: peel + pulp).

Compound	Identification	Rt (min)	PDA (λnm)	MS (APCI-) m/z
1	Lutein	18.5	445, 473	568
2	Zeaxanthin	20.2	449, 476	568
3	β-cryptoxanthin	28.4	428, 451, 478	552
4	β-citraurin-caproate	29.9	454	586
5	β-citraurin-laurate	32.8	455	614
6	β-citraurin-myristate	36.2	453	642
7	β-carotene	38.1	426, 451, 476	536
8	β-citraurin-palmitate	40.4	455	642
9	β-cryptoxanthin-laurate	44.5	428, 450, 478	734
10	β-cryptoxanthin-myristate	47.3	428, 450, 477	762
11	β-cryptoxanthin-palmitate	50.9	428, 451, 478	790

Rt: retention time; PDA: photodiode array; λnm: wavelength of maximum absorption; MS: mass spectrometry; APCI: atmospheric pressure chemical ionization.

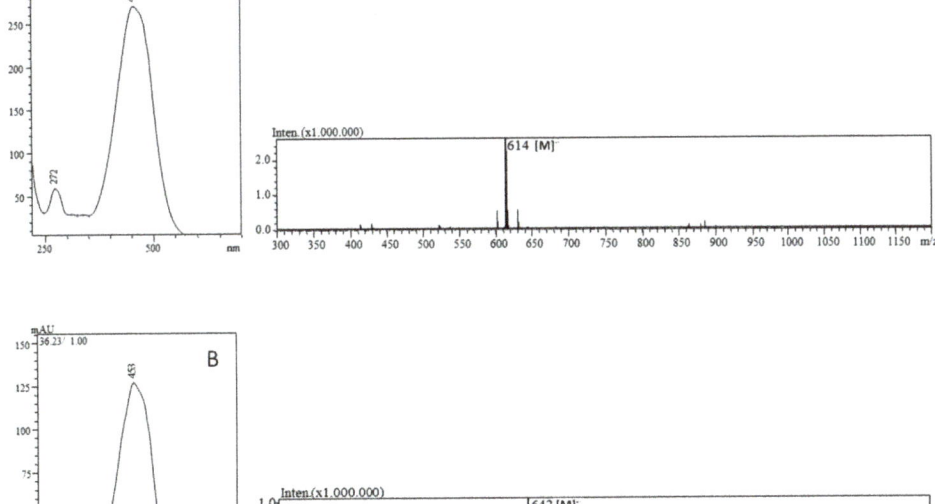

Figure 3. UV-Vis (PDA) and mass spectrum (APCI negative) of β-citraurin-laurate (A) and β-citraurin-myristate (B) of kumquat from Brazil.

Moreover, both the molecular ions [M]$^{-\bullet}$ at respectively m/z 614 and m/z 642 relative to the β-citraurin-laurate and β-citraurin-myristate esters, obtained in the negative APCI mode, are also clearly shown in the same figure. Eleven different carotenoids were identified in kumquat from Brazil; four different β-citraurin esters and three β-cryptoxanthin esters were identified.

The predominant carotenoids were β-citraurin-laurate and β-cryptoxanthin-laurate, both esterified with lauric acid. The chemical structures of these carotenoids are shown in Figure 4. Interestingly, no free β-citraurin was detected in the present study.

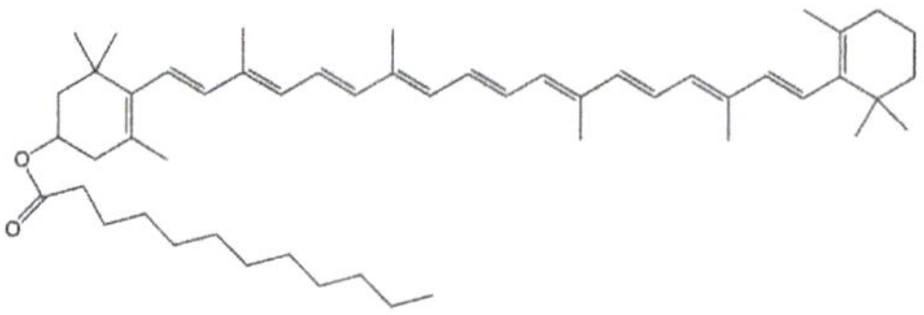

Figure 4. Chemical structures of β-cryptoxanthin-laurate and β-citraurin-laurate, main carotenoids detected in kumquat.

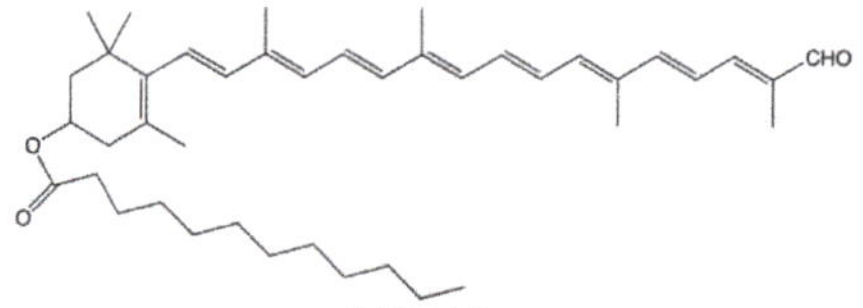

Shirra et al. [9] detected only 4 carotenoids in kumquat from Italy without saponification (β-carotene, β-cryptoxanthin, lutein and zeaxanthin), which were also detected in the present study. In contrast to our study, Agos et al. [18] reported only β-citraurin and β-cryptoxanthin in the free form in kumquat from Hungary. However, these researchers used saponification after the extraction process and did not quantify the carotenoid esters. Saponification may result in destruction or structural transformation of carotenoids [31]. Huyskens et al. [19] studied the qualitative composition of kumquat carotenoids from Israel by thin-layer chromatography and reported several carotenoids, including β-citraurin, β-cryptoxanthin, lutein, β-carotene and zeaxanthin which were, also determined in the present study. In addition, Huyskens et al. [19] reported that violaxanthin was the predominant component in kumquat from Israel, whereas in our study β-citraurin-laurate was the major component.

Esterification with saturated fatty acids improves the stability of xanthophylls such as β-citraurin and β-cryptoxanthin against heat and UV light, but does not affect their antioxidant activity [32]. During the storage and processing of the fruits the xanthophyll esters were more stable than the free xanthophyll [33]. The pigment β-citraurin is responsible for the citrus reddish color, it derives from of β-cryptoxanthin or zeaxanthin and accumulates in some citrus varieties [34,35]. In these fruits, the accumulation of β-citraurin is not a common event, it is observed only in the flavedos of some varieties during fruit ripening [34]. Frutita, a tropical fruit from Panama, showed a very high content of β-citraurin [36].

Recent studies have shown that β-cryptoxanthin and β-citraurin esterified with lauric acid, myristic acid and palmitic acid are found in the Mandarin Satsuma, [35].

The dietary intake of β-cryptoxanthin has been shown to prevent and reduce some pathologies such as: cancer, diabetes and rheumatism due to its antioxidant activity [37–39]. Breithaupt et al. [40] have verified that in chili, papaya, peach and persimmon, β-cryptoxanthin is mainly esterified with saturated fatty acids. Furthermore, the bioavailability of β-cryptoxanthin esters is comparable to the non-esterified form, since fatty acids can be effectively hydrolyzed by β-cryptoxanthin esters before intestinal absorption in the human body [40]. In several citrus varieties, the esterified form β-citraurin is found [35]. As already observed in another study [41], we have found β-cryptoxanthin and β-citraurin in both free and esterified forms.

3.2. Carotenoids Quantitative Profile of Brazilian Kumquat

In the present study, β-citraurin-laurate was the carotenoid found in the highest content in kumquat from Brazil (607.33 µg/100 g fresh matter, representing 27.80% of total carotenoids), followed by β-cryptoxanthin-laurate (552.59 µg/100 g fresh matter, representing 25.31% of total carotenoids). Thus, β-citraurin-laurate and β-cryptoxanthin-laurate represented 53.11% of the total carotenoids found. The different forms of β-citraurin (4 components esterified with fatty acids) and β-cryptoxanthin (1 component in free form and 3 in esterified form) represented 84.34% of the carotenoids found in kumquat from Brazil. Forms of β-citraurin and β-cryptoxanthin esterified with fatty acids accounted for 79.54% of the total carotenoids in kumquat. These results show the importance of the study of the intact carotenoids composition in different matrices (Table 2).

Table 2. Carotenoid content in kumquat from Brazil (peel + pulp).

Compound	Carotenoid	Content in Lyophilized Kumquat (µg/100 g) *	Content in Fresh Kumquat (µg/100 g) *	Carotenoid Composition (%)
1	Lutein	873.05 ± 93.51	144.35 ± 15.45	6.61
2	Zeaxanthin	469.55 ± 25.52	77.63 ± 4.21	3.55
3	β-Cryptoxanthin	634.93 ± 55.91	104.98 ± 9.25	4.80
4	β-Citraurin-caproate	721.81 ± 20.04	119.34 ± 3.32	5.46
5	β-Citraurin-laurate	3673.28 ± 81.78	607.33 ± 13.63	27.80
6	β-Citraurin-myristate	421.88 ± 9.52	69.75 ± 1.58	3.20
7	β-Carotene	726.96 ± 110.60	120.19 ± 18.29	5.50
8	β-Citraurin-palmitate	801.27 ± 66.35	132.48 ± 10.98	6.06
9	β-Cryptoxanthin-laurate	3342.25 ± 126.10	552.59 ± 20.75	25.31
10	β-Cryptoxanthin-myristate	927.77 ± 82.92	153.39 ± 13.71	7.02
11	β-Cryptoxanthin-palmitate	623.73 ± 46.34	103.13 ± 7.67	4.72
Total carotenes		726.96	120.19	5.50
Total free xanthophylls		1977.53	326.96	14.96
Total esterified xanthophylls		10511.99	1738.01	79.54
Total carotenoids		13,216.48	2185.16	100

* Mean of 3 repetitions ± standard deviation (SD). The carotenoid content in the fresh kumquat was calculated based on the average moisture content ($n = 3$) of the lyophilized fruit (14.08%) and the fresh fruit (81.16%).

β-carotene was found in smaller amounts (120.19 µg/100 g fresh matter), representing 5.50% of total carotenoids in kumquat, as well as free xanthophylls (β-cryptoxanthin, lutein and zeaxanthin), which represented 14.96% of total carotenoids (326.96 µg/100 g) (Table 2). β-carotene and β-cryptoxanthin possess provitamin A activity, playing a key role in human health [42]. Differently from our results, Wang et al. [20] reported that β-cryptoxanthin was the major carotenoid present in kumquat cultivated in Taiwan, while Schirra et al. [9] found lutein to be the major carotenoid present in kumquat from Italy.

Schirra et al. [9] found a lower concentration, in fresh matter, of β-carotene (33 µg/100 g), β-cryptoxanthin (26 µg/100 g), lutein (44 µg/100 g) and zeaxanthin (24 µg/100 g) in kumquat from Italy when compared to the concentrations of these carotenoids in kumquat from Brazil, reported here for the first time. Wang et al. [41] found contents in dry matter that were much lower than reported here, for lutein (9.9 µg/100 g), zeaxanthin (10.4 µg/100 g), β-cryptoxanthin (183 µg/100 g) and β-carotene (131 µg/100 g) in kumquat cultivated in Taiwan. Carotenoid composition and contents in fruits can be influenced by genetic factors, geographical regions, fruit processing, storage methods Giuffrida et al. [25] and environmental conditions Lu et al. [26], which can explain the differences found in these studies. Besides the climatic factors, the irrigation, soil conditions, fertilizers and herbicides may also affect the carotenoids accumulation [26]. Regarding the carotenoids content in food, Britton [43] have proposed the following classification ranges: low: 0–0.1 mg/100 g; moderate: 0.1–0.5 mg/100 g; high: 0.5-2 mg/100 g; very high: >2 mg/100 g. Thus, the total carotenoid content of kumquat from Brazil was very high (2185.16 µg/100 g).

4. Conclusions

In this study the native carotenoids composition in kumquat (*Fortunella margarita*) from Brazil was determined for the first time. Eleven native carotenoids in kumquat from the rural area of Minas Gerais, Brazil were successfully identified and quantified by HPLC-DAD-APCI-MS. Four carotenoids in the free form (β-carotene, β-cryptoxanthin, lutein and zeaxanthin) and 7 carotenoids in the esterified form were identified (β-citraurin-caproate, β-citraurin-laurate, β-citraurin-myristate, β-citraurin-palmitate, β-cryptoxanthin-laurate, β-cryptoxanthin-myristate, β-cryptoxanthin-palmitate). β-citraurin-laurate and β-cryptoxanthin-laurate were the most abundant native carotenoids in kumquat from Brazil. The total carotenoid content of kumquat from Brazil was high (2185.16 µg/100 g), suggesting that this fruit can contribute significantly to the ingestion of important bioactive compounds.

Author Contributions: Conceptualization, H.M.P.-S., P.C.A.; Methodology, C.S.e.S., G.X.d.P.F.; Writing-review and editing, A.S.; Supervision, G.D., and D.G.

Funding: This research received no external funding.

Acknowledgments: The authors thanks the Fundação de Amparo à Pesquisa do Estado de Minas Gerais (FAPEMIG), the Conselho Nacional de Desenvolvimento Científico e Tecnológico (CNPq), the Coordenação de Aperfeiçoamento de Ensino Superior (CAPES) and the Università degli Studi di Messina for financial support for conduction of the study.

Conflicts of Interest: The authors declare no conflict of interest.

References

1. Instituto Brasileiro de Geografia e Estatística. *Pesquisa de Orçamentos Familiares*; Ministério da Saúde: Rio de Janeiro, Brasília, 2009.
2. The Plant List. Available online: http://www.theplantlist.org/tpl/record/kew-2724150 (accessed on 15 May 2019).
3. Ogawa, K.; Kawasaki, A.; Omura, M.; Yoshida, T.; Ikoma, Y.; Yano, M. 3′,5′-Di-C-β-glucopyranosylphloretin, a flavonoid characteristic of the genus Fortunella. *Phytochemistry* **2001**, *57*, 737–742. [CrossRef]
4. Pompeu Junior, J. Rootstocks and scions in the citriculture of the São Paulo State; In Proceedings of International Congress of Citrus Nurserymen, Ribeirão Preto, Brasil, 2001; pp. 75–82.
5. Watanabe, H.S.; De Oliveira, S.L. Comercialização de frutas exóticas. *Revista Brasileira de Fruticultura* **2014**, *36*, 23–38. [CrossRef]
6. Barreca, D.; Bellocco, E.; Caristi, C.; Leuzzi, U.; Gattuso, G. Kumquat (Fortunella japonica Swingle) juice: Flavonoid distribution and antioxidant properties. *Food. Res. Int.* **2011**, *44*, 2190–2197. [CrossRef]
7. Koller, O.L. *Citricultura Catarinense*; Epagri: Florianópolis, Brazil, 2013.
8. Kawaii, S.; Tomono, Y.; Katase, E.; Ogawa, K.; Yano, M. Quantitation of flavonoid constituents in citrus fruits. *J. Agric. Food Chem.* **1999**, *47*, 3565–3571. [CrossRef]
9. Schirra, M.; Palma, A.; D'Aquino, S.; Angioni, A.; Minello, E.V.; Melis, M.; Cabras, P. Influence of postharvest hot water treatment on nutritional and functional properties of kumquat (*Fortunella japonica Lour. Swingle* Cv. Ovale) fruit. *J. Agric. Food Chem.* **2007**, *56*, 455–460. [CrossRef]
10. Liu, Y.; Liu, Y.; Liu, Y.; Liu, H.; Shang, Y. Evaluating effects of ellagic acid on the quality of kumquat fruits during storage. *Sci. Hortic.* **2018**, *227*, 244–254. [CrossRef]
11. Lou, S.N.; Ho, C.T. Phenolic compounds and biological activities of small-size citrus: Kumquat and calamondin. *J. Food Drug Anal.* **2017**, *25*, 162–175. [CrossRef] [PubMed]
12. Lou, S.N.; Lai, Y.C.; Hsu, Y.S.; Ho, C.T. Phenolic content, antioxidant activity and effective compounds of kumquat extracted by different solvents. *Food Chem.* **2016**, *197*, 1–6. [CrossRef] [PubMed]
13. Lou, S.N.; Lai, Y.C.; Huang, J.D.; Ho, C.T.; Ferng, L.H.A.; Chang, Y.C. Drying effect on flavonoid composition and antioxidant activity of immature kumquat. *Food Chem.* **2015**, *171*, 356–363. [CrossRef] [PubMed]
14. Chen, M.H.; Yang, K.M.; Huang, T.C.; Wu, M.L. Traditional small-size citrus from Taiwan: Essential oils, bioactive compounds and antioxidant capacity. *Medicines* **2017**, *4*, 28. [CrossRef]
15. Salvo, A.; Bruno, M.; La Torre, G.L.; Vadalà, R.; Mottese, A.F.; Saija, E.; Mangano, V.; Casale, K.E.; Cicero, N.; Dugo, G. Interdonato lemon from Nizza di Sicilia (Italy): Chemical composition of hexane extract of lemon peel and histochemical investigation. *Nat. Prod. Res.* **2016**, *30*, 1517–1525. [CrossRef]
16. Salvo, A.; Costa, R.; Albergamo, A.; Arrigo, S.; Rotondo, A.; La Torre, G.L.; Mangano, V.; Dugo, G. An in-depth study of the volatile variability of chinotto (Citrus myrtifolia Raf.) induced by the extraction procedure. *Eur. Food Res. Technol.* **2019**. [CrossRef]
17. Costa, R.; Salvo, A.; Rotondo, A.; Bartolomeo, G.; Pellizzeri, V.; Saija, E.; Arrigo, S.; Interdonato, M.; Trozzi, A.; Dugo, G. Combination of separation and spectroscopic analytical techniques: Application to compositional analysis of a minor citrus species. *Nat. Prod. Res.* **2018**, *32*, 2596–2602. [CrossRef] [PubMed]
18. Agócs, A.; Nagy, V.; Szabó, Z.; Márk, L.; Ohmacht, R.; Deli, J. Comparative study on the carotenoid composition of the peel and the pulp of different citrus species. *Innov. Food. Sci. Emerg. Technol.* **2007**, *8*, 390–394. [CrossRef]

19. Huyskens, S.; Timberg, R.; Gross, J. Pigment and plastid ultrastructural changes in Kumquat (*Fortunella margarita*) «Nagami» during ripening. *J. Plant. Physiol.* **1985**, *118*, 61–72. [CrossRef]
20. Wang, Y.C.; Chuang, Y.C.; Ku, Y.H. Quantitation of bioactive compounds in citrus fruits cultivated in Taiwan. *Food Chem.* **2007**, *102*, 1163–1171. [CrossRef]
21. Nisar, N.; Li, L.; Lu, S.; Khin, N.C.; Pogson, B.J. Carotenoid metabolism in plants. *Mol. Plant* **2015**, *8*, 68–82. [CrossRef] [PubMed]
22. Krinsky, N.I.; Johnson, E.J. Carotenoid actions and their relation to health and disease. *Mol. Aspects Med.* **2005**, *26*, 459–516. [CrossRef] [PubMed]
23. Fiedor, J.; Burda, K. Potential role of carotenoids as antioxidants in human health and disease. *Nutrients* **2014**, *6*, 466–488. [CrossRef]
24. Johnson, E.J.; Krinsky, N.I. Carotenoids and coronary heart disease. In *Carotenoids*; Springer: Basel, Switzerland, 2009; pp. 287–300.
25. Giuffrida, D.; Dugo, P.; Salvo, A.; Saitta, M.; Dugo, G. Free carotenoid and carotenoid ester composition in native orange juices of different varieties. *Fruits* **2010**, *65*, 277–284. [CrossRef]
26. Lu, Q.; Huang, X.; Lv, S.; Pan, S. Carotenoid profiling of red navel orange "Cara Cara" harvested from five regions in China. *Food Chem.* **2017**, *232*, 788–798. [CrossRef]
27. Donadio, L.C. *Dicionário das Frutas*; UNESP: Jaboticabal, Brazil, 2007.
28. The Angiosperm Phylogeny Group. An update of the Angiosperm Phylogeny Group classification for the orders and families of flowering plants: APG IV. *Bot. J. Linn. Soc.* **2016**, *181*, 1–20.
29. Association of Official Analytical Chemists (AOAC). *Official Methods of Analysis of the Association of Official Analytical Chemists*, 18th ed.; Association of Official Analytical Chemists: Washington, DC, USA, 2010.
30. Rodriguez-Amaya, D. *Food Carotenoids: Chemistry, Biology and Technology*; John Wiley & Sons: Hoboken, NZ, USA, 2015.
31. Amorim-Carrilho, K.; Cepeda, A.; Fente, C.; Regal, P. Review of methods for analysis of carotenoids. *Trends. Analyt. Chem.* **2014**, *56*, 49–73. [CrossRef]
32. Fu, H.; Xie, B.; Fan, G.; Ma, S.; Zhu, X.; Pan, S. Effect of esterification with fatty acid of β-cryptoxanthin on its thermal stability and antioxidant activity by chemiluminescence method. *Food Chem.* **2010**, *122*, 602–609. [CrossRef]
33. Bunea, A.; Socaciu, C.; Pintea, A. Xanthophyll esters in fruits and vegetables. *Not. Bot. Horti Agrobot. Cluj Napoca* **2014**, *42*, 310. [CrossRef]
34. Ma, G.; Zhang, L.; Iida, K.; Madono, Y.; Yungyuen, W.; Yahata, M.; Yamawaki, K.; Kato, M. Identification and quantitative analysis of β-cryptoxanthin and β-citraurin esters in Satsuma mandarin fruit during the ripening process. *Food Chem.* **2017**, *234*, 356–364. [CrossRef]
35. Ma, G.; Zhang, L.; Matsuta, A.; Matsutani, K.; Yamawaki, K.; Yahata, M.; Wahyudi, A.; Motohashi, R.; Kato, M. Enzymatic formation of β-citraurin from β-cryptoxanthin and zeaxanthin by carotenoid cleavage dioxygenase4 in the flavedo of citrus fruit. *Plant Physiol.* **2013**, *163*, 682–695. [CrossRef]
36. Murillo, E.; Giuffrida, D.; Menchaca, D.; Dugo, P.; Torre, G.; Meléndez-Martinez, A.J.; Mondello, L. Native carotenoids composition of some tropical fruits. *Food Chem.* **2013**, *140*, 825–836. [CrossRef]
37. Iskandar, A.R.; Liu, C.; Smith, D.E.; Hu, K.Q.; Choi, S.W.; Ausman, L.M.; Wang, X.D. β-Cryptoxanthin restores nicotine-reduced lung SIRT1 to normal levels and inhibits nicotine-promoted lung tumorigenesis and emphysema in A/J mice. *Cancer Prev. Res.* **2012**, *6*, 309–320. [CrossRef] [PubMed]
38. Takayanagi, K.; Morimoto, S.I.; Shirakura, Y.; Mukai, K.; Sugiyama, T.; Tokuji, Y.; Ohnishi, M. Mechanism of visceral fat reduction in Tsumura Suzuki obese, diabetes (TSOD) mice orally administered β-cryptoxanthin from Satsuma mandarin oranges (*Citrus unshiu* Marc). *J. Agric. Food Chem.* **2011**, *59*, 12342–12351. [CrossRef] [PubMed]
39. Yamaguchi, M. Role of carotenoid β-cryptoxanthin in bone homeostasis. *J. Biomed. Sci.* **2012**, *19*, 36. [CrossRef] [PubMed]
40. Breithaupt, D.E.; Weller, P.; Wolters, M.; Hahn, A. Plasma response to a single dose of dietary β-cryptoxanthin esters from papaya (*Carica papaya* L.) or non-esterified β-cryptoxanthin in adult human subjects: A comparative study. *Br. J. Nutr.* **2003**, *90*, 795–801. [CrossRef] [PubMed]
41. Wada, Y.; Matsubara, A.; Uchikata, T.; Iwasaki, Y.; Morimoto, S.; Kan, K.; Ookura, T.; Fukusaki, E.; Bamba, T. Investigation of β-cryptoxanthin fatty acid ester compositions in citrus fruits Cultivated in Japan. *Food Nutr. Sci.* **2013**, *4*, 98.

42. Mercadante, A.Z.; Rodrigues, D.B.; Petry, F.C.; Mariutti, L.R.B. Carotenoid esters in foods-A review and practical directions on analysis and occurrence. *Food Res. Int.* **2017**, *99*, 830–850. [CrossRef] [PubMed]
43. Britton, G.; Khachik, F. Carotenoids in food. In *Carotenoids*; Britton, G., Pfander, H., Liaaen-Jensen, S., Eds.; Springer: Basel, Switzerland, 2009.

© 2019 by the authors. Licensee MDPI, Basel, Switzerland. This article is an open access article distributed under the terms and conditions of the Creative Commons Attribution (CC BY) license (http://creativecommons.org/licenses/by/4.0/).

Article

Chemical Composition and Antioxidant Activity of Steam-Distilled Essential Oil and Glycosidically Bound Volatiles from *Maclura Tricuspidata* Fruit

Gyung-Rim Yong, Yoseph Asmelash Gebru, Dae-Woon Kim, Da-Ham Kim, Hyun-Ah Han, Young-Hoi Kim and Myung-Kon Kim *

Department of Food Science and Technology, Chonbuk National University, Jeonju 54896, Jeonbuk, Korea; rudfla1226@naver.com (G.-R.Y.); holden623@naver.com (Y.A.G.); eodns3344@gmail.com (D.-W.K.); dadaham@naver.com (D.-H.K.); hha208@Korea.kr (H.-A.H.); yhoi1307@hanmail.net (Y.-H.K.)
* Correspondence: kmyuko@jbnu.ac.kr; Tel.: +82-63-270-25512

Received: 12 November 2019; Accepted: 5 December 2019; Published: 9 December 2019

Abstract: Essential oil obtained from *Maclura triscuspidata* fruit has been reported to have functional properties. This study aimed at determining chemical compositions and antioxidant activities of steam-distilled essential oil (SDEO) and glycosidically bound aglycone fraction (GBAF) isolated from fully ripe *M. triscuspidata* fruit. SDEO was isolated by simultaneous steam distillation and extraction (SDE). GBAF was prepared by Amberlite XAD-2 adsorption of methanol extract, followed by methanol elution and enzymatic hydrolysis. Both fractions were analyzed by gas chromatography–mass spectrometry (GC–MS). A total of 76 constituents were identified from both oils. Apart from fatty acids and their esters, the SDEO contained *p*-cresol in the highest concentration (383.5 ± 17.7), followed by δ-cadinene (147.7 ± 7.7), β-caryophyllene (145.7 ± 10.5), β-ionone (141.0 ± 4.5), *n*-nonanal (140.3 ± 20.5), theaspirane A (121.3 ± 4.5) and theaspirane B (99.67 ± 9.05 µg/g). Thirteen carotenoid-derived compounds identified in the SDEO are being isolated from *M. triscuspidata* fruit for the first time. Out of the 22 components identified in GBAF, 14 were present only in the glycosidically bound volatiles. Antioxidant activity of the GBAF was higher than that of SDEO. These results suggest that glycosidically bound volatiles of *M. triscuspidata* fruit have a good potential as natural antioxidants.

Keywords: *Maclura triscuspidata fruit; essential oil; glycosidically bound volatiles; gas chromatography*-mass spectroscopy (GC–MS); chemical composition; antioxidant activity

---

1. Introduction

Plant-derived essential oils are complex mixtures of volatile and semi-volatile organic compounds characterized by diverse odors and chemical compositions depending on their origins. They are traditionally obtained from various plant tissues including fruits, seed, leaves, flowers, roots, woods and barks by means of hydrodistillation, steam distillation, solvent extraction or cold pressing [1,2]. Due to their organoleptic and biological properties, essential oils have been used as flavoring agents and natural preservatives in foods since ancient times [3]. More recently, essential oils and some of their isolated components are increasingly being used in various commercial products such as foods, cosmetics, perfumes, household cleaning products and hygiene products, and medicinal applications [2]. These compounds have been reported to have various biological activities including antimicrobial, antioxidant, antiviral, antiplatelet, antithrombotic, antiallergic, anti-inflammatory, antimutagenic, and anticarcinogenic properties [4–6].

Lipid oxidation causes serious problems in foods by producing unpleasant flavors, discoloration, decreasing nutritional quality and safety of foods through due to production of secondary oxidation products that have harmful effects on human health [7]. The use of essential oils as natural antioxidants is a field of growing interest because of the fact that synthetic antioxidants such as butylated hydroxyanisole (BHA) and butylated hydroxyltoluene (BHT) have been suspected of causing liver damage and carcinogenesis when used at high levels in laboratory animals [8–11]. For this reason, their use in the food industry has recently declined owing to safety concerns and consumer demand for natural products.

Maclura tricuspidata (Carr.) Bur. (formerly known as *Cudrania tricuspidata*) which belongs to the Moraceae family is a thorny tree native to East Asia including China, Japan and Korea. The leaves, root, stem and fruit of this plant have been used in traditional herbal medicines to treat jaundice, hepatitis, neuritis and inflammation in Korea [12]. Several beneficial effects of *M. tricuspidata* extracts have been reported including anticancer [13,14], anti-inflammatory [15], antioxidant [16,17], and antidiabetes effects [18]. Various bioactive compounds such as prenylated xanthones, phenolic acids and flavonoids have already been identified from its leaves, root, stem and fruit [19–21].

The ripe fruits of *Maclura tricuspidata* which have a bright red color are edible with a floral aroma and sweet taste. They have traditionally been used to prepare fresh juice, jam, wine, vinegar and fermented alcoholic beverages in Korea. Previous studies have reported that the extracts and components of *M. tricuspidata* fruits have strong antioxidant and free radical-scavenging activities in an in vitro system [22,23]. The antioxidant activity of *M. tricuspidata* fruit extract is associated with the presence of phenolic compounds such as flavonoids and phenolic acids [17,24]. We have recently identified 18 polyphenolic compounds among which five parishin derivatives (gastrodin, parishin A, B, C, E) identified for the first time in the fruit and confirmed their anti-oxidant potentials [25]. Essential oil obtained from the fruit by microwave-assisted hydrodistillation has also been reported to have antioxidant activity through 2,2-diphenyl-1-picrylhydrazyl (DPPH), nitric oxide, hydroxy and superoxide radical scavenging activities [26]. Recently, Bajpai and colleagues [26] identified 29 compounds as major constituents in the essential oil isolated from *M. tricuspidata* fruit. Although the chemical compositions and their antioxidant activities of essential oils from the stem and root of *M. tricuspidata* were elucidated [26,27], the information on the chemical composition and antioxidant activity of the essential oil of *M. tricuspidata* fruit is still very poor. Furthermore, it is known that some volatile compounds in plants are present either in a free form and glycosidically bound forms to sugar moiety [28,29]. In some plants, glycosidically bound volatiles have shown a more potent antioxidant activity than essential oils [30,31]. Nevertheless, little is known about chemical constituents and their antioxidant potentials of glycosidically bound aglycones in *M. tricuspidata* fruit. Therefore, the objective of this study was to elucidate the chemical composition of steam-distilled essential oils (SDEO), aglycone fraction and major compounds of aglycone fraction liberated from glycosidically bound volatiles (GBAF) in *M. tricuspidata* fruit and their antioxidant potentials.

2. Materials and Methods

2.1. Reagents

n-Decanol, *n*-decyl-β-D-glucopyranoside, Amberlite XAD-2 polymeric resin (20–60 mesh), butylated hydroxyanisole (BHA), butylated hydroxy toluene (BHT), ascorbic acid, 2,2-diphenyl-1-picrylhydrazyl (DPPH), 2,2'-azino-bis(3-ethylbenzothiazoline-6-sulfonic acid) diammonium salt (ABTS), 2,4,6-tri(2-pyridyl)-*s*-triazine (TPTZ) and saturated *n*-alkanes mixture (C_7–C_{30}), were purchased from Sigma-Aldrich Corp. (St. Louis, MO, USA). Authentic volatile chemicals were purchased from commercial sources (Sigma-Aldrich and Wako Pure Chemical Industries, Ltd., Osaka, Japan). The other reagents used were of analytical grade and were purchased from commercial sources.

2.2. Plant Materials

M. tricuspidata fruits were collected in late October 2017 at a fully mature stage from plants cultivated in a farm located in Milyang district, Gyeongsangnam-do, Republic of Korea. A voucher specimen has been deposited at the Herbarium of Department of Food Science and Technology, College of Agricultural Life Science, Chonbuk National University. The fruit was freeze-dried for 4 day. The samples were powdered and stored in a freezer (−20 °C) until use.

2.3. Isolation of Steam-Distilled Essential Oil

A powdered sample (100 g) and distilled water (2 L) were placed in a 3 L round flask. The essential oil was isolated by means of simultaneous steam distillation and extraction at atmospheric pressure in a modified Likens–Nickerson type apparatus using n-pentane-diethyl ether (1:1) containing n-decanol (950 µg) as an internal standard for 2 h [32]. After the isolated oil was dried over anhydrous sodium sulfate for 12 h, the solvent was concentrated to a volume of 0.5 mL using a Vigreaux column at 40 °C and thereafter was evaporated off under a stream of nitrogen. The resulting residue was redissolved in 1 mL of n-pentane-diethyl ether (1:1) and subjected to gas chromatography (GC) and GC–mass spectrometry (GC–MS) analysis.

2.4. Isolation of Free Volatiles and Glycosidically Bound Volatiles

The powdered sample (100 g) was homogenized with 300 mL of methanol for 1 min in a Waring blender. The homogenate was centrifuged at 4500× g for 20 min. The residue was homogenized with 300 mL of methanol followed by centrifugation as above. The supernatant was combined and the solvent was concentrated to remove methanol under reduced pressure at 40 °C. The residue was dissolved in 100 mL of distilled water and was passed through a previously preactivated (with methanol) Amberlite XAD-2 (20–60 mesh) adsorbent column (5 × 35 cm) at a flow rate of 3 mL/min according to a previously reported method [33]. After the column was washed with 1.5 L of distilled water, free volatiles (FV) and glycosidically bound volatile (GBV) fraction was isolated by sequentially eluting with each 1 L of n-pentane:diethyl ether (1:1) and methanol, respectively. The FV fraction was dried over anhydrous sodium sulfate for 12 h and filtered through filter paper. The filtrate was concentrated to remove solvent under reduced pressure at 40 °C. The resulting residue was redissolved in 1 mL of n-pentane-diethyl ether (1:1). The methanol eluate designated as GBV was concentrated under reduced pressure to dryness at 40 °C. After residue was redissolved in 50 mL of 0.1 M citrate-phosphate buffer (pH 4.8), the aqueous layer was washed triplicate with each 50 mL of n-pentane:diethyl ether (1:1) to remove remaining free volatiles and added n-decyl-β-D-glucopyranoside (1900 µg) as an internal standard. The GBF was hydrolyzed by *Aspergillus niger* cellulase (80 mg, 24 U as β-glucosidase) at 37 °C for 36 h with gentle shaking. The liberated aglycones were isolated by liquid-liquid extraction using ethyl acetate (50 mL × 3). After the liberated glycosidically bound aglycone fractioin (GBAF) was dried over anhydrous sodium sulfate for 12 h, the solvent was evaporated using rotary evaporator at 40 °C. The resulting residue was dissolved in ethyl acetate. The extracts prepared were stored at −20 °C until use.

2.5. Gas Chromatography (GC) and GC–Mass Spectrometry (GC–MS) Analysis

GC analysis was performed on a Hewlett-Packard model 6890 series gas chromatograph, with a flame ionization detector (FID), a split ratio of 1:30 using Agilent J&W DB-5MS fused silica capillary column (30 m × 0.32 mm, i.d., 0.25 µm film thickness, Santa Clara, CA, USA) and Agilent J&W Supelcowax 10 fused silica capillary column (30 m × 0.32 mm, i.d., 0.25 µm film thickness). The column temperatures were programmed from 50 °C to 230 °C at 2 °C/min and then kept constant at 230 °C for 20 min. The injector and detector temperatures were 250 °C, respectively. The carrier gas was nitrogen, at a flow rate of 1.0 mL/min. Peak areas were measured by electronic integration and the concentrations of volatile compounds were expressed as n-decanol equivalent (assuming response factor of all analytes was 1.0). The concentrations are to be considered only relative values as recovery after extraction and calibration factors related to the standard were not determined [34,35].

The GC–MS analysis was performed on an Agilent Technologies 7890A GC and 5975C mass selective detector operating in the EI mode at 70 eV, fitted with a DB-5MS fused silica capillary column (30 m × 0.25, i.d., 0.25 µm film thickness) and Supelcowax 10 fused silica capillary column

(30 m × 0.32 mm, i.d., 0.25 µm film thickness), respectively. Both column temperatures were programmed from 50 °C to 230 °C at 2 °C per minute and then kept constant at 230 °C for 20 min. The injector and ion source temperatures were 250 °C. The carrier gas was helium at a flow rate of 1.0 mL/min. Identification of the compounds was achieved by comparing their retention times with those of authentic standards and mass spectral data in Wiley7n,1 database (Hewlett-Packard, Palo Alto, CA, USA), and NIST (National Institute of Standards and Technology, USDA) Webbook, and reported retention indices in the literatures [36]. Retention indices of each compound was calculated by a homologous series of saturated n-alkanes (C_7–C_{30}) (concentration of 1000 µg/mL in n-hexane) under the same conditions [37]. All compounds identified based on comparisons of only mass spectral data were listed as tentatively identified.

2.6. Determination of Total Phenolic Content

Total phenol content of the sample was measured according to the method described by Chandra et al. [38] with some modifications. Briefly, 20 µL of each fraction (at concentration of 1000 µg/1 mL methanol) was mixed with 50% Folin–Ciocalteu phenol reagent (20 µL) in 96-well plates. After 5 min, 1 N sodium carbonate solution (20 µL) was added to the mixture and distilled water was added to adjust the final volume to 200 µL. After incubation at room temperature (RT) in the dark for 30 min, the absorbance of test sample against a blank was measured at 725 nm using a VersaMax enzyme-linked immunosorbent assay (ELISA) microplate reader (Molecular Devices, LLC, San Jose, CA, USA). Total phenol content was calculated based on a calibration curve of gallic acid. The results were expressed as mg gallic acid equivalent (mg GAE)/g.

2.7. Antioxidant Activity

2.7.1. Preparation of Sample

The solvent in the test samples (SDEO, FV, GBV and GBAF) were removed under a nitrogen stream. The resulting residues were dissolved in n-pentane:diethyl ether (1:1). BHA, BHT and ascorbic acid all diluted to a concentration of 1000 µg per mL in methanol were used as positive controls for the antioxidant activity assays.

2.7.2. DPPH (2,2-Diphenyl-1-Picrylhydrazyl) Free Radical-Scavenging Activity

DPPH radical scavenging activity was determined according to the method described by Thaipong et al. [39] with some modifications. For calculation of effective concentration EC_{50} value, a stock solution of DPPH was freshly prepared by dissolving 240 mg DPPH in methanol (1000 mL) and the working solution was prepared by diluting stock solution with methanol to obtain an absorbance of 1.1 ± 0.02 units at 517 nm using an ultraviolet–visible (UV–vis) spectrophotometer (Shimadzu UV-1601, Osaka, Japan). 100 µL of the samples (SDEO, FV and TBAF) and chemicals were allowed to react with 0.1M Tris-HCl buffer (900 µL) and 500 µM DPPH solution (1000 µL) for 20 min at RT in the dark. Then absorbance was taken at 517 nm using UV–vis spectrophotometer. The EC_{50} (µg/mL) were calculated from the regression curves using six different concentrations (10–100 µg/mL) of samples and chemicals. The results were expressed as EC_{50} value (µg/mL). As a blank, the test was repeated using buffer instead of samples, and the DPPH radical-scavenging activity of the extracts was calculated against a blank as follows:

$$\text{DPPH radical-scavenging activity (\%)} = (1 - A_0/A_1) \times 100$$

where A_0 and A_1 are absorbance values of the test sample and control, respectively.

2.7.3. ABTS (2,2′-Azino-Bis(3-Ethylbenzothiazoline-6-Sulfonic Acid)) Free Radical-Scavenging Activity

ABTS free radical scavenging activity was determined by the methods of Thaipong et al. [39] with some modifications. Briefly, a mixture of ABTS (7.4 mM) solution and potassium persulfate (2.6 mM) solution in 1:1 ratio was kept at room temperature for 12 h under dark condition to form ABTS cation. The solution was diluted by adding methanol to obtain an absorbance of 1.1 ± 0.02 at 734 nm. All the required solutions were freshly prepared for each assay. 100 µL of the samples and chemicals were added to 1400 µL of the diluted ABTS solution and the mixture was incubated at room temperature for 2 h in a dark. After the reaction, its absorbance was measured at wavelength of 734 nm. The results were expressed as RC_{50} value (µg/mL), and also ABTS radical scavenging activity (%) was calculated with the following equation:

$$\text{ABTS radical scavenging activity (\%)} = (1 - A_0/A_1) \times 100$$

where A_0 and A_1 are absorbance values of the test sample and control, respectively.

2.7.4. Ferric-Reducing Antioxidant Power (FRAP)

Ferric-reducing power was determined using FRAP assay [40] with some modification. The FRAP reagent was prepared by mixing 10 volume of 300 mM acetate buffer (pH 3.6) with 1 volume of 10 mM TPTZ solution in 40 mM HCl and 1 volume of 20 mM ferric chloride solution. Sample extract (75 µL) was added to 1425 µL of FRAP reagent. The reaction mixture was then incubated at RT for 30 min in a dark. The reducing power was expressed as absorbance at 593 nm and RC_{50} values (µg/mL) of FRAP were calculated from the regression lines using six different concentrations (10–100 µg/mL) in triplicate.

2.8. Statistical Analysis

All experiments were conducted in triplicate unless otherwise indicated and the results were expressed as mean ± standard deviation (SD). The statistical analysis was conducted with SPSS (ver. 10.1) for Windows and a one-way analysis of variance (ANOVA). Duncan's multiple range tests were carried out to test any significant differences among various fruit maturity stages. Values with $p < 0.05$ were considered as significantly different

3. Results and Discussion

3.1. Chemical Composition of the Steam-Distilled Essential Oil (SDEO) Fraction

The yields of total SDEO and GBAF from *M. tricuspidata* fruit were 0.03 ± 0.01% and 0.37 ± 0.03%, respectively. Table 1 shows the volatile compounds identified in the SDEO and GBAF isolated from *M. tricuspidata* fruit along with their amounts and retention indices on DB-5MS (non-polar) and DB-WAX (polar) column. A total of 55 compounds including 17 tentatively identified compounds were identified in SDEO. The compounds that were found by only DB-5MS column but not by DB-WAX column were considered as tentatively identified. The compounds were 4 alcohols, 14 aldehyde and ketones, 7 terpenoids, 13 carotenoid-derived compounds, 6 aromatic and phenolic compounds, 11 acids and 3 miscellaneous. With the exception of aliphatic acids and their esters such as palmitic acid, linoleic acid, ethyl palmitate and linoleic acid, compounds with the highest concentration in the SDEO were p-cresol (393.50 ± 17.70), followed by δ-cadinene (147.67 ± 7.50), β-caryophyllene (145.67 ± 10.50), β-ionone (141.00 ± 4.40) and n-nonanal (140.33 ± 20.50 µg/g). In particular, 10 kinds of carotenoid-derived compounds were identified in the SDEO. These compounds have been found in various plants and are known to play an important role as characteristic aroma compounds of leaves, flowers or fruits of some plants [28,41,42]. Especially, theaspirane A and theaspirane B are present in green tea, black tea, grape and corn [43], and are believed to contribute to the unique aroma of *M. tricuspidata* fruit. Their chemical structures are presented in Figure 1.

Table 1. Concentration of compounds identified in steam-distilled essential oil (SDEO) and glycosidically bound aglycone fraction (GBAF) isolated from *M. tricuspidata* fruit.

PeakNo	tR (min)	Compounds	RI [1]	RI [2]	Concentration (μg/100 g dw) [3]	
					SDEO	GBAF
		Alcohols				
1	5.363	2-Methyl-1-butanol	737	1206	3.03 ± 0.25	1036.0 ± 124.6
5	7.735	*trans*-2-Hexen-1-ol	862	1405	7.33 ± 1.53	_[5]
8	10.318	5-Methyl-2-furfuryl alcohol	956	_[4]	3.17 ± 0.76	-
20	19.585	3,4-Dimethylcyclohexanol [6]	1109	-	15.67 ± 2.08	-
		Aldehydes and ketones				
3	6.644	Furfural	819	1459	53.67 ± 6.03	-
4	7.378	*trans*-2-Hexenal	848	1201	10.33 ± 3.06	-
2	6.284	*n*-Hexanal	804	1097	3.00 ± 0.80	-
7	8.777	2-Acetyl furan	903	1493	5.33 ± 1.53	-
9	10.513	5-Methylfufural	966	1508	4.13 ± 0.81	-
10	12.102	Benzaldehyde	971	1508	6.33 ± 1.53	-
11	12.491	6-Methyl-5-hepten-2-one	989	1326	6.33 ± 1.53	-
12	13.267	1-(2-Furanyl)-3-butanone [6]	1006	-	4.03 ± 0.55	-
16	15.179	Phenylacetaldehyde	1039	1629	44.33 ± 3.51	5.33 ± 1.04
18	18.958	*n*-Nonanal	1104	1388	140.3 ± 20.5	-
23	22.487	10-Undecenal [6]	1146	-	6.67 ± 2.52	-
24	23.306	2,4-Dimethylbenzaldehyde [6]	1158	1712	7.03 ± 1.55	-
44	41.734	Genanyl acetone	1451	1860	17.33 ± 2.52	-
52	44.505	2-Tridecanone	1493	-	23.67 ± 5.51	-
		Terpenoids				
36	35.657	Ylangene	1356	1464	10.93 ± 3.10	-
37	36.379	α-Copaene	1368	1477	62.33 ± 51.47	-
41	39.05	β-Caryophyllene	1409	1565	145.7 ± 10.5	-
43	39.533	α-Bergamotene	1416	1575	5.67 ± 0.58	-
45	41.982	β-Humulene	1454	-	10.33 ± 2.52	-
53	45.905	δ-Cadinene	1517	1754	147.7 ± 7.5	-
58	50.101	Caryophyllene oxide	1588	1968	56.33 ± 3.51	-

Table 1. Cont.

PeakNo	tR (min)	Compounds	RI [1]	RI [2]	Concentration (μg/100 g dw) [3] SDEO	Concentration (μg/100 g dw) [3] GBAF
		Carotenoid-derived compounds				
14	15.079	2,2,6-Trimethylcyclohexanone [6]	1037	1300	5.57 ± 0.51	-
19	19.303	Isophorone	1119	1578	7.10 ± 1.85	-
22	21.727	4-Oxoisophorone [6]	1115	1674	5.33 ± 0.58	-
28	26.163	β-Cyclocitral	1214	1603	17.10 ± 1.85	-
29	28.870	β-Homocyclocitral [6]	1254	-	15.10 ± 0.85	-
31	31.253	Theaspirane A	1289	1482	121.3 ± 4.5	-
33	32.447	Theaspirane B	1306	1522	99.67 ± 9.02	-
42	39.454	7,8-Dihydro-α-ionone [6]	1415	1825	-	30.33 ± 2.52
48	43.389	β-Ionone	1480	1907	141.0 ± 4.4	-
49	43.637	β-Ionone epoxide	1483	1957	92.33 ± 9.71	-
55	46.692	Dihydroactinidiolide [6]	1530	2291	10.67 ± 5.51	-
59	56.486	3-Hydroxy-β-ionone [6]	1698	2646	-	160.7 ± 30.0
60	57.969	9-Hydroxymegastigma-4,6-dien-3-one (isomer #1) [6]	1705	2677	-	197.67 ± 9.45
61	58.525	4-Oxo-7,8-dihydro-β-ionol	1725	2694	-	76.00 ± 11.00
63	61.311	9-Hydroxymegastigma-4,6-dien-3-one (isomer #2) [6]	1786	2846	-	234.3 ± 24.5
		Aromatic and phenolic compounds				
15	15.292	Benzyl alcohol	1040	1864	-	883.7 ± 29.8
17	18.294	p-Cresol	1092	2074	393.5 ± 17.7	43.00 ± 7.55
21	19.694	2-Phenylethyl alcohol	1113	1892	-	58.85 ± 4.58
26	25.427	Pyrocatechol [7]	1203	2646	-	20.33 ± 5.51
30	31.225	Resorcinol	1288	-	-	57.33 ± 10.50
32	31.523	Carvacrol	1293	2213	19.37 ± 3.46	-
34	34.006	α-Methoxy-p-cresol [7]	1331	2490	-	2783.0 ± 143.0
35	34.981	p-Vinylguaiacol	1346	2181	-	17.33 ± 3.51
25	24.925	Methyl chavicol	1171	1658	66.67 ± 9.02	-
38	37.539	2,4,6-Trihydroxybenzaldehyde	1386	-	9.33 ± 1.53	-
39	38.473	p-Hydroxybenzyl alcohol [7]	1400	2952	17.67 ± 3.06	468.1 ± 30.9
40	38.977	p-Hydroxybenzaldehyde [7]	1408	2964	-	170.0 ± 19.5
46	42.529	Tyrosol [7]	1463	2969	-	68.67 ± 4.51
47	43.524	p-Methylsalicylaldehyde [7]	1478	-	43.00 ± 10.82	4088.0 ± 147.8
50	44.116	Methyl p-hydroxybenzoate [7]	1487	1969	-	289.3 ± 12.5
51	44.439	Vanillyl alcohol [7]	1492	-	-	30.67 ± 3.27
54	46.293	p-Hydroxybenzoic acid [7]	1523	-	-	20.33 ± 4.51

Table 1. *Cont.*

PeakNo	tR (min)	Compounds	RI [1]	RI [2]	Concentration (μg/100 g dw) [3]	
					SDEO	GBAF
56	46.955	Methyl caffeate [7]	1532	2593	-	31.33 ± 4.51
57	48.027	Vanillic acid [7]	1583	-	-	22.67 ± 4.04
65	64.199	Methyl ferulate	1844	-	-	92.00 ± 28.62
66	65.320	Ferulic acid [7]	1865	-	-	383.0 ± 26.6
76	79.425	*p*-(*p*-Hydroxybenzyl)phenol [6]	2166	-	-	133.1 ± 12.9
Aliphatic acids and esters						
62	61.180	Myristic acid	1775	2694	124.2 ± 10.3	-
64	61.871	Ethyl myristate	1798	2041	9.33 ± 1.53	-
67	65.807	Pentadecanoic acid	1875	2822	9.33 ± 2.52	-
68	68.480	Methyl palmitate	1928	2212	55.75 ± 6.23	-
69	71.900	Palmitic acid	1986	2953	813.1 ± 39.5	-
70	72.100	Ethyl palmitate	2002	2277	291.7 ± 29.0	-
71	76.479	Methyl linoleate	2120	2485	58.16 ± 8.23	-
72	76.776	Methyl linolenate	2119	2484	55.35 ± 10.53	-
73	79.580	Linoleic acid	2169	-	363.7 ± 39.0	-
74	79.897	Linolenic acid	2175	-	176.0 ± 22.5	-
75	80.583	Ethyl linolenate	2187	2585	9.33 ± 1.53	-
Miscellaneous						
6	7.987	*p*-Xylene	836	1279	4.17 ± 0.76	-
13	14.526	2-Acetylthiazole [6]	1027	-	4.03 ± 0.35	-
27	25.537	2,3-Dihydrobenzofuran	1205	2381	3.77 ± 0.68	316.0 ± 29.0

[1] Retention indices on DB-5MS column. [2] Retention indices on Suplecowax 10 column. [3] Values expressed as equivalents of *n*-decanol are given as mean ± standard deviation (*n* = 3). [4] Not detected or larger retention indices than 3000 in Suplecowax 10 column. [5] Not detected or less than 1.0 μg/100 g. [6] Tentatively identified based on mass spectral data only due to lack of authentic standard compound. [7] Compounds used for antioxidant activity assays.

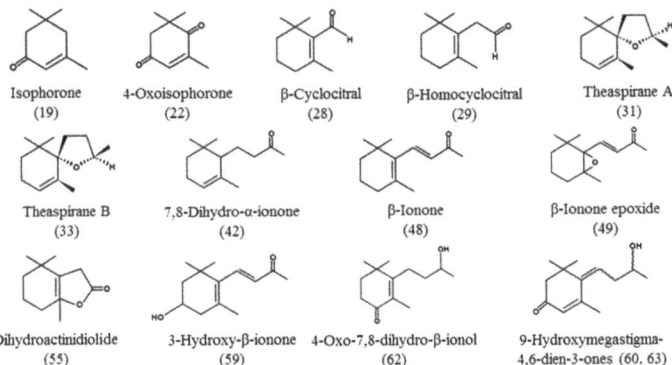

Figure 1. Chemical structures of carotenoid-derived compounds identified in steam-distilled essential oil (SDEO) and glycosidically bound aglycone fraction (GBAF) isolated from *M. tricuspidata* fruit. Numbers in brackets indicate peak numbers as listed in Table 1.

In this study, the norisoprenoid compounds, 7,8-dihydro-α-ionone, 3-hydroxy-β-ionone, 4-oxo-7,8-dihydro-β-ionol and 9-hydroxymegastigma-4,6-dien-3-one (two isomers) were not detected in fractions separated by the steam-distillation and extraction (SDE) method but in the glycosidicaly bound volatiles fraction (GBAF). These results suggest that most of the norisoprenoid compounds detected in the fruit are present in the form of glycosidic form rather than existing in free form in the maturing fruit or being formed in the process of preserving the fruits after harvesting [44]. These compounds can be derived from carotenoids by the action of related enzymes or chemical oxidation during processing or storage of *M. tricuspidata* fruit. It is considered that the carotenoid is decomposed in the process of separating volatile components by steam distillation. In particular, 3-hydroxy-β-ionone, 3,4-dihydro-α-ionone and two quantitatively detected 9-hydroxymegastigma-4,6-dien-3-one are present in glycosidic form in some plants [45,46]. In our previous study that analyzed phenolic compounds in the methanol extract of a fully matured fruit of the plant, we isolated a number of phenolic compounds including quercetin and parishin derivatives [25]. In this study, only 4-Hydroxybenzyl alcohol was able to be detected at a significant concentration suggesting most of the other phenolic compounds must have been degraded during the steam-distillation process.

To the best of our knowledge, 13 carotenoid-derived compounds (isophorone, 4-oxoisophorone, theaspiranes A, theaspiranes B, 7,8-dihydro-α-ionone, β-ionone, β-ionone epoxide, dihydroactinidiolide, 3-hydroxy-β-ionone, β-cyclocitral, β-homocyclocitral and two 9-hydroxymegastigma-4, 6-dien-3-one isomers) are being identified for the first time from *M. tricuspidata* fruit oil. These compounds are related to carotenoids [44]. *M. tricuspidata* fruit contains several carotenoids including α-carotene, β-carotene, zeaxanthin, ruboxanthin, and lutein [47]. As described in the introduction section above, Bajpai and colleagues have previously identified 29 compounds with 1,1-difluoro-4-vinylspiropentane, scyllitol, 1-phenyl-1-cyclohexylethane, diethyl phthalate and 4,4-diphenyl-5-methyl-2-cyclohexenone as major constituents in the essential oil obtained from *M. tricuspidata* fruit by microwave-assisted extraction [26]. However, most of these compounds were not detected in this study. We believe that the difference in detected components is caused by the difference in extraction method and plant samples. In the present study, we used a fresh fruit instead of a dried one.

3.2. Chemical Composition of Glycosidically Bound Aglycone Fraction (GBAF)

It is well established that volatile components in plants and foods are present in free form while some components exist in glycosidically bound forms [41,48,49]. The volatile components in the form of glycoside in association with saccharides have a hydroxyl group in the molecule and are bonded in the form of a β-glycoside. These glycosides can be hydrolyzed by β-glycosidases produced by microorganisms to produce free form of volatiles [28,33,41]. The enzyme preparation used for such a purpose are enzymes with glycosidase activities such as β-D-glucosidase, α-L-arabinopyranosidase, α-L-arabinofuranosidase and α-L-rhanosidase.

In this experiment, GBV fractions were isolated from an Amberlite XAD-2 column and then *Asp. niger* cellulase was used to release aglycones from their conjugates. Compared with the gas chromatograms of the volatile components separated by the SDE method, the number of components detected in the GBV fraction (Supplementary Figure S1) was smaller. However, it can be clearly seen that the intensities of the peaks are significantly higher in the GBV fraction. These results indicate that the overall compositions of the volatile components constituting the GBV fraction are clearly different from the volatile components present in the free form. Identities of individual compounds identified in the SDEO and GBAF are presented in Table 1.

Regarding aldehydes and ketones which belong to the oxygenated compounds, 14 components were detected in the volatile components fraction separated by the SDE method while only a small amount of phenylacetaldehyde was detected in the GBAF (Table 1). These results suggest that aldehydes and ketones present in fruits are not combined with saccharides in the form of glycosides.

In the volatile fractions separated by the SDE method, few aromatic alcohol and phenolic compounds including constituents such as *p*-cresol, estragole, 2-methyl-5-(1-methylethyl) phenol, methoxy-2-methylphenol, 2,4,6-trimethylbenzaldehyde and 2-hydroxy-4-methylbenzaldehyde were detected in lower concentrations. By contrast, in the GBAF fraction, a large amount of aromatic alcohols and phenolic compounds were detected. Among them, benzyl alcohol, 2-phenylethyl alcohol, resorcinol, α-methoxy-*p*-cresol, *p*-hydroxybenzyl alcohol, *p*-hydroxybenzaldehyde, 4-methylsalicylaldehyde, methyl *p*-hydroxybenzoate, ferulic acid, methyl caffeate, pyrocatechol, *p*-hydroxyphenylethyl alcohol, vanillyl alcohol, *p*-hydroxybenzoic acid, methyl vanillate, vanillic acid, and *p*-(*p*-hydroxybenzyl) phenol were detected only in the GBAF (Table 1). The chemical structures of the phenolic compounds detected in the GBAF are shown in Figure 2. As shown in the figure, one or more hydroxyl groups are contained in the molecular structure, and thus the β-glycoside bond is hydrolyzed by treating β-glucosidase in the presence of sugar in the form of β-glycoside in the hydroxyl group. These compounds are smaller in molecular weight and simple in structure compared to other phenolic compounds, but are widely distributed in plants and are known to contribute to various physiological activities. Interesting biological activities have been reported for tyrosol, *p*-hydroxybenzyl alcohol and *p*-hydroxybenzaldehyde including anti-oxidant activities, improving functional blood flow, preventing memory deficits, and providing protective effects on the blood–brain barrier [50–53].

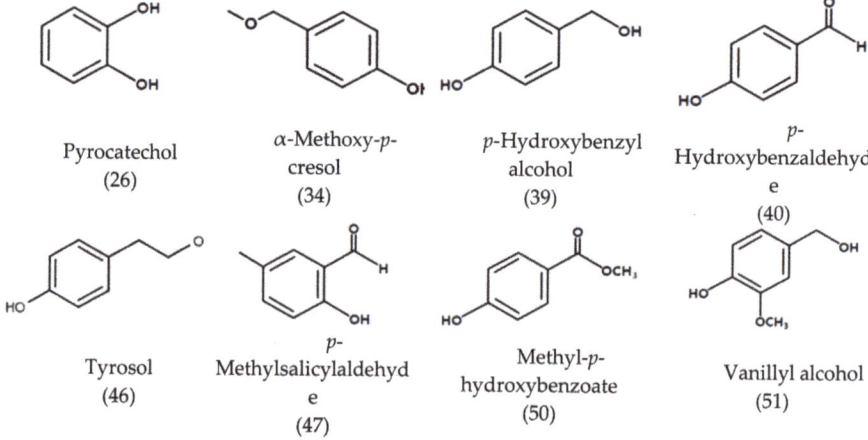

Figure 2. *Cont.*

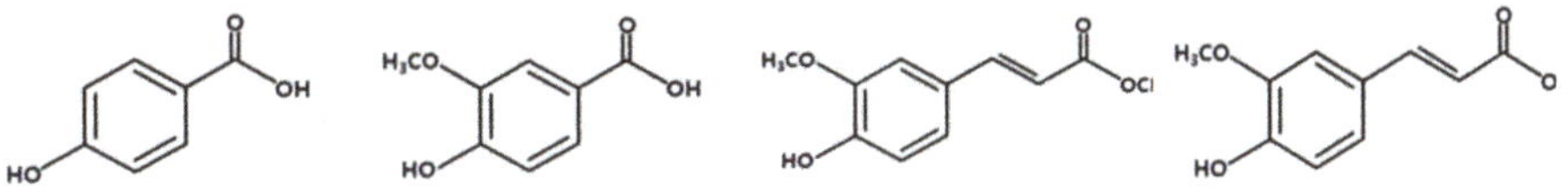

p-Hydroxybenzoic acid (54)   Vanillic acid (57)   Methyl caffeate (65)   Ferulic acid (66)

Figure 2. Chemical structures of aromatic and phenolic compounds identified in glycosidically bound aglycone fraction (GBAF) isolated from *M. tricuspidata* fruit. Numbers in brackets indicate peak numbers as listed in Table 1.

3.3. Total Phenol Contents of Fractions

The total phenol contents of the SDEO, FV and GBAF were also determined and comparisons of the results are presented in Figure 3. Among all, the highest total phenol content was obtained from the GBAF while the SDEO showed the lowest total phenol content (<10 mg/g dw). The total phenol content of the FV fraction was slightly lower than the GBAF while it was much higher than that of SDEO. The relatively higher total phenol contents in the GBAF and FV is due to the solvents used as the efficiency of the phenolics extraction depends on the type of the solvent. During isolation of the GBAF, extraction of the aglycones liberated by enzymatic hydrolysis employed a more polar solvent (ethyl acetate) while only *n*-pentane-diethyl ether (1:1) was used in the case of SDEO. It is well established that phenolic compounds are extracted more efficiently with polar solvents [54].

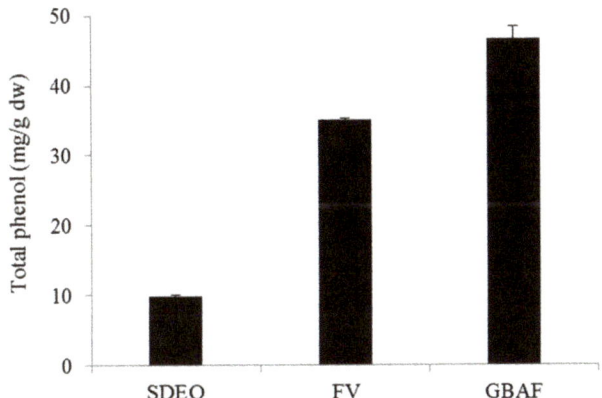

Figure 3. Total phenol contents of fractions isolated from *M. tricuspidata* fruit. SDEO, steam-distilled essential oil; FV, free volatile; GBAF, glycosidically bound aglycone fraction liberated from GBV by *Asp. niger* cellulose; GBV, glycosidically bound volatile fraction.

3.4. Antioxidant Activity of SDEO and GBAF

Antioxidant activities of fruit extracts have been characterized extensively [55]. In this study, antioxidant capacities of each fraction expressed in percent of radical (DPPH and ABTS) scavenging activities and reducing power as measured by FRAP assay, and EC_{50} as compared to the positive controls BHA and BHT, are presented in Figure 4 and Table 2. In all the antioxidant property measurement methods, the GBAF showed the highest antioxidant activity while the SDEO showed the lowest. Considering the total yields of these fractions and their respective total phenol content results described above, it can be said that there is a strong positive correlation between their concentrations and their respective antioxidant activities. Maximum antioxidant activities of the GBAF were obtained in the DPPH and FRAP methods where its activity was even higher or equivalent to those of the synthetic antioxidants BHA and BHT. While the antioxidant properties of phenolic compounds are extensively demonstrated in the literature, some of the volatile compounds exclusively detected in the GBAF might also have greatly contributed to its considerable antioxidant capacity observed in this study. It should also be noticed that the volatile aroma components detected in higher concentrations in the GBAF including p-Hydroxybenzyl alcohol, p-hydroxybenzaldehyde and tyrosol are well known to have strong biological activities [56–58]. However, while antioxidant activity estimations based on synthetic radicals are indispensable tools, many people raise concerns about their substantiation through in vivo and clinical trials which also have more safety issues [59].

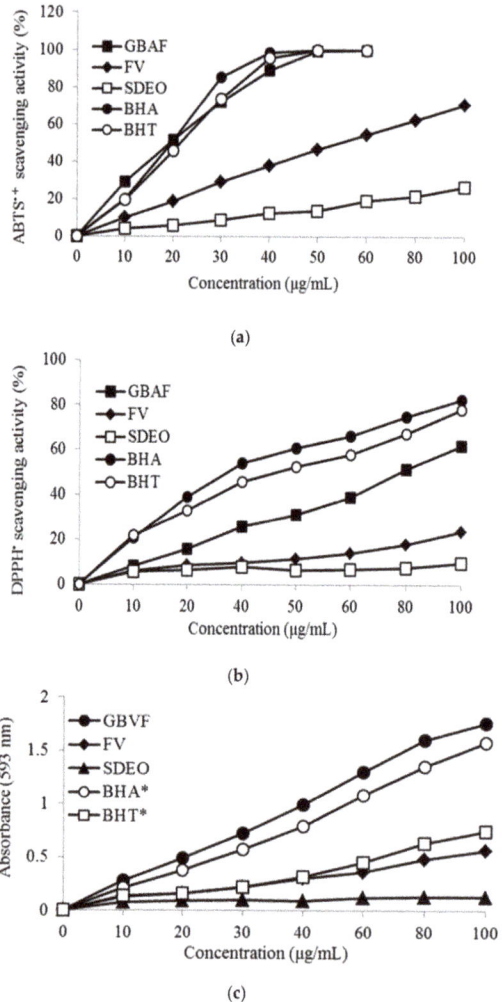

Figure 4. Antioxidant activities of steam-distilled essential oil (SDEO), free volatile (FV) and glycosidically bound aglycone fraction (GBAF) isolated from *M. tricuspidata* fruit. (**a**) 2,2-Diphenyl-1-Picrylhydrazyl (DPPH) free radical scavenging activity, (**b**) 2,2′-Azino-Bis(3-Ethylbenzothiazoline-6- Sulfonic Acid (ABTS) free radical scavenging activity; (**c**) Ferric reducing antioxidant power (FRAP). Samples, 1000 ug/mL; * Butylated hydroxyltoluene BHA, Butylated hydroxyanisole (BHT), 200 ug/mL.

Table 2. Antioxidant activity of SDEO, FV and GBAF isolated from *M. tricuspidata* fruit.

Samples	DPPH [1]	ABTS [+1]	FRAP [2]
SDEO	17,065.22 ± 146.27 [a]	1921.81 ± 49.45 [a]	10,638.56 ± 223.33 [a]
FV	2507.18 ± 24.21 [b]	660.72 ± 7.18 [b]	1963.48 ± 10.97 [b]
GBAF	835.33 ± 6.97 [d]	317.09 ± 1.99 [d]	529.6 ± 4.73 [d]
BHA	466.79 ± 7.10 [e]	89.15 ± 4.14 [e]	129.46 ± 1.61 [f]
BHT	535.75 ± 3.52 [e]	108.62 ± 1.06 [e]	331.26 ± 4.68 [e]

[1] EC_{50} (µg/mL) values were calculated from the regression lines using six different concentrations (10–100 µg/mL) in triplicate and data represent 50% scavenging activity. [2] Ferric-reducing antioxidant power (FRAP) were calculated from the regression lines using six different concentrations (10–100 µg/mL in triplicate and the values were presented by sample concentration at 0.5 of absorbance at 517 nm DPPH, 2,2-Diphenyl-1-Picrylhydrazyl; ABTS, 2,2′-Azino-Bis(3-Ethylbenzothiazoline-6-Sulfonic Acid; SDEO, steam-distilled essential oil; FV, free volatile; GBAF, glycosidically bound aglycone fraction liberated from glycosidically bound volatile fraction by *Asp. niger* cellulase. Different superscripts in the same column indicate significant differences ($p < 0.05$). [+], cation.

Even though antioxidant activity of the SDEO was found to be much lower than the other fractions, it is suggested that its observed antioxidant property is related to the compounds detected in it. Compounds such as palmitic acid, linoleic acid and *p*-cresol that were detected in relatively higher concentrations in the SDEO are not important antioxidants [60,61]. Generally, the antioxidant capacity of volatile compound fractions from *M. triscuspidata* fruit extracted with the SDE method and that of GBAF are attributed to the individual components identified. The antioxidant activities expressed in EC_{50} of some individual phenolic compounds evaluated in this study are also presented in Table 3. Based on these results, it can be suggested that as most potent bioactive compounds are glycosidically bound forms in *M. triscuspidata* fruit, enzymatic processing like fermentation can play an important role in enhancing its biological activities.

Table 3. Antioxidant activity of phenolic compounds identified in GBAF.

Compounds	EC_{50} (µg/mL)		
	DPPH [1]	ABTS [1]	FRAP [2]
Pyrocatechol	9.59 ± 1.22 [e]	67.68 ± 2.47 [jk]	74.45 ± 2.16 [jk]
α-Methoxy-*p*-cresol	1114.09 ± 114.45 [d]	59.55 ± 6.46 [jk]	3298.92 ± 126.20 [f]
p-Hydroxybenzyl alcohol	3357.55 ± 134.15 [c]	377.85 ± 4.78 [f]	2854.37 ± 43.04 [g]
p-Hydroxybenzaldehyde	1765.90 ± 364.23 [d]	1117.70 ± 7.01 [c]	7906.18 ± 60.96 [c]
Tyrosol	1331.74 ± 195.63 [d]	287.36 ± 3.70 [g]	92.64 ± 1.97 [jk]
p-Methylsalicylaldehyde	1644.14 ± 365.52 [d]	423.69 ± 3.13 [e]	19,365.27 ± 81.38 [b]
Methyl *p*-hydroxybenzoate	5241.03 ± 941.54 [b]	12,735.03 ± 47.26 [a]	6789.61 ± 82.27 [d]
Vanillyl alcohol	27.96 ± 1.65 [e]	66.98 ± 1.99 [jk]	5928.60 ± 90.87 [e]
p-Hydroxybenzoic acid	10,906.51 ± 1103.69 [a]	6921.86 ± 50.48 [b]	1116.61 ± 11.69 [h]
Vanillic acid	48.58 ± 2.50 [e]	157.22 ± 4.83 [h]	161.18 ± 4.25 [jk]
Methyl caffeate	11.92 ± 0.48 [e]	11.91 ± 1.29 [l]	7.84 ± 0.28 [k]
Ferulic acid	24.47 ± 2.59 [e]	66.39 ± 2.11 [jk]	138.98 ± 3.73 [jk]
BHA	26.10 ± 0.42 [e]	89.27 ± 4.01 [ij]	129.46 ± 1.61 [jk]
BHT	33.71 ± 1.04 [e]	108.76 ± 3.93 [i]	331.26 ± 4.68 [j]

[1] EC_{50} (µg/mL) values were calculated from the regression lines using six different concentrations (10–100 µg/mL) in triplicate and data represent 50% scavenging activity. [2] FRAP were calculated from the regression lines curve using six different concentrations (10–100 µg/mL) of authentic standards in triplicate and the values were presented by sample concentration at 0.5 of absorbance at 517 nm. Different superscripts in the same column indicate significant differences ($p < 0.05$).

3.5. Antioxidant Activity of Individual Phenolic Compounds in GBAF

In order to evaluate the antioxidant activities of individual compounds, EC_{50} of 12 compounds identified in the GBAF was determined and the results are presented in Table 3. In all the three assay methods, methyl caffeate displayed by far the strongest antioxidant activity expressed in EC_{50}. Pyrocatechol also showed the highest DPPH scavenging activity and was even higher than the synthetic antioxidants BHA and BHT. As can be seen from Table 3, several phenolic compounds including pyrocatechol, vanillyl alcohol, methyl caffeate and ferulic acid have shown antioxidant potencies higher than that of positive controls. Ferulic acid and methyl caffeate, the two compounds that showed the highest DPPH-scavenging activities in this study, have been previously reported to have antioxidant activities expressed in EC_{50} of DPPH scavenging activity of 22 and 10.64 µg/mL for ferulic acid methyl caffeate, respectively [50,62].

Therefore, it can be assumed that these compounds have greatly contributed to the overall higher antioxidant activity observed in the GBAF. As described above, only a few phenolic compounds were detected in lower concentrations in the SDEO fraction. Considering this, proper processing techniques are required before application of *M. triscuspidata* fruit for its biological activity. While processing techniques such as specific enzymatic treatments can help release some compounds, processing methods like fermentation with microorganisms may give more efficient results. A previous study has demonstrated an increase in the levels of phenolic compounds such as kaempferol and quercetin after lactobacillus-mediated fermentation of *M. triscuspidata* leaf [63].

4. Conclusions

This study explored the chemical compositions and antioxidant activities of steam-distilled essential oil (SDEO) and glycosidically bound aglycone fraction (GBAF) extracts from fully ripe *M. triscuspidata* fruit. Thirteen carotenoid-derived compounds are being isolated for the first time in *M. triscuspidata* fruit. These compounds have been associated with a variety of organoleptic properties in other plants. A number of bioactive compounds were exclusively identified in the GBAF. It can be suggested that the relatively higher antioxidant activity observed in this particular fraction compared to the SDEO fraction is mainly associated with these exclusive compounds. Therefore, enzymatic treatments of fruits suh as *M. triscuspidata* can significantly enhance functional properties by releasing glycosidically bound bioactive components.

Supplementary Materials: The following are available online at http://www.mdpi.com/2304-8158/8/12/659/s1, Figure S1: Gas chromatograms of the volatile components detected in (A) steam distilled essential oil (SDEO); (B) glycosidically bound aglycone fraction (GBAF) isolated from Maclura triscuspidata.

Author Contributions: Conceptualization & Methodology, M.-K.K. and Y.-H.K.; Investigation & Resources, M.-K.K. and Y.-H.K.; Funding Acquisition, M.-K.K.; Formal analysis, G.-R.Y., D.-W.K., D.-H.K. and H.-A.H.; Writing—Original Draft Preparation, Y.-H.K.; Writing—Review and Editing, Y.A.G. All authors read and approved the final manuscript.

Funding: This work was carried out with the support of the Cooperative Research Program for Agricultural Science & Technology Development, grant number PJ012588022019, National Academy of Agricultural Science, Rural Development Administration, Republic of Korea.

Conflicts of Interest: The authors declare no conflict of interest. The funders had no role in the design of the study; in the collection, analyses, or interpretation of data; in the writing of the manuscript; or in the decision to publish the results.

References

1. Raut, J.S.; Karuppayil, S.M. A status review on the medicinal properties of essential oils. *Ind. Crops Prod.* **2014**, *62*, 250–264. [CrossRef]
2. Shaaban, H.A.E.; El-Ghorab, A.H.; Shibamoto, T. Bioactivity of essential oils and their volatile aroma components: Review. *J. Essent. Oil Res.* **2012**, *24*, 203–212. [CrossRef]
3. Sendra, E. Essential Oils in Foods: From Ancient Times to the 21st Century. *Foods* **2016**, *5*, 43. [CrossRef]
4. Edris, A.E. Pharmaceutical and therapeutic Potentials of essential oils and their individual volatile constituents: A review. *Phyther. Res.* **2007**, *21*, 308–323. [CrossRef]
5. Adorjan, B.; Buchbauer, G. Biological properties of essential oils: An updated review. *Flavour Fragr. J.* **2010**, *25*, 407–426. [CrossRef]
6. Wang, H.F.; Yih, K.H.; Yang, C.H.; Huang, K.F. Anti-oxidant activity and major chemical component analyses of twenty-six commercially available essential oils. *J. Food Drug Anal.* **2017**, *25*, 881–889. [CrossRef]

7. Alamed, J.; Chaiyasit, W.; McClements, D.J.; Decker, E.A. Relationships between Free Radical Scavenging and Antioxidant Activity in Foods. *J. Agric. Food Chem.* **2009**, *57*, 2969–2976. [CrossRef]
8. Williams, G.M.; Iatropoulos, M.J.; Whysner, J. Safety assessment of butylated hydroxyanisole and butylated hydroxytoluene as antioxidant food additives. *Food Chem. Toxicol.* **1999**, *37*, 1027–1038. [CrossRef]
9. Bakkali, F.; Averbeck, S.; Averbeck, D.; Idaomar, M. Biological effects of essential oils—A review. *Food Chem. Toxicol.* **2008**, *46*, 446–475. [CrossRef]
10. McClements, D.J.; Decker, E.A.; Weiss, J. Emulsion-Based Delivery Systems for Lipophilic Bioactive Components. *J. Food Sci.* **2007**, *72*, R109–R124. [CrossRef]
11. Shahidi, F.; Ambigaipalan, P. Phenolics and polyphenolics in foods, beverages and spices: Antioxidant activity and health effects—A review. *J. Funct. Foods* **2015**, *18*, 820–897. [CrossRef]
12. Hiep, N.T.; Kwon, J.; Kim, D.W.; Hong, S.; Guo, Y.; Hwang, B.Y.; Kim, N.; Mar, W.; Lee, D. Neuroprotective constituents from the fruits of Maclura tricuspidata. *Tetrahedron* **2017**, *73*, 2747–2759. [CrossRef]
13. Seo, W.G.; Pae, H.O.; Oh, G.S.; Chai, K.Y.; Yun, Y.G.; Chung, H.T.; Jang, K.K.; Kwon, T.O. Ethyl acetate extract of the stem bark of cudrania tricuspidata induces apoptosis in human leukemia HL-60 cells. *Am. J. Chin. Med.* **2001**, *29*, 313–320. [CrossRef]
14. Kwon, S.-B.; Kim, M.-J.; Yang, J.M.; Lee, H.P.; Hong, J.T.; Jeong, H.S.; Kim, E.S.; Yoon, D.Y. Cudrania tricuspidata Stem Extract Induces Apoptosis via the Extrinsic Pathway in SiHa Cervical Cancer Cells. *PLoS ONE* **2016**, *11*, e0150235. [CrossRef]
15. Chang, S.H.; Jung, E.J.; Lim, D.G.; Oyungerel, B.; Lim, K.I.; Her, E.; Choi, W.S.; Jun, M.H.; Choi, K.D.; Han, D.J.; et al. Anti-inflammatory action of *Cudrania tricuspidata* on spleen cell and T lymphocyte proliferation. *J. Pharm. Pharmacol.* **2008**, *60*, 1221–1226. [CrossRef]
16. Kang, D.-H.; Kim, J.-W.; Youn, K.-S. Antioxidant Activities of Extracts from Fermented Mulberry (*Cudrania tricuspidata*) Fruit and Inhibitory Actions on Elastase and Tyrosinase. *Korean J. Food Preserv.* **2011**, *18*, 236–243. [CrossRef]
17. Jeong, C.-H.; Choi, G.-N.; Kim, J.-H.; Kwak, J.-H.; Heo, H.-J.; Shim, K.-H.; Cho, B.-R.; Bae, Y.-I.; Choi, J.-S. In vitro Antioxidative Activities and Phenolic Composition of Hot Water Extract from Different Parts of *Cudrania tricuspidata*. *Prev. Nutr. Food Sci.* **2009**, *14*, 283–289. [CrossRef]
18. Kim, D.H.; Lee, S.; Chung, Y.W.; Kim, B.M.; Kim, H.; Kim, K.; Yang, K.M. Antiobesity and Antidiabetes Effects of a *Cudrania tricuspidata* Hydrophilic Extract Presenting PTP1B Inhibitory Potential. *Biomed Res. Int.* **2016**. [CrossRef]
19. Han, X.H.; Hong, S.S.; Jin, Q.; Li, D.; Kim, H.K.; Lee, J.; Kwon, S.H.; Lee, D.; Lee, C.K.; Lee, M.K.; et al. Prenylated and Benzylated Flavonoids from the Fruits of *Cudrania tricuspidata*. *J. Nat. Prod.* **2009**, *72*, 164–167. [CrossRef]
20. Hwang, J.H.; Hong, S.S.; Han, X.H.; Hwang, J.S.; Lee, D.; Lee, H.; Yun, Y.P.; Kim, Y.; Ro, J.S.; Hwang, B.Y. Prenylated Xanthones from the Root Bark of *Cudrania tricuspidata*. *J. Nat. Prod.* **2007**, *70*, 1207–1209. [CrossRef]
21. Xin, L.T.; Yue, S.J.; Fan, Y.C.; Wu, J.S.; Yan, D.; Guan, H.S.; Wang, C.Y. *Cudrania tricuspidata*: An updated review on ethnomedicine, phytochemistry and pharmacology. *RSC Adv.* **2017**, *7*, 31807–31832. [CrossRef]
22. Lee, Y.J.; Kim, S.; Lee, S.J.; Ham, I.; Whang, W.K. Antioxidant activities of new flavonoids from *Cudrania tricuspidata* root bark. *Arch. Pharm. Res.* **2009**, *32*, 195–200. [CrossRef]
23. Song, S.-H.; Ki, S.; Park, D.-H.; Moon, H.S.; Lee, C.D.; Yoon, I.S.; Cho, S.S. Quantitative Analysis, Extraction Optimization, and Biological Evaluation of *Cudrania tricuspidata* Leaf and Fruit Extracts. *Molecules* **2017**, *22*, 1489. [CrossRef]
24. Shin, G.R.; Lee, S.; Lee, S.; Do, S.G.; Shin, E.; Lee, C.H. Maturity stage-specific metabolite profiling of *Cudrania tricuspidata* and its correlation with antioxidant activity. *Ind. Crops Prod.* **2015**, *70*, 322–331. [CrossRef]
25. Kim, D.-W.; Lee, W.-J.; Asmelash Gebru, Y.; Choi, H.S.; Yeo, S.H.; Jeong, Y.J.; Kim, S.; Kim, Y.H.; Kim, M.K. Comparison of Bioactive Compounds and Antioxidant Activities of Maclura tricuspidata Fruit Extracts at Different Maturity Stages. *Molecules* **2019**, *24*, 567. [CrossRef]
26. Bajpai, V.K.; Sharma, A.; Baek, K.H. Antibacterial mode of action of *Cudrania tricuspidata* fruit essential oil, affecting membrane permeability and surface characteristics of food-borne pathogens. *Food Control* **2013**, *32*, 582–590. [CrossRef]
27. Bajpai, V.K.; Baek, K. Antioxidant efficacy, lipid peroxidation inhibition and phenolic content of essential oil of fruits of cudrania tricuspidata. *Bangladesh J. Bot.* **2017**, *46*, 1015–102046.
28. Stahl-Biskup, E.; Intert, F.; Holthuijzen, J.; Stengele, M.; Schulz, G. Glycosidically bound volatiles—A review 1986–1991. *Flavour Fragr. J.* **1993**, *8*, 61–80. [CrossRef]

29. Winterhalter, P.; Skouroumounis, G.K. Glycoconjugated aroma compounds: Occurrence, role and biotechnological transformation. *Adv. Biochem. Eng. Biotechnol.* **1997**, *55*, 73–105. [CrossRef]
30. Politeo, O.; Jukic, M.; Milos, M. Chemical composition and antioxidant capacity of free volatile aglycones from basil (*Ocimum basilicum* L.) compared with its essential oil. *Food Chem.* **2007**, *101*, 379–385. [CrossRef]
31. Maric, S.; Jukic, M.; Katalinic, V.; Milos, M. Comparison of Chemical Composition and Free Radical Scavenging Ability of Glycosidically Bound and Free Volatiles from Bosnian Pine (Pinus heldreichii Christ. var. leucodermis). *Molecules* **2007**, *12*, 283–289. [CrossRef]
32. Schultz, T.H.; Flath, R.A.; Mon, T.R.; Eggling, S.B.; Teranishi, R. Isolation of Volatile Components from a Model System. *J. Agric. Food Chem.* **1977**, *25*, 446–449. [CrossRef]
33. Gunata, Y.Z.; Bayonove, C.L.; Baumes, R.L.; Cordonnier, R.E. The aroma of grapes I. Extraction and determination of free and glycosidically bound fractions of some grape aroma components. *J. Chromatogr. A* **1985**, *331*, 83–90. [CrossRef]
34. Aubert, C.; Ambid, C.; Baumes, R.; Günata, Z. Investigation of Bound Aroma Constituents of Yellow-Fleshed Nectarines (*Prunus persica* L. Cv. Springbright). Changes in Bound Aroma Profile during Maturation. *J. Agric. Food Chem.* **2003**, *51*, 6280–6286. [CrossRef]
35. Aubert, C.; Günata, Z.; Ambid, C.; Baumes, R. Changes in Physicochemical Characteristics and Volatile Constituents of Yellow- and White-Fleshed Nectarines during Maturation and Artificial Ripening. *J. Agric. Food Chem.* **2003**, *51*, 3083–3091. [CrossRef]
36. Babushok, V.I.; Linstrom, P.J.; Zenkevich, I.G. Retention Indices for Frequently Reported Compounds of Plant Essential Oils. *J. Phys. Chem. Ref. Data* **2011**, *40*. [CrossRef]
37. Van Den Dool, H.; Kratz, P.D. A generalization of the retention index system including linear temperature programmed gas-liquid partition chromatography. *J. Chromatogr.* **1963**, *11*, 463–471. [CrossRef]
38. Assessment of total phenolic and flavonoid content, antioxidant properties, and yield of aeroponically and conventionally grown leafy vegetables and fruit crops: A comparative study. Available online: http://www.hindawin.com/journals/ecam/2014/253875/abs/ (accessed on 23 March 2014).
39. Thaipong, K.; Boonprakob, U.; Crosby, K.; Cisneros-Zevallos, L.; Hawkins Byrne, D. Comparison of ABTS, DPPH, FRAP, and ORAC assays for estimating antioxidant activity from guava fruit extracts. *J. Food Compos. Anal.* **2006**, *19*, 669–675. [CrossRef]
40. Benzie, I.F.F.; Strain, J.J. The ferric reducing ability of plasma (FRAP) as a measure of "antioxidant power": The FRAP assay. *Anal. Biochem.* **1996**, *239*, 70–76. [CrossRef]
41. Strauss, C.R.; Gooley, P.R.; Wilson, B.; Williams, P.J. Application of droplet countercurrent chromatography to the analysis of conjugated forms of terpenoids, phenols, and other constituents of grape juice. *J. Agric. Food Chem.* **1987**, *35*, 519–524. [CrossRef]
42. Gunata, Y.Z.; Dugelay, I.; Sapis, J.C.; Baumes, R.; Bayonove, C. Role of enzymes in the use of the flavor potential from grape glycosides in winemaking. In *Progress in Flavor Precursor Studies*; 1994.
43. Synthesis and enantiodifferentiation of isomeric theaspiranes. Available online: https://doi.org/10.1021/jf00019a022 (accessed on 1 July 1992).
44. Zelena, K.; Hardebusch, B.; Hülsdau, B.; Berger, R.G.; Zorn, H. Generation of Norisoprenoid Flavors from Carotenoids by Fungal Peroxidases. *J. Agric. Food Chem.* **2009**, *57*, 9951–9955. [CrossRef]
45. Cai, Y.; Zheng, H.; Ding, S.; Kropachev, K.; Schwaid, A.G.; Tang, Y.; Mu, H.; Wang, S.; Geacintov, N.E.; Zhang, Y.; et al. Free energy profiles of base flipping in intercalative polycyclic aromatic hydrocarbon-damaged DNA duplexes: Energetic and structural relationships to nucleotide excision repair susceptibility. *Chem. Res. Toxicol.* **2013**, *26*, 1115–1125. [CrossRef]
46. Winterhalter, P.; Rouseff, R.L. (Eds.) *Carotenoid-Derived Aroma Compounds*; American Chemical Society: Washington, DC, USA, 2001; Volume 802. [CrossRef]
47. Novruzov, E.N.; Agamirov, U.M. Carotenoids of *Cudrania tricuspidata* fruit. *Chem. Nat. Compd.* **2002**, *38*, 468–469. [CrossRef]
48. Schwab, W.; Davidovich-Rikanati, R.; Lewinsohn, E. Biosynthesis of plant-derived flavor compounds. *Plant J.* **2008**, *54*, 712–732. [CrossRef]
49. Adedeji, J.; Hartman, T.G.; Lech, J.; Ho, C.T. Characterization of glycosidically bound aroma compounds in the African mango (*Mangifera indica* L.). *J. Agric. Food Chem.* **1992**, *40*, 659–661. [CrossRef]
50. Kicel, A.; Wolbiś, M. Study on the phenolic constituents of the flowers and leaves of *Trifolium repens* L. *Nat. Prod. Res.* **2012**, *26*, 2050–2054. [CrossRef]

51. Zhou, D.Y.; Sun, Y.X.; Shahidi, F. Preparation and antioxidant activity of tyrosol and hydroxytyrosol esters. *J. Funct. Foods.* **2017**, *37*, 66–73. [CrossRef]
52. Cho, B.R.; Ryu, D.R.; Lee, K.S.; Lee, D.K.; Bae, S.; Kang, D.G.; Ke, Q.; Singh, S.S.; Ha, K.S.; Kwon, Y.G.; et al. P-Hydroxybenzyl alcohol-containing biodegradable nanoparticle improves functional blood flow through angiogenesis in a mouse model of hindlimb ischemia. *Biomaterials* **2015**, *53*, 679–687. [CrossRef]
53. Zhu, Y.P.; Li, X.; Du, Y.; Zhang, L.; Ran, L.; Zhou, N.N. Protective effect and mechanism of p-hydroxybenzaldehyde on blood-brain barrier. *Zhongguo Zhongyao Zazhi* **2018**, *43*, 1021–1027. [CrossRef]
54. Chirinos, R.; Rogez, H.; Campos, D.; Pedreschi, R.; Larondelle, Y. Optimization of extraction conditions of antioxidant phenolic compounds from mashua (Tropaeolum tuberosum Ruíz & Pavón) tubers. *Sep. Purif. Technol.* **2007**, *55*, 217–225. [CrossRef]
55. Durazzo, A.; Lucarini, M.; Novellino, E.; Daliu, P.; Santini, A. Fruit-based juices: Focus on antioxidant properties—Study approach and update. *Phyther. Res.* **2019**, *33*. [CrossRef]
56. Parada, F.; Duque, C.; Fujimoto, Y. Free and Bound Volatile Composition and Characterization of Some Glucoconjugates as Aroma Precursors in Melón de Olor Fruit Pulp (*Sicana o dorifera*). *J. Agric. Food Chem.* **2000**, *48*, 6200–6204. [CrossRef]
57. Choi, J.; Yeo, S.; Kim, M.; Lee, H.; Kim, S. *p*-Hydroxybenzyl alcohol inhibits four obesity-related enzymes in vitro. *J. Biochem. Mol. Toxicol.* **2018**, *32*, e22223. [CrossRef]
58. Coelho, E.; Genisheva, Z.; Oliveira, J.M.; Teixeira, J.A.; Domingues, L. Vinegar production from fruit concentrates: Effect on volatile composition and antioxidant activity. *J. Food Sci. Technol.* **2017**, *54*, 4112–4122. [CrossRef]
59. Daliu, P.; Santini, A.; Novellino, E. From pharmaceuticals to nutraceuticals: Bridging disease prevention and management. *Expert Rev. Clin. Pharmacol.* **2019**, *12*, 1–7. [CrossRef]
60. Kashanian, S.; Ezzati Nazhad Dolatabadi, J. In vitro studies on calf thymus DNA interaction and 2-tert-butyl-4-methylphenol food additive. *Eur. Food Res. Technol.* **2010**, *230*, 821–825. [CrossRef]
61. Fagali, N.; Catalá, A. Antioxidant activity of conjugated linoleic acid isomers, linoleic acid and its methyl ester determined by photoemission and DPPH{radical dot} techniques. *Biophys. Chem.* **2008**, *137*, 56–62. [CrossRef]
62. Mishra, K.; Ojha, H.; Chaudhury, N.K. Estimation of antiradical properties of antioxidants using DPPH-assay: A critical review and results. *Food Chem.* **2012**, *130*, 1036–1043. [CrossRef]
63. Lee, Y.; Oh, J.; Jeong, Y.S. Lactobacillus plantarum-mediated conversion of flavonoid glycosides into flavonols, quercetin, and kaempferol in *Cudrania tricuspidata* leaves. *Food Sci. Biotechnol.* **2015**, *24*, 1817–1821. [CrossRef]

 © 2019 by the authors. Licensee MDPI, Basel, Switzerland. This article is an open access article distributed under the terms and conditions of the Creative Commons Attribution (CC BY) license (http://creativecommons.org/licenses/by/4.0/).

Article

Oleic Acid Is not the Only Relevant Mono-Unsaturated Fatty Ester in Olive Oil

Archimede Rotondo [1,*], Giovanna Loredana La Torre [1], Giacomo Dugo [1], Nicola Cicero [1], Antonello Santini [2] and Andrea Salvo [3]

[1] Department of Biomedical and Dental Sciences and Morpho-functional Imaging, University of Messina, Polo Universitario Annunziata, Viale Annunziata, 98168 Messina, Italy; llatorre@unime.it (G.L.L.T.); dugog@unime.it (G.D.); ncicero@unime.it (N.C.)
[2] Department of Pharmacy, University of Napoli Federico II, via D. Montesano 49, 80131 Napoli, Italy; asantini@unina.it
[3] Department of Chemistry and Drug Technology, University of Roma La Sapienza, via P.le A. Moro 5, 00185 Roma, Italy; andrea.salvo@uniroma1.it
* Correspondence: arotondo@unime.it; Tel.: 39-090-676-6890

Received: 26 February 2020; Accepted: 19 March 2020; Published: 26 March 2020

Abstract: (1) Background: Extra-virgin olive oil (EVOO) is a precious and universally studied food matrix. Recently, the quantitative chemical composition was investigated by an innovative processing method for the nuclear magnetic resonance (NMR) experiments called Multi-Assignment Recovered Analysis (MARA)-NMR. (2) Methods: Any EVOO 13-carbon NMR (^{13}C-NMR) profile displayed inconsistent signals. This mismatch was resolved by comparing NMR data to the official gas-chromatographic flame ionization detection (GC-FID) experiments: the analyses concerned many EVOOs but also the "exotic" *Capparis spinosa* oil (CSO). (3) Results: NMR and GC-FID evidenced the overwhelming presence of *cis*-vaccenic esters in the CSO and, more importantly, *cis*-vaccenic ^{13}C-NMR resonances unequivocally matched the misunderstood ^{13}C-NMR signals of EVOOs. The updated assignment revealed the unexpected relevant presence of *cis*-vaccenic ester (around 3%) in EVOOs; it was neglected, so far, because routine and official GC-FID profiles did not resolve oleic and *cis*-vaccenic signals leading to the total quantification of both monounsaturated fatty esters. (4) Conclusions: The rebuilt MARA-NMR and GC-FID interpretations consistently show a meaningful presence of *cis*-vaccenic esters in EVOOs, whose content could be a discrimination factor featuring specific cultivar or geographical origin. The study paves the way toward new quantification panels and scientific research concerning vegetable oils.

Keywords: *cis*-vaccenic; monounsaturated fatty; glycerols; NMR analysis; olive oil; *Capparis spinosa*; ^{13}C-NMR; MARA-NMR

---

1. Introduction

Extra-virgin olive oil (EVOO) comes from the supernatant phase of juice obtained after cold pressing of *Olea europaea* fruits and is the fundamental dressing of any Mediterranean dish. It is considered the liquid gold in food trading because of its crucial role in the healthy way of life model called "Mediterranean Diet" [1]. Many scientific studies reveal that the chemical composition of EVOO is a perfect balance leading to countless benefits for humans [2–5]. The positive biological activities are reasonably due to the suitable presence of vegetable sterols [6], liposoluble polyphenols [7] and other anti-oxidant hydrocarbons [8] joined to the most abundant presence of mono-unsaturated tri-acyl-glycerol esters among the vegetable oils. Albeit the oleic ester in EVOOs is considered the overwhelming main mono-unsaturated fatty ester so far, this work casts another important mono-unsaturated fat potentially playing important biological roles. The wide impact of EVOOs composition accounts for the constantly updated European Regulation stating the chemical and taste features, limits and official analytical techniques recognized for olive oil trade [9,10]. In the last decades, the traditional food analysis was shocked by the nuclear magnetic resonance (NMR) as alternative quantitative (qNMR) approach [11] flanking the officially recognized separation techniques. The nondestructive NMR spectroscopy allows the *in-situ* detection of several chemical species without the requirement of a real physical separation [8,12,13]; moreover qNMR is feasible directly or through a clever data throughput [14,15]. The definite advantages of the NMR analyses are: a) minimal sample treatment [8], b) simultaneous detection of a great amount of data [16], c) reduction of systematic errors controlled by the intrinsic instrumental stability, d) constant and direct dependence between signal integration and quantitative values because of the constant nuclear magnetic momentum for the measured nuclei [17]. Criticism toward NMR concerned mainly sensitivity; however, it actually depends on the machine, on sample type, on used solvent, on observed nuclei and on specific experimental runs; this is the reason it should be evaluated from case to case [18]. After several years of research on EVOOs composition, Rotondo et al. have developed a Multi-Assignment Recovered Analysis (MARA-NMR) involving multi-nuclear ^{1}H and $\{^1H\}^{13}$C-NMR experiments processed by an accustomed processing "MARA" algorithm [19]. This method successfully and quickly achieves the quantification of many components in EVOOs samples through high-resolution spectroscopy at 500 MHz (500 MHz HR-NMR). On the other hand, the "first" MARA-NMR scheme did not take into account some ^{13}C-NMR resonances whose intensity was significant, but these were associated to EVOO minor components (theoretically negligible and contributing for less than 1%) and, for these studies, the best fitting goodness never reached the expected convergence. Since the official method for the quantitative determination of glyceryl fatty esters consists in the gas-chromatographic analysis of the corresponding methyl esters using the gas-chromatographic flame ionization detection (GC-FID) [20], this work focused on the data comparison between NMR and GC-FID on oils in order to solve inconsistent "leftovers" from literature. The paper evidences the neat importance of *cis*-vaccenic fatty ester in EVOOs as its content is around 3%; however, it was neglected so far because, according the official method, it is included in the level of oleic ester.

2. Materials and Methods

2.1. Materials and Samples

Deuterated chloroform with a small amount of Tetra-Methyl-Silane (TMS), used as internal reference, was purchased at reagent grade from Cambridge Isotope Laboratories (CIL) Inc. Extra-virgin olive oils were samples from awarded cultivars of different provenience representing top level food in the seasons 2014–2015. These samples were kindly given by producers in order to carry out scientific projects belonging to the BIOOIL program, aiming to improve knowledge about top quality products.

Some seeds were isolated from *Capparis spinosa* fruits (known in Sicily as "cucunci"). Afterward seeds were dried in oven at 30 °C for 2 h. The matter was grinded in a mortar until the formation of a raw powder. This matter (20 g) was extracted in 100 mL of hexane, sonicated for 30 min at 30 °C and stirred overnight. The solution was then filtrated, and the hexane removed from the solution by using, at first, the rotating evaporator and later N_2 flow over the sample. Finally, cucunci's seed oil (CSO) was recovered (yield 15% *w/w*).

2.2. GC-FID Analysis for the Comparative Tests

Fatty acids methyl esters (FAMEs) analysis was performed according to European Union (EU) Regulations [10]. It consists of the hydrolysis of tri-acyl-glycerides and cold transesterification with a methanol KOH solution; in particular, the methyl esters were prepared by vigorously shaking solution of the oil in heptane (0.1 g in 2 mL) with 0.2 mL of the methanolic KOH solution. The resulting solution was then injected into a gas chromatograph DANI MASTER GC-FID (Milan, Italy), equipped with a fused silica capillary column Phenomenex Zebron ZB-WAX (polar phase in polyethylene glycol) with a length of 30 m, internal diameter of 0.25 mm and film thickness of 0.25 µm. Helium was used as a carrier gas at a column flow rate of 1.2 mL/min, with a split ratio of 1:100. The temperature of the injector (split/splitless) and detector was of 220 °C and 240 °C, respectively. The oven was programmed as follows: initial temperature at 130 °C, final temperature at 200 °C (10 min) with an increase of 3 °C/min. The fatty acid methyl esters were identified by comparing the retention times with those of standard compounds. The relative percentage area of the fatty acids was obtained using the following relationship: %FAX = [AX/AT] × 100, where FAX stands for fatty acids to quantify, AX is the area of the methyl-esters and AT is the total area of the identified peaks in the chromatogram [21]. This analytical strategy is chosen for data comparison because it is officially recognized for the fatty esters quantification, on another hand the reader should be aware that the hydrolysis-esterification step is always tedious, laborious and time consuming, decreasing accuracy and precision. This is the reason why, lately, alternative analytical chromatographic methods have also been proposed [22], still showing limitations.

2.3. Sample Preparation for NMR

Sample preparation follows the same procedure successfully used by our group several years ago [7,8,12,19]. Briefly, all the $CDCl_3$ solutions were kept homologous by mixing 122 µL of oil and 478 µL of deuterated chloroform ($CDCl_3$) into a 5 mm test-tube (EVOO or CSO in a 13.5% weight ratio). In this study we used the same EVOOs studied in Reference 19; however, these were dissolved as different samples and the experiments were repeated in light of the new assignments. Tubes were immediately sealed to prevent solvent evaporation; it would affect the sample concentration influencing the chemical shift of many signals, especially the unsaturated and carbonyl ^{13}C signals. These samples were readily used for the NMR scheduled analysis so that outcomes could be suitably processed and compared to each other.

2.4. NMR Analysis

All the samples were analyzed at a constant temperature of 298 K on a 500 MHz Avance III NMR spectrometer endowed with a gradient assisted probe (SMARTprobe, Faellanden, Switzerland). The shimming procedure was carried out until the field homogeneity was assessed by less than 1.5 Hz of half-height line-width for the TMS signal.

The 1D ^1H and $^{13}C\{^1H\}$ NMR spectra were run at 499.74 and 125.73 MHz, respectively. This research exploited the analytical procedure including two experiments: a) the standard ^1H experiment with 64 scans; b) the standard ^{13}C NMR experiment with 32 scans. The entire procedure takes around 30 min of experimental time for any sample including preparation. Hard pulse for the maximum sensitivity (90° pulse), was calibrated and constantly checked for ^1H throughout the samples being always 8.2 ± 0.1 µs at −6 dB. ^1H-NMR experiments (type A) were run with a spectral width of 12 ppm, 64 scans, 10 s of acquisition time and 5 s of recycle delay in order to overcome problems coming from the differences in the proton relaxation times. For the same reason the ^{13}C spectra (type B) were acquired with the 90° hard pulse (11.2 ± 0.3 us at 6 dB) with 32 scans, 5 s of acquisition time and 20 s for the time delay. Thanks to the MARA-NMR algorithm, these experimental elements were conveyed together for the overall quantitative evaluation.

2.5. NMR Processing and Data Treatment

All the spectra were processed through three main software programs (ACDLab/NMR 2012 (Toronto, Ontario, Canada), MestreNova 6.6.2 (Galicia, Spain), Topspin 4.0.5 (Bruker, Milan, Italy) and using several procedures for the coherent alignment, spectral phasing, calibration, base-line correction and integration procedure. The best processing choices are here reported regardless the many other adoptable procedures. Topspin processed data were selected with manual phase-correction, parametric base-line correction with an implemented polynomial curve (for example, for experiment I *absd 16* command). Calibration of experiment A was performed on the methyl group of the β-sitosterol signal to (δ_H = 0.738 ppm) with the TMS always being (δ = 0.0 ± 0.005 ppm); for ^{13}C calibration of experiment B the divinyl- methylene group of the linoleate glycerols (L11; δ_{13C} = 25.6614 ppm) was used always keeping the known TMS ^{13}C signal to δ_{13C} = 0.0 ± 0.05 ppm. The TMS calibration would not really change the results; here, the calibration over internal signals is preferred because these are less dependent on random conditions as explained elsewhere [19].

The serial integration of 100 regions for all the A-type experiments, and of 90 regions for experiment B profiles, provided a pretty big matrix whose columns were the 40 studied samples EVOO and rows represented 190 homologous integrations (see Supplementary Materials). Every column of this matrix was processed by the mentioned MARA algorithm [19]; this theoretical architecture is modified according to the original knowledge and assignments concerning *cis*-vaccenic esters (V). The experimental coherences simply confirm the presence of a relevant amount of V, also improving consistency assessed by low best fitting goodness (ρ) values. The extended procedure outputs up to 20 quantitative parameters [7,8,19] (Table 1) but this manuscript focuses on the 11 quantitative parameters showing sound precision and important significance (Table 2). The data validation and experimental error is evaluated through reproducibility (several samplings) and repeatability (analyses in different days of the same sample).

Table 1. Abbreviations used to indicate quantitative values.

tri-acyl-glycerol percent	TG%
1,2 di-acyl-glycerol percent	1,2-DG%
1,3 di-acyl-glycerol percent	1,3-DG%
squalene molecular%	SQ$_{mol}$%
linolenate esters %	Ln%
linoleates esters %	L%
oleic esters %	O%
palmitoleic esters %	PO%
cis-vaccenic esters %	V%
palmitate esters %	P%
stearate esters	S%
linolenate esters % in internal glyceril position	Lni%
linoleates esters % in internal glyceril position	Li%
oleic esters % in internal glyceril position	Oi%
palmitoleic esters % in internal glyceril position	POi%
cis-vaccenic esters % in internal glyceril position	Vi%
palmitate esters % in internal glyceril position	Pi%
sterarate esters % in internal glyceril position	Si%
β-sitosterol + avenasterol + camposterol in molecular ppm	VSTR
cyclo arthenol and other cyclosterols in molecular ppm	CYSR

Table 2. Quantitative data and relative deviation for 11 main variables (whose code is reported in Table 1), as measured through Multi-Assignment Recovered Analysis-Nuclear Magnetic Resonance (MARA-NMR) processing method working on mono dimensional ^{1}H and ^{13}C-NMR experiments for 33 samples. Standard deviations were measured through 9 different experiments on 3 identical samples analyzed on three different days.

Sample	TG%	1,2-DG%	1,3-DG%	SQ$_{mol}$%	Ln%	L%	O%	PO%	V%	P%	S%
S_1	96.7 ± 0.1	1.19 ± 0.06	2.1 ± 0.1	1.7 ± 0.1	0.59 ± 0.03	10.2 ± 0.1	63.9 ± 0.5	1.3 ± 0.3	2.8 ± 0.3	19.1 ± 0.3	2.2 ± 0.2
S_2	97.4 ± 0.1	1.30 ± 0.07	1.34 ± 0.09	2.2 ± 0.1	0.57 ± 0.03	12.5 ± 0.1	61.8 ± 0.4	0.9 ± 0.2	2.5 ± 0.2	19.7 ± 0.3	1.9 ± 0.2
S_3	97.7 ± 0.1	1.65 ± 0.09	0.68 ± 0.05	1.3 ± 0.1	0.51 ± 0.03	12.6 ± 0.1	59.4 ± 0.4	0.8 ± 0.2	3.7 ± 0.3	21.1 ± 0.3	2.0 ± 0.2
S_4	97.4 ± 0.1	1.39 ± 0.07	1.21 ± 0.08	0.8 ± 0.0	0.63 ± 0.03	15.4 ± 0.1	51.9 ± 0.4	1.1 ± 0.2	4.9 ± 0.4	23.7 ± 0.3	2.4 ± 0.2
S_5	97.2 ± 0.1	2.1 ± 0.1	0.70 ± 0.05	2.9 ± 0.2	0.67 ± 0.04	5.7 ± 0.1	72.8 ± 0.5	0.7 ± 0.1	1.8 ± 0.2	16.0 ± 0.2	2.4 ± 0.2
S_6	97.3 ± 0.1	1.70 ± 0.09	0.97 ± 0.07	2.7 ± 0.2	0.65 ± 0.03	2.6 ± 0.0	73.9 ± 0.5	0.7 ± 0.1	3.5 ± 0.3	16.7 ± 0.2	1.9 ± 0.2
S_7	97.3 ± 0.1	2.0 ± 0.1	0.77 ± 0.05	2.5 ± 0.2	0.61 ± 0.03	3.1 ± 0.0	74.2 ± 0.5	0.8 ± 0.2	2.7 ± 0.2	16.1 ± 0.2	2.5 ± 0.2
S_8	97.7 ± 0.1	1.59 ± 0.08	0.74 ± 0.05	2.2 ± 0.1	0.61 ± 0.03	6.4 ± 0.1	70.8 ± 0.5	0.9 ± 0.2	3.6 ± 0.3	15.5 ± 0.2	2.1 ± 0.2
S_9	96.8 ± 0.1	1.78 ± 0.09	1.4 ± 0.1	1.3 ± 0.1	0.44 ± 0.02	10.2 ± 0.1	60.0 ± 0.4	1.1 ± 0.2	5.0 ± 0.4	21.7 ± 0.3	1.6 ± 0.1
S_10	96.9 ± 0.1	1.30 ± 0.07	1.8 ± 0.1	2.3 ± 0.1	0.68 ± 0.04	7.6 ± 0.1	65.2 ± 0.5	1.1 ± 0.2	3.4 ± 0.3	20.0 ± 0.3	2.1 ± 0.2
S_11	97.4 ± 0.1	1.21 ± 0.06	1.4 ± 0.1	2.9 ± 0.2	0.61 ± 0.03	5.9 ± 0.1	67.2 ± 0.5	1.2 ± 0.3	3.6 ± 0.3	19.2 ± 0.3	2.4 ± 0.2
S_12	97.6 ± 0.1	1.32 ± 0.07	1.04 ± 0.07	3.1 ± 0.2	0.64 ± 0.03	7.7 ± 0.1	65.4 ± 0.5	2.5 ± 0.5	4.6 ± 0.4	17.4 ± 0.3	1.9 ± 0.2
S_13	97.7 ± 0.1	1.47 ± 0.08	0.88 ± 0.06	2.8 ± 0.2	0.79 ± 0.04	8.3 ± 0.1	67.1 ± 0.5	1.0 ± 0.2	2.4 ± 0.2	18.3 ± 0.3	2.0 ± 0.2
S_14	97.8 ± 0.1	1.40 ± 0.07	0.80 ± 0.06	3.4 ± 0.2	0.64 ± 0.03	4.5 ± 0.0	71.3 ± 0.5	1.2 ± 0.2	2.5 ± 0.2	17.7 ± 0.3	2.3 ± 0.2
S_15	97.8 ± 0.1	1.14 ± 0.06	1.07 ± 0.07	4.0 ± 0.3	0.60 ± 0.03	8.3 ± 0.1	64.0 ± 0.5	0.7 ± 0.1	3.1 ± 0.3	20.7 ± 0.3	2.7 ± 0.2
S_16	97.8 ± 0.1	1.48 ± 0.08	0.72 ± 0.05	3.1 ± 0.2	0.58 ± 0.03	6.2 ± 0.1	70.2 ± 0.5	0.6 ± 0.1	3.0 ± 0.3	17.5 ± 0.3	2.0 ± 0.2
S_17	97.4 ± 0.1	1.53 ± 0.08	1.08 ± 0.07	2.1 ± 0.1	0.61 ± 0.03	6.4 ± 0.1	68.4 ± 0.5	1.2 ± 0.3	3.3 ± 0.3	17.9 ± 0.3	2.2 ± 0.2
S_18	97.4 ± 0.1	1.22 ± 0.06	1.4 ± 0.1	2.3 ± 0.1	0.58 ± 0.03	6.7 ± 0.1	68.5 ± 0.5	0.8 ± 0.2	2.5 ± 0.2	19.1 ± 0.3	1.7 ± 0.1
S_19	97.6 ± 0.1	1.43 ± 0.07	0.98 ± 0.07	3.5 ± 0.2	0.79 ± 0.04	7.5 ± 0.1	65.6 ± 0.5	0.7 ± 0.2	2.8 ± 0.3	20.7 ± 0.3	1.9 ± 0.2
S_20	97.2 ± 0.1	1.59 ± 0.08	1.17 ± 0.08	2.0 ± 0.1	0.63 ± 0.03	9.3 ± 0.1	64.8 ± 0.5	1.1 ± 0.2	3.7 ± 0.3	19.0 ± 0.3	1.5 ± 0.1
S_21	97.2 ± 0.1	1.70 ± 0.09	1.10 ± 0.08	3.0 ± 0.2	0.70 ± 0.04	7.6 ± 0.1	66.2 ± 0.5	0.6 ± 0.1	3.1 ± 0.3	19.9 ± 0.3	1.9 ± 0.2
S_22	97.7 ± 0.1	1.49 ± 0.08	0.78 ± 0.05	3.6 ± 0.2	0.73 ± 0.04	7.5 ± 0.1	64.8 ± 0.5	1.3 ± 0.3	3.4 ± 0.3	20.7 ± 0.3	1.7 ± 0.1
S_23	98.0 ± 0.1	1.63 ± 0.09	0.37 ± 0.03	3.4 ± 0.2	0.76 ± 0.04	6.9 ± 0.1	65.4 ± 0.5	1.4 ± 0.3	2.8 ± 0.3	20.1 ± 0.3	2.5 ± 0.2
S_24	97.6 ± 0.1	1.58 ± 0.08	0.85 ± 0.06	3.4 ± 0.2	0.73 ± 0.04	7.9 ± 0.1	65.9 ± 0.5	1.1 ± 0.2	2.5 ± 0.2	20.0 ± 0.3	2.0 ± 0.2
S_25	98.0 ± 0.1	1.21 ± 0.06	0.78 ± 0.05	2.4 ± 0.2	0.58 ± 0.03	7.2 ± 0.1	66.9 ± 0.5	0.6 ± 0.1	3.6 ± 0.3	19.2 ± 0.3	2.0 ± 0.2
S_26	97.3 ± 0.1	1.27 ± 0.07	1.4 ± 0.1	2.0 ± 0.1	0.68 ± 0.04	4.4 ± 0.0	73.3 ± 0.5	1.3 ± 0.3	3.0 ± 0.3	15.6 ± 0.2	1.8 ± 0.1
S_27	97.4 ± 0.1	1.48 ± 0.08	1.13 ± 0.08	2.5 ± 0.2	0.64 ± 0.03	4.4 ± 0.0	72.2 ± 0.5	0.9 ± 0.2	2.6 ± 0.2	16.9 ± 0.2	2.3 ± 0.2
S_28	97.8 ± 0.1	1.31 ± 0.07	0.85 ± 0.06	2.5 ± 0.2	0.64 ± 0.03	7.1 ± 0.1	73.9 ± 0.5	1.3 ± 0.3	1.5 ± 0.1	13.7 ± 0.2	1.8 ± 0.1
S_29	96.9 ± 0.1	1.45 ± 0.08	1.6 ± 0.1	2.7 ± 0.2	0.65 ± 0.04	5.2 ± 0.0	70.5 ± 0.5	1.4 ± 0.3	2.6 ± 0.2	17.1 ± 0.2	2.4 ± 0.2
S_30	97.9 ± 0.1	1.22 ± 0.06	0.89 ± 0.06	1.9 ± 0.1	0.64 ± 0.03	4.8 ± 0.0	73.1 ± 0.5	0.9 ± 0.2	2.9 ± 0.3	15.4 ± 0.2	2.3 ± 0.2
S_31	96.9 ± 0.1	2.5 ± 0.1	0.50 ± 0.03	2.6 ± 0.2	0.56 ± 0.03	6.3 ± 0.1	64.5 ± 0.5	0.5 ± 0.1	5.2 ± 0.5	21.5 ± 0.3	1.4 ± 0.1
S_32	96.0 ± 0.1	3.0 ± 0.2	0.99 ± 0.07	3.1 ± 0.2	0.69 ± 0.04	9.4 ± 0.1	61.8 ± 0.4	1.0 ± 0.2	4.4 ± 0.4	21.1 ± 0.3	1.8 ± 0.1
S_33	95.8 ± 0.1	2.3 ± 0.1	1.9 ± 0.1	2.2 ± 0.1	0.66 ± 0.04	8.7 ± 0.1	65.6 ± 0.5	1.0 ± 0.2	3.9 ± 0.4	18.3 ± 0.3	1.9 ± 0.1

2.6. Mathematical Background of MARA-NMR and Updates

The used algorithm MARA-NMR was invented in this laboratory, exploiting the very simple idea that all NMR signals rise from active nuclei that belong to compounds and contribute according to: a) relative concentration, b) number of resonating nuclei, c) possible overlaps with homologous nuclei maybe belonging to other compounds [19]. If this theoretical statement and a suitable assignment is correct, the experimental profile should perfectly match our theoretical reconstruction. As explained in the original paper [18] experimental data are not ideal data-points, however we have designed this algorithm able to optimize quantitative parameters in order to minimize the overall deviations between experimental and theoretical outcomes enclosed in the function ρ which is the best-fitting goodness.

$$\rho = \sum_{xj=x1}^{xf} \omega_{xj} \left(\frac{\gamma_{xj} I_{xj}}{I_{ref}} - \frac{\sum_{i=a}^{n} N°NUC_i * C_i}{N°NUC_{ref} * C_{ref}} \right)^2. \tag{1}$$

The intensity of any signal in the spectrum I_{xj} respect to a reference signal I_{ref} should even out the relative concentration (C_i against C_{ref}) of the magnetically active nuclei NUC_i actually assigned to that signal. Coefficients ω and γ are empirical parameters able to reduce experimental deviation improving the algorithm; theoretically speaking the best fitting goodness ρ should be 0 but in the real world we accept low values. The introduction of 18 new assignments for the *cis*-vaccenic ester, by enhancing just one quantitative parameter referred to the "new" component greatly lowered the best fitting goodness giving the proof of concept about the assignment. The 20 quantitative parameters are derived by 11 expressions derived from A experiments and 65 expressions derived from B experiments put together in the same expression as equation (1) containing 76 xj members and 20 i compounds. In order to preserve the quantitative proportion of ^{13}C integrations, despite the uneven nOe relayed on total decoupled carbon nuclei, adopted equations in the sum (1) are divided in blocks of nuclei with the same chemical environment (methyl terminal carbons, methylene inner chain carbons, vynil-methylene, etc.). It is demonstrated that MARA-NMR keeps the quantitative information as reported in Supplementary Materials and in Reference [19].

3. Results

Figure 1 shows the chemical moieties, related abbreviations and the adopted labelling scheme; Figure 2 reports the GC-FID profile referring to the CSO extracted in our laboratory and Figure 3 represents the ^{13}C-NMR profile of EVOO and CSO in the unsaturated region (127–131 ppm) along with the relative assignment witnessing the presence of the *cis*-vaccenic ester. As easily foreseeable, other NMR spectral regions also clearly showed *cis*-vaccenic resonances; however, a total assignment of 18 ^{13}C carbon atoms was challenged by the many overlaps. Previous pioneering studies pointed out the challenging quantitative decoding of the mono-unsaturated fatty esters mixture in EVOOs [23]. Specifically, other minor mono-unsaturated fatty esters (MUFE) were taken into account; beyond the oleic (O) are also considered cis-vaccenic (V), eicosenoic (E) and palmitoleic (PO) [24,25]. On the other hand, data coming from known EVOOs compositions, limit the quantitative contribution of E and PO below 1% [9] and it is consistently witnessed by the lack of defined resonances in the regions where these esters should not have overlap with other similar constructs. The Multiple Assignment Recovered Analysis (MARA-NMR) takes advantage of any spectral section also overcoming the overlap issues hampering, so far, the independent quantification of mono-unsaturated fatty esters. Specifically, in this case, MARA-NMR processing definitely led to the detection and quantification of the V esters (consistently all over the recorded spectral span). Among the 20 variables feed out from MARA-NMR whose code is reported in Table 1, we herein have restricted our considerations to the most meaningful 11 variables reported in Table 2 along with the relative standard deviation.

With respect to the other studies [24,25] the new information remarkably smooths discrepancies between ^{1}H and ^{13}C-NMR as the mono-unsaturated fatty esters contribution in ^{1}H-NMR matches the contribution of O and V esters, which actually should be also somewhat enhanced by the minor PO and E esters' contribution. Because of the tricky GC-FID resolution between V and O, also referred to in the European regulation (which suggests to report the whole V+O contribution), it is not always possible to compare GC and NMR data. However, the new available data, display the best fitting so far obtainable (Figure 4) concerning the measurements of mono-unsaturated (O + V + PO), saturated (P + S), di-unsaturated (L) and tri-unsaturated (Ln) fatty esters. The average V contribution is around 3% and it is consistent with previous NMR [23] and GC-FID [26] analyses; on the other hand, we think that MARA-NMR is the most versatile method suitable for serial processing of several samples and data. We think that this remarkable parameter in EVOOs cannot be ignored, since it is not constant by shifting from sample to sample, therefore it could assess specific features of different food products. The V quantification is not a marker for this study according to Table 2; however, it will trigger many important statistical considerations.

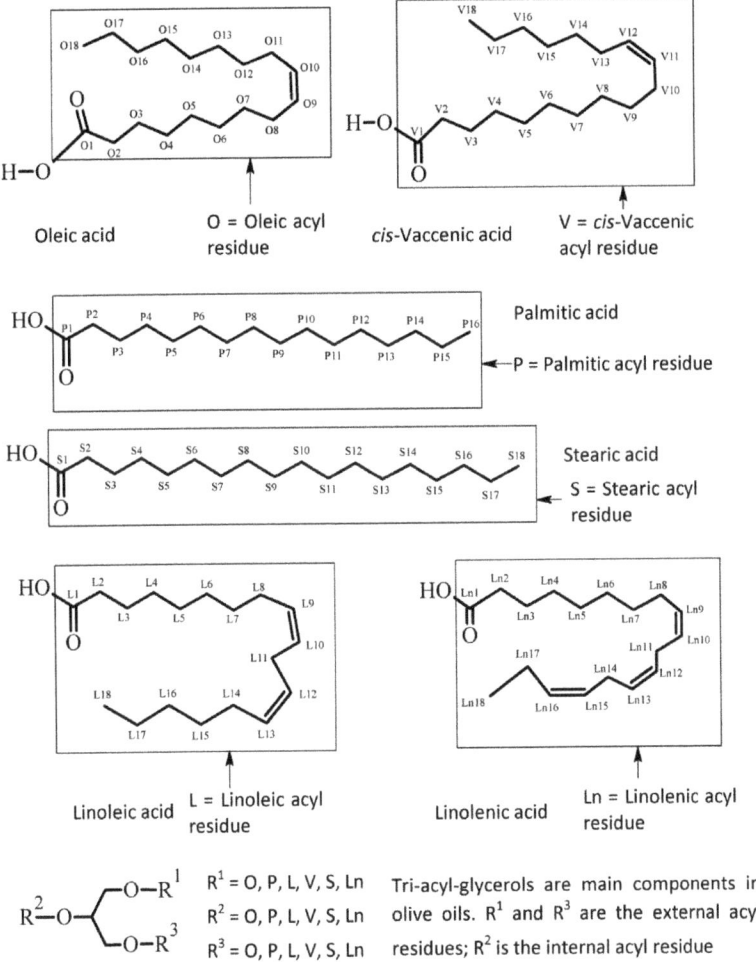

Figure 1. Chemical scheme of the fatty esters commonly found in olive oils with relative abbreviation. Usually these acyl residues are esters of the glycerol moiety. The labelling scheme of carbon atoms is adopted in this paper for assignments and discussion.

This enlightened an important piece of information concerning the *cis*-vaccenic ester as main compound in CSO but also as relevant ester contributing to the EVOO mixture. This last element was incredibly ignored so far. Table S1 (Supplementary Materials) reports the extended panel of 20 quantitative variables considered in the study for 33 samples (see details in Supplementary Materials). These values are obtained by MARA-NMR—a post-processing algorithm working over the two experiments A and B type.

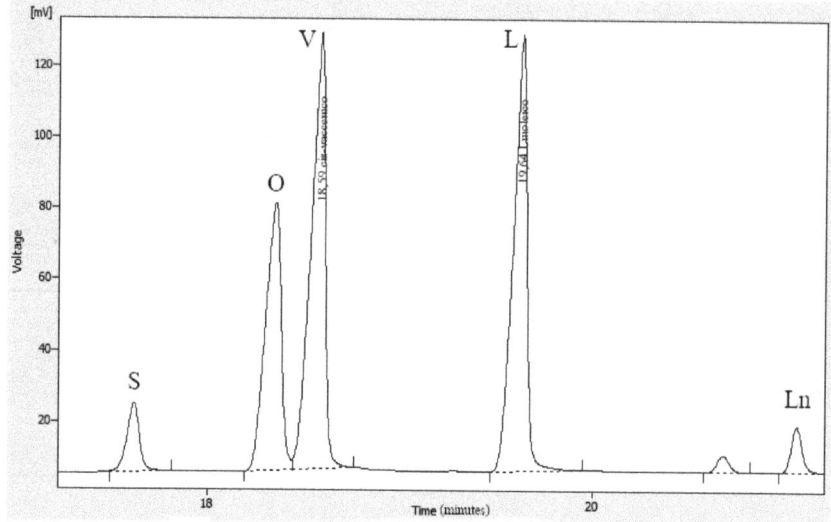

Figure 2. Expanded region of interest in the gas-chromatographic flame ionization detection (GC-FID) profile for *Capparis spinosa* oil; oleic (O) and *cis*-vaccenic (V) methyl esters are resolved for the quantification. In the case of extra-virgin oil the O peak is around 20 times more than V. Other labelled signals are linolenic (Ln), linoleic (L) and stearic (S) esters

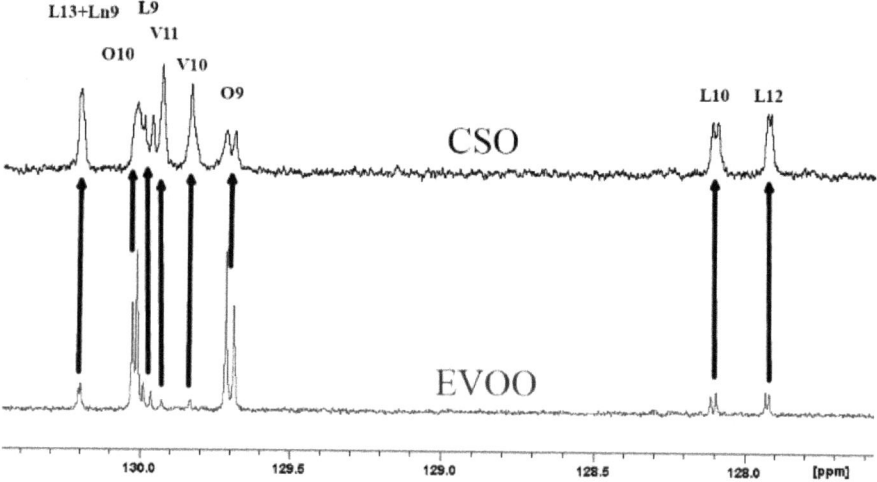

Figure 3. ^{13}C-NMR profiles for olive oil (EVOO) in gray and capparis seed oil (CSO) in black. All the assignments for oleic (O), linoleic (L) linolenic (Ln) and *cis*-vaccenic (V), with the number representing carbon atom position respect to the 1 carboxyl position, are pretty known and coherent with quantitative and literature data.

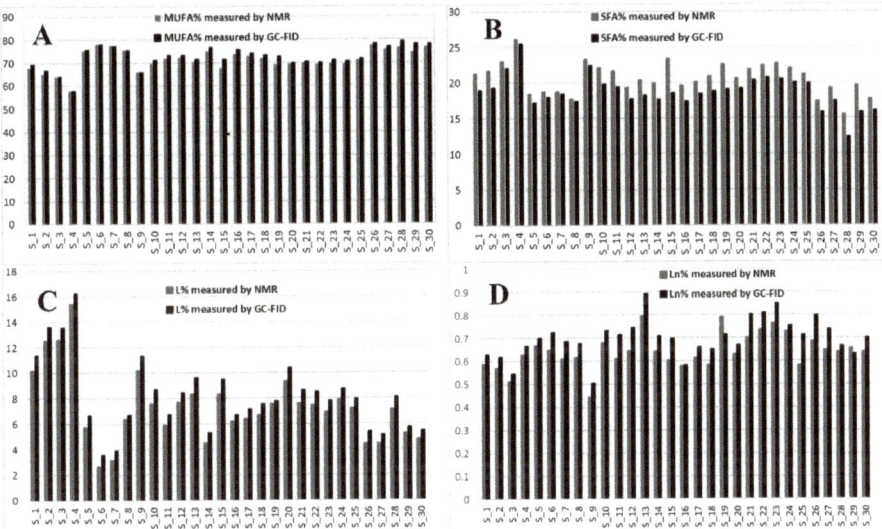

Figure 4. Comparison between MARA-NMR and GC-FID measured quantitative parameters referred to: (**A**) mono-unsaturated (MUFA), (**B**) saturated (SFA), (**C**) Linoleic (L) and (**D**) Linolenic (Ln) esters in relative percent ratio.

4. Discussion

The previously reported assignments for EVOOs ^{13}C-NMR definitely accounted for five fatty esters in the following quantitative order: oleic (O), palmitic (P), linoleic (L), stearic (S) and linolenic (Ln) [27,28]. Some other tentative assignments concerned mono-unsaturated fatty esters like palmitoleic (PO) and 11-eicosenoic (E) constructs [29]. Despite the wide availability of NMR reports [30], none of these clearly explained the systematic presence of unknown resonances (in our processing batch at 129.92, 129.82, 31.82, 22.69 ppm and others) which account for a relevant quantitative contribution (around 3%, Figure 3). Scientific hesitancy probably owes to the general opinion that the total amount of other fatty esters is limited to less than 1% of EVOOs. This idea was questioning the NMR technique itself as possible analytical method but the serendipitous extraction of the *Capparis spinosa* oil (CSO) allowed us to solve this inconsistency because of the remarkable presence of *cis*-vaccenic (V) esters. The comparison between NMR and GC-FID analyses of CSO consistently confirmed the main presence of the V ester with a minor contribution of the O ester. The analogous analytical approach executed over several EVOO samples made us realize that the detected mono-unsaturated fatty esters were again O and V but in a reversed quantitative proportion respect to the CSO. Against this background, the main NMR resonances attributed to V in other peculiar food matter [31–33] (as also the reported CSO sample) were matching the EVOO signals as reported in Figure 3. It definitely gave us the chance to include the V component in the EVOO quantitative panel according to the ^{13}C-NMR resonances afore mentioned. The V remarkable presence is not just a production side product as we did not observe the presence of *trans* isomers (resonances downfield respect 5.40 ppm in the ^{1}H-NMR and relative other singlets in the ^{13}C-NMR). Once again these results confirm the stability and sound presence of the *cis* form of unsaturated esters in spite of the minor thermodynamic stability. In order to perform the updated comprehensive quantitative NMR analysis of EVOO we have adopted an accustomed procedure based on MARA-NMR. Although it is not the first analytical comparison between GC and NMR [34,35], the novel MARA-NMR strategy suitably refined according to the new information led to a very good fitting (Figure 4). The whole outcome is reported in Tables (Tables 1 and 2, and Tables S1 and S2 in Supplementary Data). In order to get consistent data, we have chosen to compare the percent presence of L and Ln as detected, whereas the saturated fatty esters (SFA%) were considered as the sum S+P and the mono-unsaturated fatty esters (MUFA) were considered as O+V+PO. We think it is actually an important parallel evaluation whose general trend shows a very good fitting also kept with samples showing sensibly different proportions. Finally, by properly considering all the mono-unsaturated fatty esters, the MUFA% estimation reached an unprecedented very good matching. On the other hand, the slight systematic overestimation of GC-FID respect to the NMR for L% and Ln% and underestimation of SFA% deserves to be elucidated with further studies requiring standard mixtures similar to EVOO, which is a tri-acyl-glycerol mixture. At the moment, these substrates are not available but work is in progress to develop further information. Although it is not the first case of V detection and also quantification [36], the EVOOs routine quantifications barely evidence the resolution for O-V peaks; this is clearly shown in the GC picture of the European Regulation 2013 [9]. Our observations also demonstrated that new GC-FID columns keep a better (affordable) resolution, whereas routine instruments adopted for serial records easily present the V peak as O shoulder. Fortunately, recorded ^{13}C-NMR provide the missing information about the V fraction (not really taken into account so far) for any EVOO sample (Figure 3). According to our opinion, future studies could take advantage from a "powered" MARA-NMR working over sensitivity-enhanced ^{13}C-NMR profile (optimized scans); these could push further the frontiers of quick qNMR in EVOOs by enabling the independent quantification of fatty esters in the 2- internal position of glycerides but also the improved quantification of other minor components (see Supplementary Materials). This contribution also opens the way toward new studies concerning sensory attributes, geographical origin and beneficial effects [37] of EVOO as fundamental functional food with the major presence of glycerol esters [38].

5. Conclusions

This study definitely assesses the constant and relevant presence in olive oils of a not-oleic mono-unsaturated fatty ester called *cis*-vaccenic ester. It resolves the literature controversies concerning the assignment of some ^{13}C-NMR resonances but, more importantly, it brings back the expected coherency between NMR and chromatography data. The serendipitous comparison of GC-FID and NMR profiles for the "exotic" *Capparis spinosa* oil evidenced the overwhelming amount of *cis*-vaccenic ester in this matrix but also unambiguously confirmed ^{13}C-NMR assignments also validated in olive oil. By reconsidering the NMR and GC-FID of olive oils, it turned out the surprising quantitative contribution (around 3%) of *cis*-vaccenic ester. The official GC method does not always perform the required resolution to resolve and quantify oleic and *cis*-vaccenic esters and this is leading to the undistinguished quantification of both mono-unsaturated fatty esters. It opens up great potential for any technique able to clearly resolve *cis*-vaccenic moieties (just like ^{13}C-NMR) in the study of extra-virgin olive oils.

Supplementary Materials: The following are available online at http://www.mdpi.com/2304-8158/9/4/384/s1, Table S1: Analyzed samples coming from awarded BIOOIL competition 2014. The used code is connected to the provenance and to the known cultivar. Table S2: Quantitative data and relative deviation for 20 main variables, as measured through MARA-NMR processing method working on mono dimensional ^{1}H and ^{13}C-NMR experiments for 33 samples. Table S3: General scheme of MARA-NMR referred just to the first sample. There are several blocks: namely ^{1}H-NMR integrations with assignment (100 entries), DPFGSE ^{1}H-NMR integrations (17 entries, not used in this study), ^{13}C-NMR integrations (90 entries) along with some sum of integrals belonging to the same spectral block. Where possible, assignments are performed respecting the chemical position indicated also in other studies about the NMR of olive oil compounds (see Figure 1), for the fatty esters the abbreviation is followed by a number indicating the distance from the carboxyl position (generally from 1 to 18). These first rows will be used in the following equations according to the style of (1) (see main text) conveyed as square sum in the raw called RHO. The RHO value is minimized playing around with the quantitative variables so that the theoretical outcome is best-fitting the real (independent) variables, Figure S1. Stack-plot of eight olive oils coming from Sicily. The reported assignment follows the labeling used in the main manuscript (scheme 1). The expanded regions around 22 and 32 ppm show the clear presence of cis-vaccenic acid signals useful for the quantification within MARA-NMR quantification. The aromatic region is already reported in Figure 2 of the main text.

Author Contributions: Conceptualization, A.R. and A.S. (Andrea Salvo); methodology, A.R.; software, A.R.; validation, A.R., G.L.L.T. and A.S. (Andrea Salvo); formal analysis, A.R., G.L.L.T.; investigation, A.R.; data curation, A.R., A.S. (Andrea Salvo) and G.L.L.T.; writing—original draft preparation, A.R.; writing—review and editing, A.R and G.L.L.T.; visualization and supervision, A.R., A.S. (Antonello Santini), A.S. (Andrea Salvo) and G.D.;

background, N.C. and A.S. (Antonello Santini). All authors have read and agreed to the published version of the manuscript.

Funding: This research received no external funding.

Acknowledgments: Once again we thank the institution "University of Messina," which, in spite of poor means and the difficult situation of the south Italian research, does not give up.

Conflicts of Interest: The authors declare no conflict of interest.

References

1. Trichopolou, A.; Vasilopoulou, E. Mediterranean diet and longevity. *Br. J. Nutr.* **2000**, *84*, S205–S209. [CrossRef] [PubMed]
2. Visioli, F.; Bernardini, E. Extra virgin olive oil's polyphenols: Biological activities. *Curr. Pharm. Des.* **2011**, *17*, 786–804. [CrossRef] [PubMed]
3. Frankel, E.N. Nutritional and biological properties of extra virgin olive oil. *J. Agric. Food Chem.* **2011**, *59*, 785–792. [CrossRef] [PubMed]
4. Pérez-Jiménez, F.; Ruano, J.; Perez-Martinez, P.; Lopez-Segura, F.; Lopez-Miranda, J. The influence of olive oil on human health: Not a question of fat alone. *Mol. Nutr. Food Res.* **2007**, *51*, 1199–1208. [CrossRef] [PubMed]
5. Schwingshackl, L.; Christoph, M.; Hoffmann, G. Effects of Olive Oil on Markers of Inflammation and Endothelial Function—A Systematic Review and Meta-Analysis. *Nutrients* **2015**, *7*, 7651–7675. [CrossRef] [PubMed]
6. Azadmard-Damirchi, S.; Dutta, P.C. Phytosterol Classes in Olive Oils and their Analysis by Common Chromatographic Method. In *Olives and Olive Oil in Health and Disease Prevention*; Victor, R., Preedy, V.R., Watson, R.R., Eds.; Academic Press: Cambridge, MA, USA, 2010; Chapter 27; pp. 249–257.
7. Klikarova, J.; Rotondo, A.; Cacciola, F.; Ceslova, L.; Dugo, P.; Mondello, L.; Rigano, F. The Phenolic Fraction of Italian Extra Virgin Olive Oils: Elucidation through Combined Liquid Chromatography and NMR Approaches. *Food Anal. Methods* **2019**, *12*, 1759–1770. [CrossRef]
8. Rotondo, A.; Salvo, A.; Gallo, V.; Rastrelli, L.; Dugo, G. Quick unreferenced NMR quantification of Squalene in vegetable oils. *Eur. J. Lipid Sci. Technol.* **2019**, *119*, 1700151. [CrossRef]
9. EU. Commission Implementing Regulation (EU) No 1348/2013 of 16 December 2013 amending Regulation (EEC) No 2568/91 on the characteristics of olive oil and olive-residue oil and on the relevant methods of analysis. *Off. J. Eur. Union* **2013**, *L 338*, 31–67.
10. EU. Commission Delegated Regulation (EU) No 2015/1830 of 8 July 2015 amending Regulation (EEC) No 2568/91 on the characteristics of olive oil and olive-residue oil and on the relevant methods of analysis. *Off. J. Eur. Union* **2015**, *L 266*, 9–13.
11. Mannina, L.; Sobolev, A.P.; Segre, A. NMR. Olive oil as seen by NMR and chemometrics. *Spectroscopy* **2003**, *15*, 6–14.
12. Simmler, C.; Napolitano, J.G.; Mc Alpine, J.B.; Chen, S.-N.; Pauli, G.F. Universal quantitative NMR analysis of complex natural samples. *Curr. Opin. Biotechnol.* **2014**, *25*, 51–59. [CrossRef]
13. Salvo, A.; Rotondo, A.; La Torre, G.L.; Cicero, N.; Dugo, G. Determination of 1,2/1,3-diglycerides in Sicilian extra-virgin olive oils by 1H-NMR over a one-year storage period. *Nat. Prod. Res.* **2017**, *31*, 822–828. [CrossRef]
14. Bharti, S.K.; Roi, R. Quantitative 1H NMR spectroscopy. *Trends Anal. Chem.* **2012**, *35*, 5–26. [CrossRef]
15. Monakhova, Y.B.; Tsikin, A.M.; Kuballa, T.; Lachenmeiera, D.W.; Mushtakovab, S.P. Independent component analysis (ICA) algorithms for improved spectral deconvolution of overlapped signals in ^{1}H NMR analysis: Application to foods and related products. *Magn. Reson. Chem.* **2014**, *52*, 231–240. [CrossRef] [PubMed]
16. Ruiz-Aracama, A.; Goicoechea, E.; Guillén, M.D. Direct study of minor extra-virgin olive oil components without any sample modification. ^{1}H NMR multisupression experiment: A powerful tool. *Food Chem.* **2017**, *228*, 301–314. [CrossRef] [PubMed]
17. Mannina, L.; Sobolev, A.P. High resolution NMR characterization of olive oils in terms of quality, authenticity and geographical origin. *Magn. Reson. Chem.* **2011**, *49*, S3–S11. [CrossRef] [PubMed]
18. Salvo, A.; Rotondo, A.; Mangano, V.; Grimaldi, V.; Stillitano, I.; D'Ursi, A.M.; Dugo, G.; Rastrelli, L. High-resolution magic angle spinning nuclear magnetic resonance (HR-MAS-NMR) as quick and direct insight of almonds. *Nat. Prod. Res.* **2019**, *34*, 71–77. [CrossRef]

19. Rotondo, A.; Mannina, L.; Salvo, A. Multiple Assignment Recovered Analysis (MARA) NMR for a Direct Food Labeling, the Case Study of Olive Oils. *Food Anal. Methods* **2019**, *12*, 1238–1245. [CrossRef]
20. Christie, W.W. Preparation of Ester Derivatives of Fatty Acids for Chromatographic Analysis. *Adv. Lipid Methodol.* **1993**, *2*, 69–111.
21. Naccari, C.; Rando, R.; Salvo, A.; Donato, D.; Bartolomeo, G.; Mangano, V.; Lo Turco, V.; Dugo, G. Study on the composition and quality of several sicilian EVOOs (harvesting year 2015). *Rivista Italiana Delle Sostanze Grasse* **2017**, *94*, 231–237.
22. Zhang, H.; Wang, Z.; Liu, O. Development and validation of a GC–FID method for quantitative analysis of oleic acid and related fatty acids. *J. Pharm. Anal.* **2015**, *5*, 223–230. [CrossRef] [PubMed]
23. Scano, P.; Casu, M.; Lai, A.; Saba, G.; Dessi, M.A.; Deiana, M.; Corongiu, F.P.; Bandino, G. Recognition and quantitation of cis-vaccenic and eicosenoic fatty acids in olive oils by 13C nuclear magnetic resonance spectroscopy. *Lipids* **1999**, *34*, 757–759. [CrossRef] [PubMed]
24. Barison, A.; Pereira da Silva, C.W.; Ramos Campos, F.; Simonelli, F.; Lenz, C.A.; Ferreira, A.G. A simple methodology for the determination of fatty acid composition in edible oils through ^{1}H NMR spectroscopy. *Magn. Reson. Chem.* **2010**, *48*, 642–650. [CrossRef] [PubMed]
25. Knothe, G.; Kenar, J.A. Determination of the fatty acid profile by 1H-NMR spectroscopy. *Eur. J. Lipid Sci. Technol.* **2004**, *106*, 88–96. [CrossRef]
26. Boudour-Benrachou, N.; Plard, J.; Pinatel, C.; Artaud, J.; Dupuy, N. Fatty acid compositions of olive oils from six cultivars from East and South-Western Algeria. *Adv. Food Technol. Nutr. Sci. Open J.* **2017**, *3*, 1–5. [CrossRef]
27. Retief, L.; McKenzie, J.M.; Koch, K.R. A novel approach to the rapid assignment of 13C NMR spectra of major components of vegetable oils such as avocado, mango kernel andmacadamia nut oils. *Magn. Reson. Chem.* **2009**, *47*, 771–781. [CrossRef]
28. Chira, N.A.; Nicolescu, A.; Stan, R.; Rosca, S. Fatty Acid Composition of Vegetable Oils Determined from 13C-NMR Spectra. *Rev. Chim.* **2016**, *67*, 1257–1263.
29. Vlahov, G.; Schiavone, C.; Simone, N. Quantitative ^{13}C NMR method using the DEPT pulse sequence for the determination of the geographical origin (DOP) of olive oils. *Magn. Reson. Chem.* **2001**, *39*, 689–695. [CrossRef]
30. Zamora, R.; Alba, V.; Hidalgo, F.J. Use of High-Resolution ^{13}C Nuclear Magnetic Resonance Spectroscopy for the Screening of Virgin Olive Oils. *J. Am. Oil Chem. Soc.* **2001**, *78*, 89–94. [CrossRef]
31. Ng, S.; Koh, H.F. Detection of *cis-Vaccenic* Acid in Palm Oil by 13C NMR Spectroscopy. *Lipids* **1988**, *23*, 140–143. [CrossRef]
32. Kuznetsova, E.I.; Pchelkin, V.P.; Tsydendambaev, V.D.; Vereshchagin, A.G. Distribution of unusual fatty acids in the mesocarp triacylglycerols of maturing sea buckthorn fruits. *Russ. J. Plant Physiol.* **2010**, *57*, 852–858. [CrossRef]
33. Vlahov, G.; Chepkwony, P.K.; Ndalut, P.K. 13C NMR Characterization of Triacylglycerols of *Moringa oleifera* Seed Oil: An "Oleic-Vaccenic Acid" Oil. *J. Agric. Food Chem.* **2002**, *50*, 970–975. [CrossRef] [PubMed]
34. Barding, G.A.; Béni, S.; Fukao, T.; Bailey-Serres, J.; Larive, C.K. Comparison of GC-MS and NMR for Metabolite Profiling of Rice Subjected to Submergence Stress. *J. Proteome Res.* **2012**, *12*, 898–909. [CrossRef] [PubMed]
35. Romano, R.; Giordano, A.; Le Grottaglie, L.; Manzo, N.; Paduano, A.; Sacchi, R.; Santini, A. Volatile compounds in intermittent frying by gas chromatography and nuclear magnetic resonance. *Eur. J. Lipid Sci. Technol.* **2013**, *115*, 764–773. [CrossRef]
36. Shibaharaa, A.; Yamamoto, K.; Nakayama, T.; Kajimotoa, G. *cis*-Vaccenic Acid in Pulp Lipids of Commonly Available Fruits. *J. Am. Oil Chem. Soc.* **1987**, *64*, 397–398. [CrossRef]
37. Circi, S.; Capitani, D.; Randazzo, A.; Ingallina, C.; Mannina, L.; Sobolev, A. Panel test and chemical analyses of commercial olive oils: A comparative study. *Chem. Biol. Technol. Agric.* **2017**, *4*, 1–10. [CrossRef]
38. Naviglio, D.; Romano, R.; Pizzolongo, F.; Santini, A.; Vito, A.D.; Schiavo, L.; Nota, G.; Musso, S.S. Rapid determination of esterified glycerol and glycerides in triglyceride fats and oils by means of periodate method after transesterification. *Food Chem.* **2007**, *102*, 399–405. [CrossRef]

© 2020 by the authors. Licensee MDPI, Basel, Switzerland. This article is an open access article distributed under the terms and conditions of the Creative Commons Attribution (CC BY) license (http://creativecommons.org/licenses/by/4.0/).

Article

Bioactivity-Guided Fractionation and NMR-Based Identification of the Immunomodulatory Isoflavone from the Roots of *Uraria crinita* (L.) Desv. ex DC

Ping-Chen Tu [1], Chih-Ju Chan [2], Yi-Chen Liu [3], Yueh-Hsiung Kuo [2,4,5], Ming-Kuem Lin [2,*] and Meng-Shiou Lee [2,*]

1. Program for Cancer Biology and Drug Discovery, China Medical University and Academia Sinica, Taichung 404, Taiwan; pingchen.tu@gmail.com
2. Department of Chinese Pharmaceutical Sciences and Chinese Medicine Resources, China Medical University, Taichung 404, Taiwan; ruby24301@gmail.com (C.-J.C.); kuoyh@mail.cmu.edu.tw (Y.-H.K.)
3. Institute of Biomedical Science and Rong Hsing Research Center for Translational Medicine, National Chung-Hsing University, Taichung 402, Taiwan; sealioler@gmail.com
4. Department of Biotechnology, Asia University, Taichung 413, Taiwan
5. Chinese Medicine Research Center, China Medical University, Taichung 404, Taiwan
* Correspondence: linmk@mail.cmu.edu.tw (M.-K.L.); leemengshiou@mail.cmu.edu.tw (M.-S.L.); Tel.: +886-4-2205-3366 (M.-K.L.); +886-4-2205-3366 (M.-S.L.)

Received: 7 August 2019; Accepted: 10 October 2019; Published: 3 November 2019

Abstract: *Uraria crinita* is used as a functional food ingredient. Little is known about the association between its immunomodulatory activity and its metabolites. We applied a precise strategy for screening metabolites using immunomodulatory fractions from a *U. crinata* root methanolic extract (UCME) in combination with bioactivity-guided fractionation and NMR-based identification. The fractions from UCME were evaluated in terms of their inhibitory activity against the production of pro-inflammatory cytokines (IL-6 and TNF-α) by lipopolysaccharide (LPS)-stimulated mouse bone marrow-derived dendritic cells (BMDC). The role of the isoflavone genistein was indicated by the ^{1}H NMR profiling of immunomodulatory subfractions (D-4 and D-5) and supported by the result that genistein-knockout subfractions (D-4 w/o and D-5 w/o) had a lower inhibitory activity compared to genistein-containing subfractions. This study suggests that genistein contributes to the immunomodulatory activity of UCME and will help in the standardization of functional food.

Keywords: *Uraria crinita*; isoflavone; genistein; NMR-based identification; dendritic cells

1. Introduction

Uraria crinita (L.) Desv. ex DC. (UC) belongs to the family of Leguminosae. It is a popular and commercially important medicinal plant, distributed widely in Taiwan and cultivated mostly in Mingjian Township, Nantou (approximately 20–60 hectares per year). UC roots, also known as "Taiwanese ginseng" due to the similar potency and aroma of its decoction and ginseng, have been traditionally used to coordinate the gastrointestinal system, thanks to their detumescent and antipyretic effects, indicating immunomodulatory activity [1]. UC roots are used as dietary supplements for treating childhood skeletal dysplasia. UC roots have therefore been developed as valuable and commercial functional food in Taiwan. This herb has been shown to have antioxidant [2] and antidiabetic activities [3] and the potential to stimulate bone formation and regeneration [4]. Previous phytochemical investigations on UC roots led to the isolation of fatty acids, steroids, triterpenoids, phenolics, lignans, flavonoids, and isoflavonoids [2,4–6]. However, little is known about the association between the immunomodulatory activity and the metabolites in this herb.

Dendritic cells (DCs), acting as antigen-presenting cells (APCs), are the major leukocytes, with a critical role in regulating adaptive immune responses [7]. Immature DCs, characterized by a high antigen uptake ability and poor antigen-presenting function, reside in the peripheral tissues, where they regularly uptake and process self-antigens and maintain self-tolerance [7]. Upon activation, immature DCs undergo maturation and migrate to adjacent lymph nodes or to the lymph organs, after the recognition of pathogen-associated molecular patterns and damage-associated molecular patterns by pattern recognition receptors, mostly Toll-like receptors (TLRs) [8]. This process is accompanied by the upregulation of the expression of major histocompatibility complex (MHC) class II molecules and several co-stimulatory molecules (CD40, CD80, and CD86) on the surface of cells [9]. Mature DCs generate more pro-inflammatory cytokines (TNF-α, IL-6, and IL-12) required for T cell activation and have the ability to present antigens to T cells, linking the innate and adaptive immune systems [9]. Therefore, targeting DCs is a promising strategy for immunomodulation. Medicinal herbs, which modulate the function of DCs, can potentially be developed into botanical drugs for treating immune disorders.

The promising potential of the herbal industry can only be achieved through the standardization of the composition of herbal products and the assurance of proper quality control [10]. However, the variable constituents of herbal products, owing to genetic, cultural, and environmental factors, have made the quality of herbal medicines difficult to control. Nuclear magnetic resonance (NMR) has been demonstrated as one of the best analytical approaches to identifying metabolites and providing both qualitative and quantitative information [11]. While NMR suffers from a relatively poor sensitivity compared to mass spectrometry (MS), it has some unique advantages over MS-based approaches, including a non-destructive nature, high robustness, high reproducibility, high reliability, and powerful ability to provide structural information for unknown metabolites [12]. NMR spectroscopy could offer structural elucidation, achieved by the chemical shift, multiplicity, coupling constant, and integration of (primary or secondary) metabolite signals in crude extracts. Additionally, the ^{1}H NMR signal intensity, proportional to the relative number of protons, could provide useful information about the quantity of the different metabolites in herbal extracts. Therefore, NMR has been successfully used for fingerprinting and the metabolite discrimination of herbs [13,14]. Currently, NMR-based metabolomic analysis is widely used in studies of nutrients, environment, plant physiology, drug metabolism, toxicology, as well as for diagnoses and for the quality control of herbal products [10,15,16].

In this study, a combination of bioactivity-guided fractionation and NMR-based identification led to the elucidation of the central role of the immunomodulatory isoflavone genistein present in UC root methanolic extract (UCME) against the activation and maturation of lipopolysaccharide (LPS)-stimulated DCs. This strategy could prevent unnecessary time-consuming isolation procedures and provide a rapid tool for the identification of the active ingredients of herbs. Our findings suggest that UC roots can be applied as an immunosuppressive functional food, of which genistein can be a chemical marker for quality control. The standardization of UC roots could therefore be improved.

2. Results and Discussion

2.1. The EtOAc-Soluble Fraction from UCME Inhibited LPS-Stimulated DC Activation

The immunomodulatory effects of UC roots on DCs have not been reported. Here, the Gram-negative bacterial endotoxin LPS was used to stimulate bone marrow-derived dendritic cell (BMDC) activation, as a model for investigating the immunomodulatory effects of UC roots on DCs. First, the immunomodulatory effects of UCME and its EtOAc-, n-BuOH-, and H$_2$O-soluble fractions were evaluated against the production of pro-inflammatory cytokines, including TNF-α and IL-6, which is a hallmark of DC activation. As shown in Figure 1A, LPS-stimulated BMDC activation was suppressed by UCME, and UCME ability to inhibit DC activation was mainly associated with its EtOAc-soluble fraction. In addition, the treatment with UCME and its various partitioned fractions, at concentrations below 100 µg/mL, did not exhibit any cytotoxicity in BMDC (data not shown). In summary, our results revealed that the EtOAc-soluble fraction of UCME may contain immunomodulatory phytochemicals which attenuate the activity of DCs.

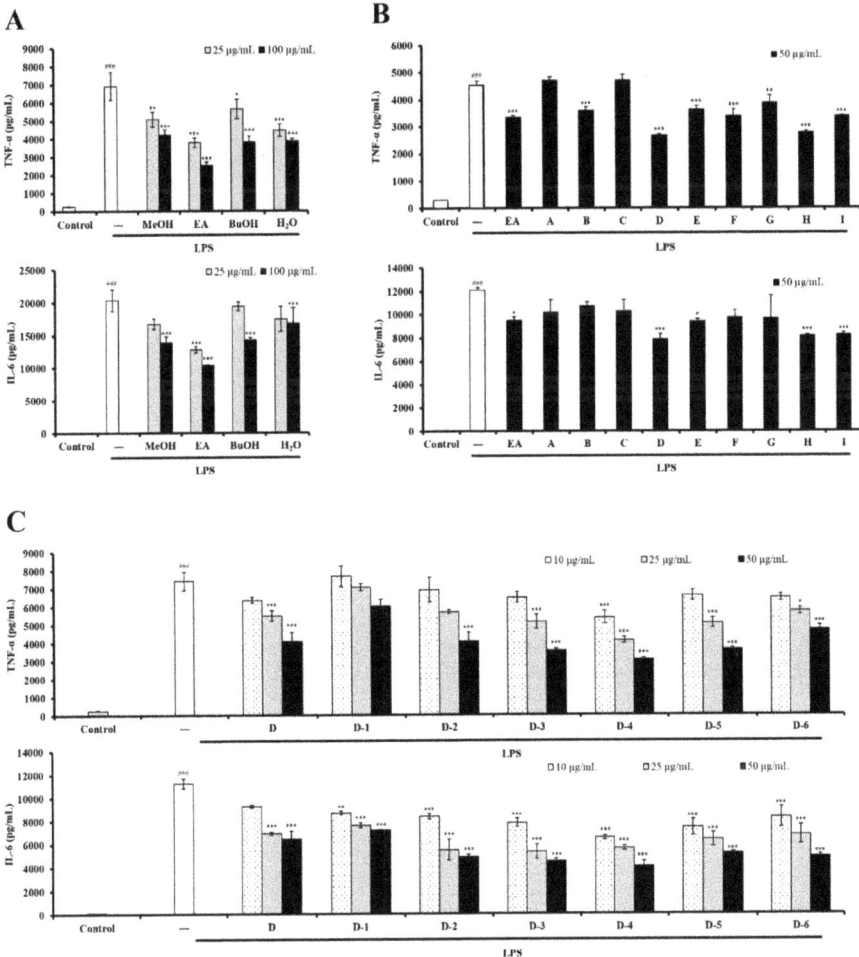

Figure 1. The effects of UC methanolic extract and of its EtOAc-, n-BuOH-, H$_2$O-soluble fractions and subfractions of the EtOAc-soluble fraction on pro-inflammatory cytokine production in LPS-stimulated dendritic cells (DCs). DCs were untreated or treated with LPS (100 ng/mL, white bar). (**A**) Methanolic extract and EtOAc, n-BuOH, and H$_2$O subfractions (25 µg/mL, gray bar, or 100 µg/mL, black bar) were used. (**B**) EtOAc and Fr. A to I subfractions (50 µg/mL, black bar) were used. (**C**) Fr. D and D-1 to D-6 subfractions (10 µg/mL, hatch bar, 25 µg/mL, gray bar, or 50 µg/mL, black bar) were used. Supernatants were collected 6 h after the treatment. The production of cytokines (TNF-α and IL-6) was measured by ELISA. The data shown are the mean ± SD of three independent experiments; ### $p < 0.001$; * $p < 0.05$; ** $p < 0.01$; *** $p < 0.001$ (Scheffe's test) for comparisons of the treated and untreated LPS-stimulated DC samples.

2.2. Bioactivity-Guided Fractionation and NMR-Based Identification of the EtOAc-Soluble Fraction of UCME

The EtOAc-soluble fraction of UCME was subjected to silica gel column chromatography (EtOAc/n-hexane/MeOH, gradient), and each collected fraction was analyzed by thin-layer chromatography (TLC) and assign to one of nine main fractions (Fr. A to I, Figure 2).

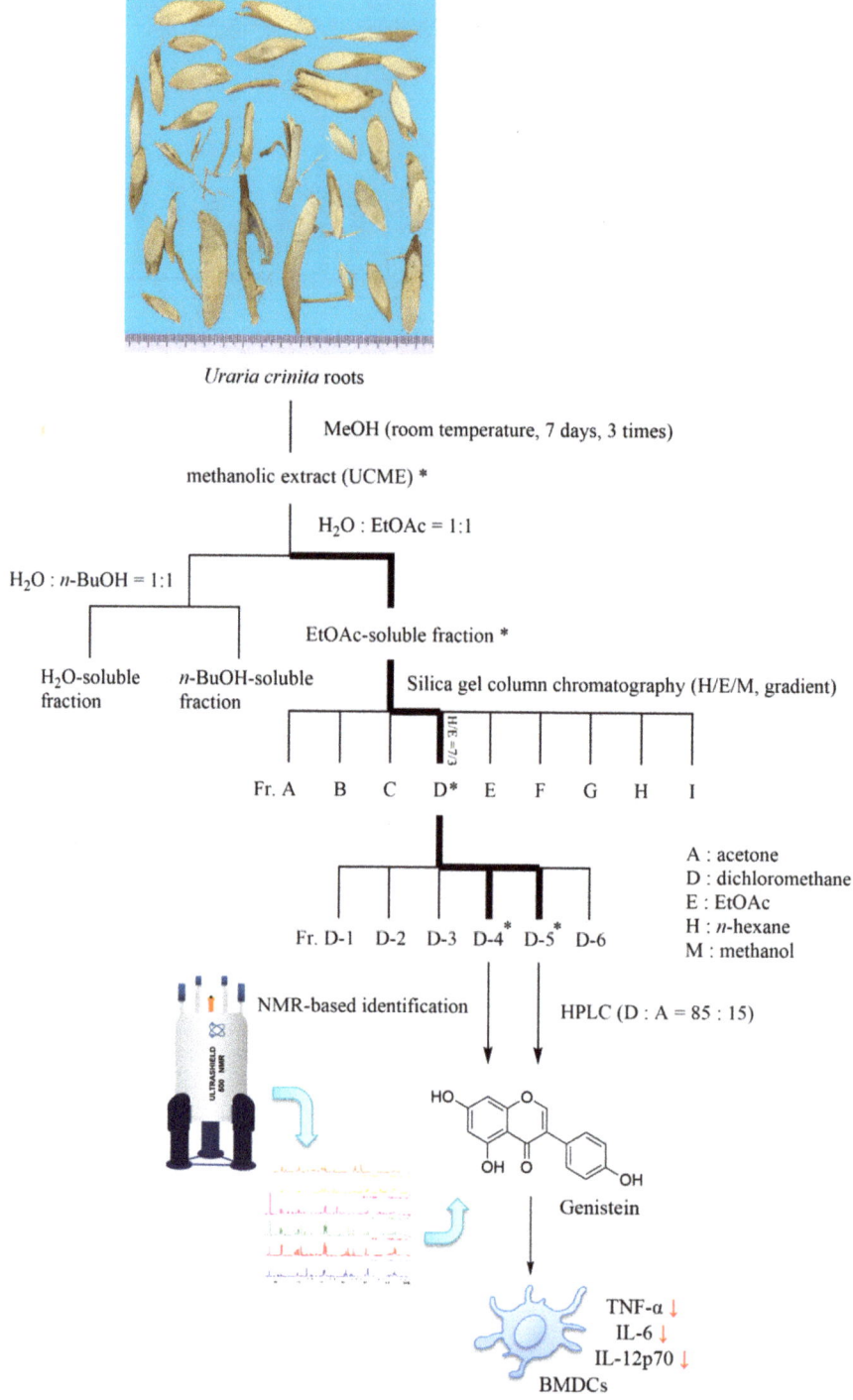

Figure 2. Bioactivity-guided fractionation and NMR-based identification of genistein from the roots of *Uraria crinita* (UC). * indicates the most potent subfractions or constituents against pro-inflammatory cytokine production in lipopolysaccharide (LPS)-stimulated DCs. UCME: UC root methanolic extract, BMDCs: bone marrow-derived dendritic cells.

Among them, fraction D significantly inhibited the production of TNF-α and IL-6 in BMDC (Figure 1B). Furthermore, the subfractions D-4 and D-5 from fraction D indicated the most potent inhibitory effects against DC activation (Figure 1C).

In order to elucidate the association between bioactivity and metabolites in subfractions D-1 to D-6, ^{1}H NMR spectroscopy was conducted (Figure 3). This could offer structural elucidation, achieved by the chemical shift, multiplicity, coupling constant, and integration of metabolite signals in the mixtures.

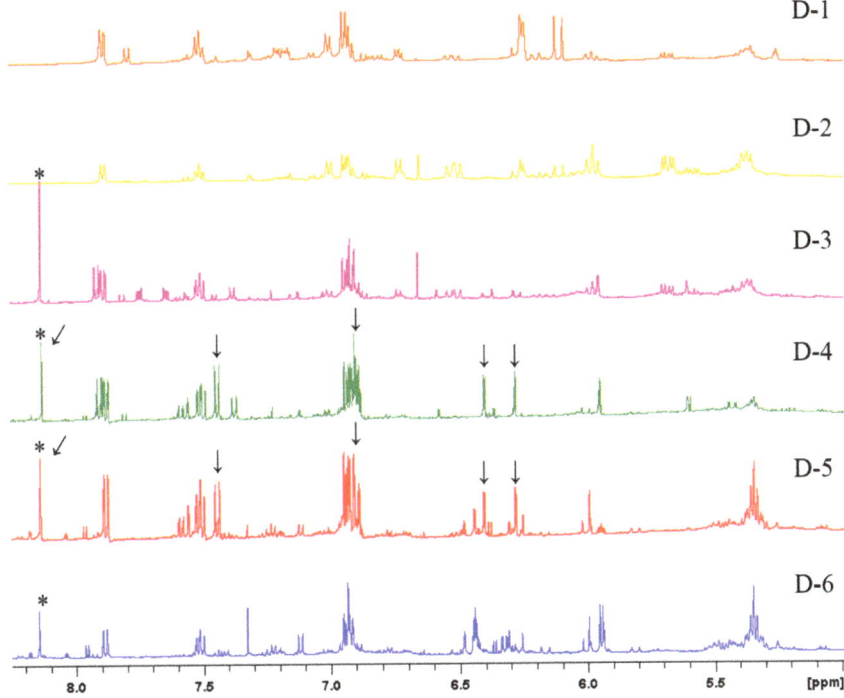

Figure 3. Selected ^{1}H NMR profiling (acetone-d_6, 500 MHz) of subfractions D-1 to D-6. * indicates the characteristic singlet signals, δ_H 8.13, of isoflavones; ↓ indicates the ^{1}H spectral data of genistein: δ_H 7.44 (2H, d, J = 8.7 Hz), 6.88–6.94 (overlaps), 6.40 (1H, d, J = 2.2 Hz), and 6.28 (1H, d, J = 2.2 Hz).

According to the ^{1}H NMR profiling of subfractions D-1 to D-6, the characteristic singlet signals (δ_H 8.13) indicated the presence of isoflavonoids in D-3, D-4, D-5, and D-6. The inhibitory activity of subfractions D-4 and D-5 was then related to the additional signals [δ_H 7.44 (2H, d, J = 8.7 Hz), 6.88–6.94 (overlaps), 6.40 (1H, d, J = 2.2 Hz), and 6.28 (1H, d, J = 2.2 Hz)], which were not visible for subfractions D-3 and D-6. A pair of aromatic protons with a *meta* coupling (J = 2.2 Hz) indicated the presence of a 1,3,4,5-tetrasubstituted aromatic ring. An aromatic proton signal at δ_H 7.44 (2H, d, J = 8.7 Hz) revealed that the other proton signal might be overlapping at δ_H 6.88–6.94 ppm, suggestive of a 1,4-disubstituted aromatic ring. To determine the overlapping peaks at δ_H 6.88–6.94 ppm, the ^{13}C NMR and 2D NMR experiments (Figures S1–S3), including heteronuclear single-quantum correlation (HSQC) and heteronuclear multiple-bond correlation (HMBC), were then conducted. On the basis of these results, genistein was identified [17] from the genistein-containing subfractions (D-4 and D-5). Bioactive fractions, together with the active ingredient, were rapidly obtained from the combination of bioactivity-guided fractionation and NMR-based identification of the UCME EtOAc-soluble fraction.

2.3. The Role of Genistein in Modulating LPS-Stimulated DC Activation of UCME

To clarify the role of genistein in modulating LPS-stimulated DC activation, the genistein-containing subfractions D-4 and D-5 were further isolated by normal-phase semipreparative HPLC to isolate genistein (24 mg/1.136 kg dry material) and genistein-knockout subfractions (D-4 w/o and D-5 w/o, Figure S4). The yield of genistein was 15 times higher than that achieved in a previous study [6]. These subfractions were then evaluated in terms of their inhibitory activity by measuring cytokine production in LPS-stimulated DCs. As shown in Figure 4A, the genistein-containing subfractions (D-4 and D-5) significantly suppressed the production of cytokines (IL-6 and TNF-α), while the genistein-knockout subfractions D-4 w/o and D-5 w/o showed lower inhibitory activity. As illustrated in Figure 4B, concerning TNF-α production, the inhibition percentage of the genistein-knockout subfraction D-4 w/o (9% at 25 µg/mL and 18% at 50 µg/mL) was dramatically lower than that of the genistein-containing subfraction D-4 (35% at 25 µg/mL and 55% at 50 µg/mL).

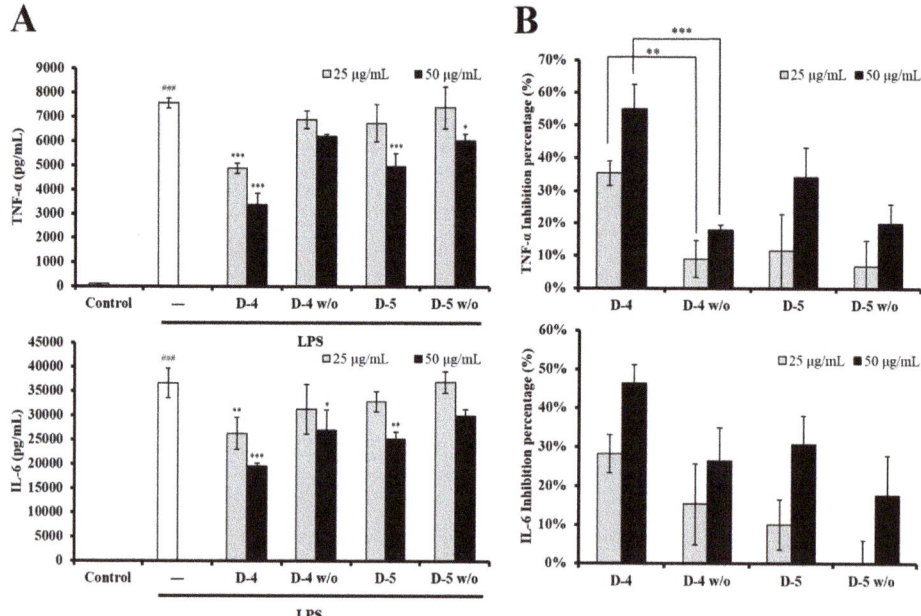

Figure 4. The effects of the subfractions, containing or not containing genistein, on the production of pro-inflammatory cytokines in LPS-stimulated DCs. (**A**) DCs were untreated or treated with LPS (100 ng/mL, white bar), LPS + genistein-containing subfractions, or LPS + genistein-knockout subfractions (25 µg/mL, gray bar or 50 µg/mL, black bar), as indicated. Supernatants were collected 6 h after the treatment. The production of cytokines (TNF-α and IL-6) was measured by ELISA. The data shown are the mean ± SD of three independent experiments. ### $p < 0.001$; * $p < 0.05$; ** $p < 0.01$; *** $p < 0.001$ (Scheffe's test) for comparisons of the treated and untreated LPS-stimulated DCs. (**B**) The inhibition percentage (%) of the subfractions, with and without genistein, of cytokine production was derived from the data in (A).

In addition to the identification of genistein, the fractionation and HPLC isolation of UCME identified eight compounds, including an isoflavone (lupinalbin A [18]), three phenolic acids (p-hydroxybenzoic acid [19], salicylic acid [20], and vanillic acid [19]), two fatty acids (monomethyl succinate [21] and 1,10-decanedioic acid [22]), and two steroids (a mixture of β-sitosterol and stigmasterol [23]). Their structures (Figure 5) were identified by comparing their spectroscopic data with data in the literature. Among the isolates, lupinalbin A, p-hydroxybenzoic acid, vanillic acid, monomethyl succinate, and monomethyl succinate were isolated from this herb for the first time. All isolates were also assessed in terms of their inhibitory effect on DC activation. However, only the analogue of genistein, lupinalbin A, showed a moderate inhibitory activity against LPS-stimulated DC activation (Figures S5 and S6).

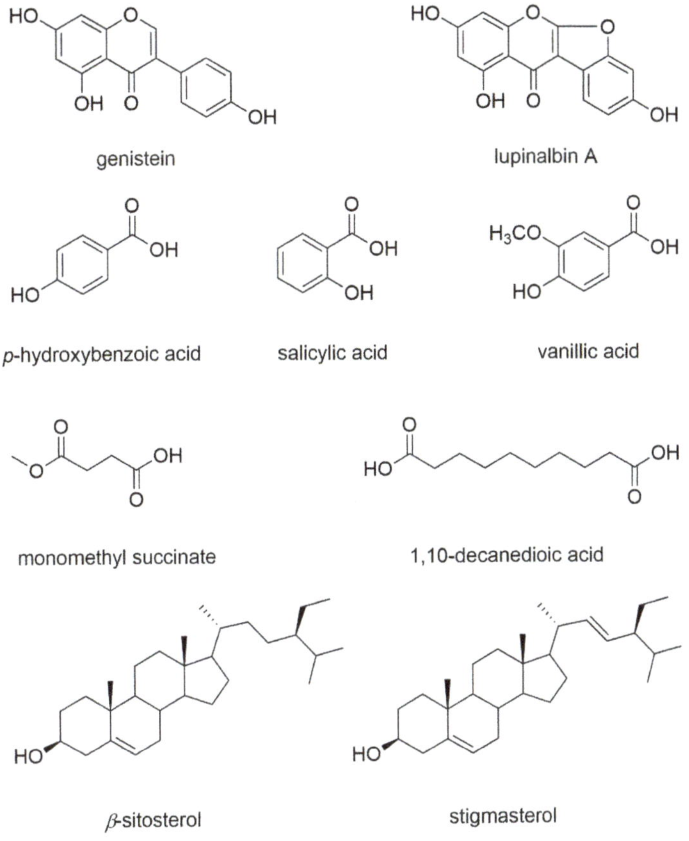

Figure 5. Chemical structures of the compounds identified in this study.

^{1}H NMR signal intensity is absolutely proportional to the relative number of protons. Therefore, the relative quantity of the different metabolites could easily be observed in the ^{1}H NMR spectra (Figure 3). The major metabolites (identified by signals with a stronger intensity) in subfractions D-4 and D-5 are shown in Table 1. Interestingly, the ^{1}H NMR profiling of subfractions D-4 and D-5 revealed the presence of other unknown genistein derivatives, which might contribute slightly to the immunomodulatory effects, because of their low abundance or poor activity. Additionally, the well-known bioactive isoflavonoids, including daidzein, formononetin, equol, and glycitein, were not present as major metabolites in fraction D. Therefore, genistein was suggested as having the central role in the modulation by UCME of LPS-stimulated DC activation.

2.4. LPS-Stimulated DC Maturation was Impaired by Genistein at Non-Cytotoxic Concentrations

To exclude the possibility that genistein caused cytotoxicity, the cell viability of DCs was determined via the CCK8 assay. As shown in Figure 6, genistein induced a significant level of DC death at 40 µM. Thus, the significant inhibition of cytokine production by 40 µM genistein should be attributed to its cytotoxicity in DCs. However, the LPS-stimulated production of pro-inflammatory cytokines

(TNF-α, IL-6, and IL-12) was suppressed by genistein below 20 µM, suggesting that genistein possesses an immunosuppressive activity. Therefore, we selected concentrations below 20 µM of genistein for further assessment of DC maturation. The complex process of DC maturation is accompanied by the upregulation of the expression of MHC class II molecules and three major co-stimulatory molecules (CD40, CD80, and CD86) on the surface of DCs. As shown in Figure 7, LPS stimulation upregulated the expression of MHC class II and also of the co-stimulatory molecules (CD40, CD80, and CD86) in DCs, while genistein treatment significantly decreased the expression levels of all these molecules. These data indicated that genistein indeed impaired LPS-stimulated DC maturation at non-cytotoxic concentrations.

Table 1. Chemical shifts (δ_H) and assignment of major compounds present in the D-4 and D-5 subfractions.

Compounds	δ_H (mult, J in Hz)
Genistein	13.01 (s), 8.13 (s), 7.44 (d, 8.7), 6.88–6.94 [1], 6.40 (d, 2.2), 6.28 (d, 2.2)
p-Hydroxybenzoic acid [2]	7.90 (d, 8.9), 6.88–6.94 [1]
Salicylic acid	7.88 (dd, 7.6, 1.7), 7.55 (ddd, 8.9, 7.6. 1.8), 6.88–6.94 [1]
Vanillic acid	7.58 (dd, 8.2, 2.0), 6.88–6.94 [1], 3.89 (s)

[1] Overlapping peaks at 6.88–6.94 ppm. [2] Only in the D-4 subfraction.

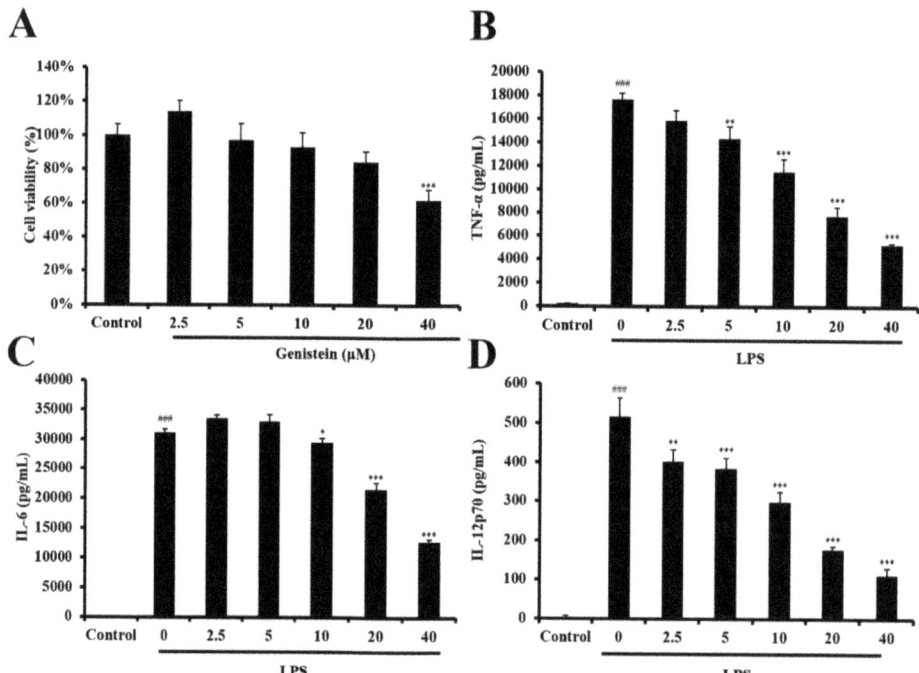

Figure 6. The effect of genistein on cell viability (A) and the production of pro-inflammatory cytokines in LPS-stimulated DCs (B, C, and D). (A) DCs were treated with genistein at various concentrations for 24 hours, and then their viability was measured by the CCK-8 assay (Sigma). DCs were untreated or treated with LPS (100 ng/mL) and LPS + genistein (2.5, 5, 10, 20, and 40 µM), as indicated. Supernatants were collected 6 h after treatment. The production of the cytokines TNF-α (B), IL-6 (C), and IL-12p70 (D) was measured by ELISA. The data shown are the mean ± SD of three independent experiments. ### $p < 0.001$; * $p < 0.05$; ** $p < 0.01$; *** $p < 0.001$ (Scheffe's test) for comparisons of the genistein-treated and untreated LPS-stimulated DCs.

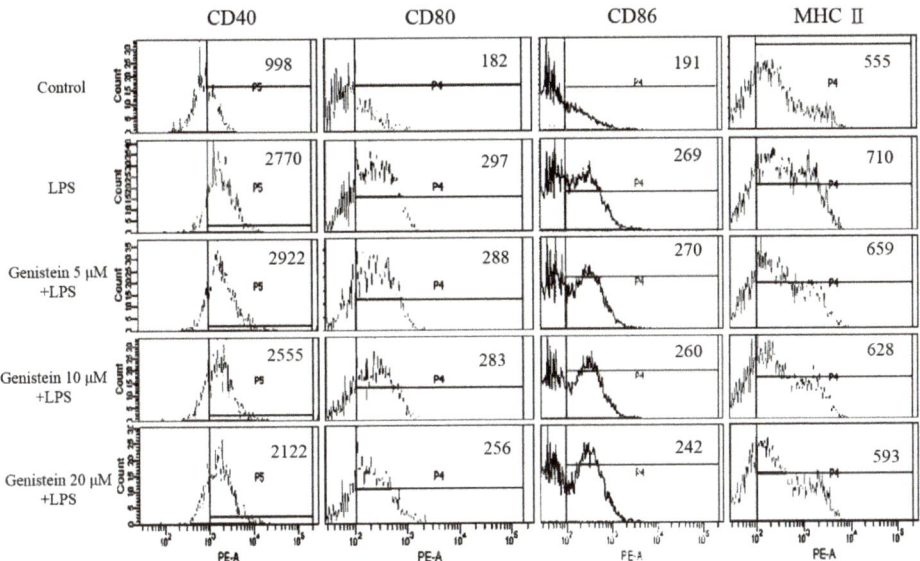

Figure 7. Genistein-mediated suppression of a maturation-associated surface marker on LPS-stimulated DCs. DCs were untreated or treated with LPS or LPS + genistein for 24 h. The suppression of major histocompatibility complex (MHC) class II and of co-stimulatory molecules (CD40, CD80, and CD86) was analyzed by flow cytometry. All of the data shown were gated on CD11c$^+$ cells. All results are representative of three independent experiments.

Isoflavones, including genistein, daidzein, and glycitein, are generally found in leguminous plants and have been reported as antioxidants and immunosuppressant agents, capable of suppressing the allergic sensitization to peanuts by regulating human monocyte-derived dendritic cell function [24]. The majority of the dietary isoflavonoids are present in inactive glycosides forms (e.g., genistin) and then converted to active aglycone forms (e.g., genistein) by the bacterial microbiote in the digestive tract. Thus, DCs have direct access to dietary antigens and are therefore poised to uptake isoflavones directly from the lumen [24]. Previously, genistein was shown to have promising activities, such as neuroprotective effects by improving hippocampus neuronal cell viability and proliferation in vitro [25], antioxidant capacity by regulating β-oxidation and energy metabolism in vivo [26], and anti-inflammatory effects by inhibiting the ERK pathway [27] and NF-κB-dependent gene expression in TLR4-stimulated DCs [28].

In this study, the association between genistein and the immunomodulatory effect of UCME was carefully elucidated through the combination of bioactivity-guided fractionation and NMR-based identification. An ^{1}H NMR-based metabolomics approach was applied to partially purified subfractions D-1 to D-6 from UCME. ^{1}H NMR profiling suggested the presence of genistein in the D-4 and D-5 subfractions, which exhibited a stronger inhibitory activity against cytokine production in LPS-stimulated DCs. Genistein was therefore supposed to be a possible marker of the immunosuppressive activity of UCME. This suggestion was supported by the results of the genistein-knockout subfractions, which showed a much lower inhibitory activity. Moreover, HPLC isolation of the D-4 and D-5 subfractions provided other compounds with poor inhibitory activity. On the basis of this evidence, genistein could be used as a chemical marker for the quality control of the potentially immunosuppressive functional food UC roots. A literature survey disclosed that the ^{1}H NMR-based metabolomics approach for screening bioactive secondary metabolites has not been widely used [29,30]. Our study provides a powerful tool for discovering the active ingredients in immunosuppressive functional foods. This approach can be applied for investigating the active ingredients of herbal products.

3. Materials and Methods

3.1. General Experimental Procedures

Column chromatography (CC) was performed on Silica gel 60 (40–63 µm, Merck, Darmstadt, Germany). Thin-layer chromatography (TLC) was performed on silica gel 60 F254 plates (200 µm, Merck). High-performance liquid chromatography (HPLC) was performed, using Keystone Spherisorb silica (5 µm, 250 × 10 mm), on a Knauer Smartline 2400 refractive index (RI) detector and a Knauer Smartline 100 pump. The NMR experiments were performed on a Bruker DRX-500 NMR spectrometer (Bruker, Rheinstetten, Germany). Flow cytometry was conducted using a BD FACSCanto II Flow Cytometer (BD Biosciences, CA, USA).

3.2. Sample Preparation and Isolation

UC was purchased from Mingjian Township, Nantou, Taiwan. The procedure of extraction and isolation is summarized in Figure 2. Briefly, the roots (1.136 kg) were pulverized into a fine powder and extracted with methanol (12 L). The supernatant was collected and concentrated under reduced pressure to obtain the methanolic extract (UCME, 80 g). A portion of the residue (60 g) was suspended in H_2O and sequentially fractionated with EtOAc and n-BuOH. The EtOAc-soluble fraction (9.6 g) was then subjected to silica-gel column chromatography (150 g, 70–230 mesh), using a gradient solvent system of n-hexane, EtOAc, and MeOH as a mobile phase. Each fraction, from which a sample was collected for the immunomodulating assessment, was analyzed by thin-layer chromatography (TLC) and assigned to one of 9 main fractions (Fr.A to Fr.I). A mixture of β-sitosterol and stigmasterol was obtained from fraction C (n-hexane/EtOAc = 9/1). The subfractions of fraction D (n-hexane/EtOAc = 7/3) were further analyzed by ^{1}H NMR, indicating the presence of genistein in the D-4 and D-5 subfractions. The genistein-knockout subfractions (D-4 w/o and D-5 w/o) were obtained by HPLC and evaluated in terms of their immunomodulating activity.

3.3. HPLC Conditions Used for the D-4 and D-5 Subfractions

D-4 and D-5 were isolated by semipreparative HPLC (dichloromethane/acetone = 85/15, flow rate = 3 mL/min) to obtain the genistein-knockout subfractions D-4 w/o, D-5 w/o and genistein (24.0 mg, t_R = 8.0 min). The genistein-knockout subfractions D-4 w/o and D-5 w/o were further isolated by semipreparative HPLC (n-hexane/acetone = 2/1, flowrate = 3 mL/min) to obtain salicylic acid (t_R = 8.3 min), lupinalbin A (t_R = 11.0 min), 1,10-decanedioic acid (t_R = 11.3 min), monomethyl succinate (t_R = 13.2 min), p-hydroxybenzoic acid (t_R = 16.7 min), and vanillic acid (t_R = 18.5 min).

3.4. NMR Analysis

The samples were dissolved in the deuterated solvent acetone-d_6 and put into a 5 mm NMR tube. All experiments were performed on a Bruker DRX-500 NMR spectrometer (Bruker, Rheinstetten, Germany), operating at a frequency of 500 MHz for ^{1}H NMR observation, and 125 MHz for ^{13}C NMR observation (at room temperature). The 2D NMR experiments included heteronuclear single-quantum correlation (HSQC) and heteronuclear multiple-bond correlation (HMBC). NMR spectra were carefully processed with the TOPSPIN2.1®software (Bruker). The spectra recorded in acetone-d_6 were referenced to the solvent signal at $δ_H$ 2.05 ppm and $δ_C$ 29.92 ppm.

3.5. Preparation of BMDC

The ICR mice used in this study were obtained from the National Laboratory Animal Center (NLAC, Taipei, Taiwan). The mouse bone marrow-derived DCs were prepared as described previously [31]. Bone marrow cells were isolated from tibias and femurs and then seeded on 6-well plates (Corning) in 4 mL/well RPMI 1640 medium (Thermo), with 10% FBS and 10 ng/mL recombinant mouse GM-CSF and IL-4 (Peprotech). On day 3 and 5, a 2 mL/well fresh medium containing 10 ng/mL GM-CSF and IL-4 was added. On day 7, BMDCs (> 80% CD11c$^+$ cells) were harvested and used for all experiments.

3.6. Measurement of Cytokine Production

Cytokine production was measured using an enzyme-linked immunosorbent assay (ELISA), as described previously [31]. The DCs were treated with 100 ng/mL lipopolysaccharide (LPS) (Sigma) or LPS + sample for 6 h for TNF-α, IL-6, and IL-12p70 determination. The production of cytokines was measured using the ELISA kit (eBioscience).

3.7. Cytotoxicity Assessment

DCs were treated with genistein (2.5, 5, 10, 20, and 40 µM) for 24 h. The cells were then measured in terms of their cell viability by the CCK-8 assay (Sigma), according to standard protocols, as described previously [32]. Triplicate treatments were performed for each sample in all experiments.

3.8. Analysis of DC Maturation

Maturation was determined by measuring the upregulation of MHC class II and three co-stimulatory molecules (CD40, CD80, and CD86), as described previously [31,32]. DCs were untreated or treated with LPS (100 ng/mL) or LPS + genistein (5, 10, and 20 µM) for 24 h. Cell aggregation was examined by microscopy (40×). Then, the cells were stained with monoclonal antibodies (mAbs), specific to mouse CD11c, MHC class II, CD40, CD80, and CD86 (Biolegend), and analyzed by flow cytometry. The fluorescence intensity of MHC class II, CD40, CD80, and CD86 was determined, following gating with a forward side scatter (FSC) and CD11c$^+$ expression. The change in the mean fluorescence intensity (MFI) from LPS alone to LPS + genistein was indicated.

3.9. Data Analysis

The significance of the suppressions was determined using one-way ANOVA, followed by Scheffe's test. A value of * $p < 0.05$ was considered significant. Values of ** $p < 0.01$ and *** $p < 0.001$ were considered highly significant.

4. Conclusions

In the present study, we assessed the effect of UC roots on the immune function of DCs and found that the immunomodulatory effect of UCME was mainly associated with its EtOAc-soluble fraction. After one-step chromatography, genistein was rapidly identified by the ^{1}H NMR profiling of the immunomodulatory subfractions (D-4 and D-5). The central role of genistein in the immunomodulatory activity of UC roots was supported by the result that the genistein-knockout subfractions (D-4 w/o and D-5 w/o) had a lower inhibitory activity.

Importantly, this work elucidates a rapid strategy for identifying immunomodulatory phytochemicals, distinguishing the chemical marker(s) for quality control and providing a rationale for the traditional immunomodulatory use of UC roots. The findings indicated that UC roots can potentially be used as an immunosuppressive functional food, of which genistein can be a chemical marker for quality control. In conclusion, our strategy will help in the standardization of *U. crinita* roots, a famous Taiwanese functional food.

Supplementary Materials: The following are available online at http://www.mdpi.com/2304-8158/8/11/543/s1, Figure S1: Selected ^{13}C NMR spectrum (acetone-d_6, 125 MHz) of subfraction D-4; Figure S2: Selected HSQC spectrum (acetone-d_6) of subfraction D-4; Figure S3: Selected HMBC spectrum (acetone-d_6) of subfraction D-4; Figure S4: Chromatogram of genistein-containing subfractions D-4 and D-5; Figure S5: The effects of the compounds LA (lupinalbin A), MS (*p*-hydroxybenzoic acid), HA (*p*-hydroxybenzoic acid), DDA (*p*-hydroxybenzoic acid), and ST (a mixture of β-sitosterol and stigmasterol) on the production of pro-inflammatory cytokines in LPS-stimulated DCs; Figure S6: The effects of the compounds SA (salicylic acid) and VA (vanillic acid) on the production of pro-inflammatory cytokines in LPS-stimulated DCs.

Author Contributions: Conceptualization, M.-K.L. and M.-S.L.; methodology, Y.-H.K. and M.-K.L.; investigation, P.-C.T., C.-J.C., and Y.-C.L.; resources, Y.-H.K., M.-S.L., and M.-K.L.; writing—original draft preparation, P.-C.T. and C.-J.C.; writing—review and editing, P.-C.T., M.-K.L., and M.-S.L.; supervision, M.-K.L. and M.-S.L.

Funding: This research was funded by grants from China Medical University and the Ministry Science and Technology of Taiwan, ROC (grant number, CMU 107-5-49 and MOST 107-2313-B-039-001).

Conflicts of Interest: The authors declare no conflict of interest.

Abbreviations

CCK-8	Cell-Counting Kit-8
CD	co-stimulatory molecules
ERK	extracellular-regulated protein kinases
GM-CSF	granulocyte-macrophage colony-stimulating factor
HSQC	heteronuclear single-quantum correlation
HMBC	heteronuclear multiple-bond correlation
IL	interleukin
MHC	major histocompatibility complex molecules
MS	mass spectrometry
NF-κB	nuclear factor kappa-light-chain-enhancer of activated B cells
NMR	nuclear magnetic resonance spectroscopy
TNF-α	tumor necrosis factor alpha
TLRs	Toll-like receptors

References

1. Liu, S.Y.; Liou, P.C.; Wang, J.Y.; Shyu, Y.T.; Hu, M.F.; Chang, Y.M.; Shieh, J.I. Production and electrophoretic analysis of medical plants in Taiwan. In *Proceeding of a Symposium on Development and Utilization of Resources of Medicinal Plants in Taiwan*; Tu, C.C., Lu, H.S., Liu, S.Y., Eds.; Taiwan Agricultural Research Institute: Taichung, Taiwan, 1995; pp. 149–188.
2. Yen, G.C.; Lai, H.H.; Chou, H.Y. Nitric oxide-scavenging and antioxidant effects of *Uraria crinita* root. *Food Chem.* **2001**, *74*, 471–478. [CrossRef]
3. Liu, X.P.; Cao, Y.; Kong, H.Y.; Zhu, W.F.; Wang, G.; Zhang, J.J.; Qiu, Y.C.; Pang, J.X. Antihyperglycemic and antihyperlipidemic effect of *Uraria crinita* water extract in diabetic mice induced by STZ and food. *J. Med. Plants Res.* **2010**, *4*, 370–374.
4. Mao, Y.W.; Lin, R.D.; Hung, H.C.; Lee, M.H. Stimulation of osteogenic activity in human osteoblast cells by edible *Uraria crinita*. *J. Agric. Food Chem.* **2014**, *62*, 5581–5588. [CrossRef] [PubMed]
5. Okawa, M.; Akahoshi, R.; Kawasaki, K.; Nakano, D.; Tsuchihashi, R.; Kinjo, J.; Nohara, T. Two new triterpene glycosides in the roots of *Uraria crinita*. *Chem Pharm Bull*. **2019**, *67*, 159–162. [CrossRef]
6. Wang, Y.Y.; Zhang, X.Q.; Gong, L.M.; Ruan, H.L.; Pi, H.F.; Zhang, Y.H. Studies on chemical constituents in roots of *Uraria crinita*. *Chin. Pharm. J.* **2009**, *44*, 1217–1220.
7. Rossi, M.; Young, J.W. Human dendritic cells: Potent antigen-presenting cells at the crossroads of innate and adaptive immunity. *J. Immunol.* **2005**, *175*, 1373–1381. [CrossRef]
8. Blanco, P.; Palucka, A.K.; Pascual, V.; Banchereau, J. Dendritic cells and cytokines in human inflammatory and autoimmune diseases. *Cytokine Growth Factor Rev.* **2008**, *19*, 41–52. [CrossRef]
9. Kim, S.K.; Yun, C.H.; Han, S.H. Induction of dendritic cell maturation and activation by a potential adjuvant, 2-hydroxypropyl-β-Cyclodextrin. *Front. Immunol.* **2016**, *7*, 435. [CrossRef]
10. Heyman, H.M.; Meyer, J.J.M. NMR-based metabolomics as a quality control tool for herbal products. *S. Afr. J. Bot.* **2012**, *82*, 21–32. [CrossRef]
11. van der Kooy, F.; Maltese, F.; Choi, Y.H.; Kim, H.K.; Verpoorte, R. Quality control of herbal material and phytopharmaceuticals with MS and NMR based metabolic fingerprinting. *Planta Med.* **2009**, *75*, 763–775. [CrossRef]

12. Kumar Bharti, S.; Roy, R. Metabolite identification in NMR-based metabolomics. *Curr. Metabol.* **2014**, *2*, 163–173. [CrossRef]
13. Hussin, M.; Abdul Hamid, A.; Abas, F.; Ramli, N.S.; Jaafar, A.H.; Roowi, S.; Majid, N.A.; Pak Dek, M.S. NMR-based metabolomics profiling for radical scavenging and anti-aging properties of selected herbs. *Molecules* **2019**, *24*, 3208. [CrossRef] [PubMed]
14. Mediani, A.; Abas, F.; Maulidiani, M.; Khatib, A.; Tan, C.P.; Ismail, I.S.; Shaari, K.; Ismail, A. Characterization of metabolite profile in *Phyllanthus niruri* and correlation with bioactivity elucidated by nuclear magnetic resonance based metabolomics. *Molecules* **2017**, *22*, 902. [CrossRef] [PubMed]
15. Pontes, J.G.M.; Brasil, A.J.M.; Cruz, G.C.F.; de Souza, R.N.; Tasic, L. NMR-based metabolomics strategies: Plants, animals and humans. *Anal. Methods* **2017**, *9*, 1078–1096. [CrossRef]
16. Zhang, L.; Hatzakis, E.; Patterson, A.D. NMR-based metabolomics and its application in drug metabolism and cancer research. *Curr. Pharmacol. Rep.* **2016**, *2*, 231–240. [CrossRef]
17. Ma, Q.; Liu, Y.; Zhan, R.; Chen, Y. A new isoflavanone from the trunk of *Horsfieldia pandurifolia*. *Nat. Prod. Res.* **2016**, *30*, 131–137. [CrossRef]
18. Tsukayama, M.; Oda, A.; Kawamura, Y.; Nishiuchi, M.; Yamashita, K. Facile synthesis of polyhydroxycoumaronochromones with quinones: Synthesis of alkylpolyhydroxy- and alkoxycoumaronochromones from 2′-hydroxyisoflavones. *Tetrahedron Lett.* **2001**, *42*, 6163–6166. [CrossRef]
19. Ni, J.C.; Shi, J.T.; Tan, Q.W.; Chen, Q.J. Phenylpropionamides, piperidine, and phenolic derivatives from the fruit of *Ailanthus altissima*. *Molecules* **2017**, *22*, 2017. [CrossRef]
20. Zhang, Y.H.; Yu, J.Q. Pd(II)-catalyzed hydroxylation of arenes with 1 atm of O_2 or air. *J. Am. Chem. Soc.* **2009**, *131*, 14654–14655. [CrossRef]
21. Wu, D.W.; Liang, Q.L.; Zhang, X.L.; Jiang, Z.; Fan, X.H.; Yue, W.; Wu, Q.N. New isocoumarin and stilbenoid derivatives from the tubers of *Sparganium stoloniferum* (Buch.-Ham.). *Nat. Prod. Res.* **2017**, *31*, 131–137. [CrossRef]
22. Schievano, E.; Morelato, E.; Facchin, C.; Mammi, S. Characterization of markers of botanical origin and other compounds extracted from unifloral honeys. *J. Agric. Food Chem.* **2013**, *61*, 1747–1755. [CrossRef] [PubMed]
23. Liu, X.; Yin, C.; Cao, Y.; Zhou, J.; Wu, T.; Cheng, Z. Chemical constituents from *Gueldenstaedtia verna* and their anti-inflammatory activity. *Nat. Prod. Res.* **2018**, *32*, 1145–1149. [CrossRef] [PubMed]
24. Masilamani, M.; Wei, J.; Bhatt, S.; Paul, M.; Yakir, S.; Sampson, H.A. Soybean isoflavones regulate dendritic cell function and suppress allergic sensitization to peanut. *J. Allergy Clin. Immunol.* **2011**, *128*, 1242–1250. [CrossRef] [PubMed]
25. Pan, M.; Han, H.; Zhong, C.; Geng, Q. Effects of genistein and daidzein on hippocampus neuronal cell proliferation and BDNF expression in H19-7 neural cell line. *J. Nutr. Health Aging* **2012**, *16*, 389–394. [CrossRef] [PubMed]
26. Lv, Z.; Xing, K.; Li, G.; Liu, D.; Guo, Y. Dietary genistein alleviates lipid metabolism disorder and inflammatory response in laying hens with fatty liver syndrome. *Front. Physiol.* **2018**, *9*, 1493. [CrossRef]
27. Kim, D.H.; Jung, W.S.; Kim, M.E.; Lee, H.W.; Youn, H.Y.; Seon, J.K.; Lee, H.N.; Lee, J.S. Genistein inhibits proinflammatory cytokines in human mast cell activation through the inhibition of the ERK pathway. *Int. J. Mol. Med.* **2014**, *34*, 1669–1674. [CrossRef]
28. Dijsselbloem, N.; Goriely, S.; Albarani, V.; Gerlo, S.; Francoz, S.; Marine, J.C.; Goldman, M.; Haegeman, G.; Vanden Berghe, W. A critical role for p53 in the control of NF-KB-dependent gene expression in TLR4-stimulated dendritic cells exposed to genistein. *J. Immunol.* **2007**, *178*, 5048–5057. [CrossRef]
29. Graziani, V.; Scognamiglio, M.; Belli, V.; Esposito, A.; D'Abrosca, B.; Chambery, A.; Russo, R.; Panella, M.; Russo, A.; Ciardiello, F.; et al. Metabolomic approach for a rapid identification of natural products with cytotoxic activity against human colorectal cancer cells. *Sci. Rep.* **2018**, *8*, 5309. [CrossRef]
30. Freire, R.T.; Bero, J.; Beaufay, C.; Selegato, D.M.; Coqueiro, A.; Choi, Y.H.; Quetin-Leclercq, J. Identification of antiplasmodial triterpenes from *Keetia* species using NMR-based metabolic profiling. *Metabolomics* **2019**, *15*, 27. [CrossRef]

31. Lin, M.K.; Yu, Y.L.; Chen, K.C.; Chang, W.T.; Lee, M.S.; Yang, M.J.; Cheng, H.C.; Liu, C.H.; Chen Dz, C.; Chu, C.L. Kaempferol from Semen cuscutae attenuates the immune function of dendritic cells. *Immunobiology* **2011**, *216*, 1103–1109. [CrossRef]
32. Lin, C.H.; Lin, S.H.; Lin, C.C.; Liu, Y.C.; Chen, C.J.; Chu, C.L.; Huang, H.C.; Lin, M.K. Inhibitory effect of clove methanolic extract and eugenol on dendritic cell functions. *J. Funct. Foods* **2016**, *27*, 439–447. [CrossRef]

© 2019 by the authors. Licensee MDPI, Basel, Switzerland. This article is an open access article distributed under the terms and conditions of the Creative Commons Attribution (CC BY) license (http://creativecommons.org/licenses/by/4.0/).

Article

The Essential Oil and Hydrolats from *Myristica fragrans* Seeds with Magnesium Aluminometasilicate as Excipient: Antioxidant, Antibacterial, and Anti-inflammatory Activity

Inga Matulyte [1,2], Aiste Jekabsone [2], Lina Jankauskaite [2,3], Paulina Zavistanaviciute [4], Vytaute Sakiene [4], Elena Bartkiene [4], Modestas Ruzauskas [5], Dalia M. Kopustinskiene [2], Antonello Santini [6,*] and Jurga Bernatoniene [1,2]

1. Department of Drug Technology and Social Pharmacy, Lithuanian University of Health Sciences, LT-50161 Kaunas, Lithuania; inga.matulyte@lsmuni.lt (I.M.); jurga.bernatoniene@lsmuni.lt (J.B.)
2. Institute of Pharmaceutical Technologies, Medical Academy, Lithuanian University of Health Sciences, LT-50161 Kaunas, Lithuania; aiste.jekabsone@lsmuni.lt (A.J.); lina.jankauskaite@lsmuni.lt (L.J.); daliamarija.kopustinskiene@lsmuni.lt (D.M.K.)
3. Department of Pediatrics, Lithuanian University of Health Sciences Hospital Kauno Klinikos, LT-50161 Kaunas, Lithuania
4. Department of Food Safety and Quality, Lithuanian University of Health Sciences, LT-47181 Kaunas, Lithuania; paulina.zavistanaviciute@lsmuni.lt (P.Z.); vytaute.sakiene@lsmuni.lt (V.S.); elena.bartkiene@lsmuni.lt (E.B.)
5. Institute of Microbiology and Virology, Lithuanian University of Health Sciences, LT-47181 Kaunas, Lithuania; modestas.ruzauskas@lsmuni.lt
6. Department of Pharmacy, University of Napoli Federico II, Via D. Montesano 49, 80131 Napoli, Italy
* Correspondence: asantini@unina.it

Received: 10 December 2019; Accepted: 30 December 2019; Published: 2 January 2020

Abstract: Nutmeg (*Myristica fragrans*) essential oil has antimicrobial, antiseptic, antiparasitic, anti-inflammatory, and antioxidant properties. We have recently demonstrated that hydrodistillation of nutmeg essential oil by applying magnesium aluminometasilicate as an excipient significantly increases both the content and amount of bioactive substances in the oil and hydrolats. In this study, we aimed to compare the antioxidant, antimicrobial, and anti-inflammatory activity of hydrolats and essential oil obtained by hydrodistillation in the presence and absence of magnesium aluminometasilicate as an excipient. The 2,2-diphenyl-1-picrylhydrazyl (DPPH) radical scavenging method revealed that magnesium aluminometasilicate did not significantly improved antioxidant activity of both essential oil and hydrolat. Antibacterial efficiency was evaluated by monitoring growth of 15 bacterial strains treated by a range of dilutions of the essential oil and the hydrolats. Essential oil with an excipient completely inhibited the growth of *E. faecalis*, *S. mutans* (referent), and *P. multocida*, whereas the pure oil was only efficient against the latter strain. Finally, the anti-inflammatory properties of the substances were assessed in a fibroblast cell culture treated with viral dsRNA mimetic Poly I:C. The essential oil with an excipient protected cells against Poly I:C-induced necrosis more efficiently compared to pure essential oil. Also, both the oil and the hydrolats with aluminometasilicate were more efficient in preventing IL-6 release in the presence of Poly I:C. Our results show that the use of magnesium aluminometasilicate as an excipient might change and in some cases improve the biological activities of nutmeg essential oil and hydrolats.

Keywords: nutmeg; essential oil; antioxidant activity; antibacterial activity; poly I:C-induced inflammation; fibroblasts; magnesium aluminometasilicate

1. Introduction

Since ancient times, *Myristica fragrans* (nutmeg) seeds have been used as a food spice, flavoring agent, a natural remedy for headaches and fever [1]. Nutmeg seeds have essential and fatty oils, resins, wax, and other components [2]. Nutmeg's essential oil has antimicrobial, antiseptic, antiparasitic, anti-inflammatory, and antioxidant properties [1,3]. The concentration of essential oil in nutmeg seeds is about 5–15% [4], and its major components are terpene hydrocarbons (sabinene, pinene, camphene, *p*-cymene, phellandrene, terpinene, limonene, and myrcene altogether make up 60% to 80% of the oil), oxygenated terpenes (linalool, geraniol, and terpineol, which make up approximately 5% to 15%) and aromatic ethers (myristicin, elemicin, safrole, eugenol, and eugenol derivatives, together constituting 15 to 20%) [5–8]. The toxicity of nutmeg seeds at high doses has been reported, mainly due to myristicin oil and elemicin, causing tachycardia, nausea, vomiting, agitation, and hallucinations. However, these effects are related to the abuse of the spice and are not observed at usual low concentrations [9]. There are many studies on the beneficial effects of nutmeg seed and various nutmeg seed extracts. One of the most prominent biological activities of the nutmeg preparations is antibacterial. Nutmeg seed lignans exert antimicrobial activity on *Bacillus subtilis*, *Staphylococcus aureus*, and *Shigella dysenteriae* [10]. Ethanol and acetone extracts of nutmeg crust have strong antibacterial activity against gram-positive bacteria *Staphylococcus aureus* [5]. Ethyl acetate extracts of flesh of the nutmeg fruit have inhibitory potential against both gram-positive and gram-negative bacteria with a minimum inhibitory concentration (MIC) ranging from 0.625 to 1.25 mg/mL [11]. Used for the preservation of sweets, nutmeg methanol extracts inhibit growth of *Staphylococcus aureus*, *Aspergillus niger*, *Saccharomyces cerevisiae*, and *Escherichia coli* at MIC between 250 and 300 mg/mL [12]. However, there are only a few studies on the biological activity of nutmeg essential oil. Takikawa et al. showed a higher antibacterial effect of essential nutmeg oil on pathogenic compared to non-pathogenic strains of *Escherichia coli* [13]. Furthermore, nutmeg essential oil decreased the growth and survival of *Yersinia enterocolitica* and *Listeria monocytogenes* in broth culture [14].

Nutmeg oil preparations are also known for their antioxidant capacity. Using the 2,2-diphenyl-1-picrylhydrazyl (DPPH) free radical scavenging assay, Piaru et al. reported a significant antioxidant activity of nutmeg oil [15]. The antioxidant properties are often related to the alleviation of inflammation. Nutmeg oil diminished chronic inflammation and pain through the inhibition of COX-2 expression and substance P release in vivo [16]. In another study, nutmeg oil suppressed reactive oxygen species (ROS) production in human neutrophils stimulated by PMA (phorbol 12-myristate 13-acetate) [17] and mildly inhibited phagocytosis in human neutrophils [18]. However, there is no published research on the effect of nutmeg seed essential oil on virus-triggered inflammatory response.

Hydrodistillation is a popular method used for the preparation of essential oils. However, hydrodistillation with excipients is not widely used—we have found just three studies applying this method so far [19–21]. Therefore, we have applied magnesium aluminometasilicate in hydrodistillation as the new excipient and have tested its effects on the nutmeg essential oil yield and its composition [22]. Aluminometasilicate is widely used as a disintegrator in the manufacturing of tablets. Furthermore, this compound is non-toxic and inexpensive, as the price is ~300 eur for 25 kg. Magnesium aluminometasilicate has significantly increased both the yield and composition of some chemical compounds (sabinene, α-pinene, and limonene). The use of the excipient also increased the essential oil yield by about 61% (hydrodistillation with water—the yield is 0.79 ± 0.04 g, using 1% excipient—1.29 ± 0.05 g; the nutmeg quantity was 15 g, the water content was 300 mL) [22].

The increased amount of active substances suggests that oil preparations with aluminometasilicate might have stronger biological activities. Therefore, in this study we compared the antioxidant, antimicrobial, and anti-inflammatory properties of *Myristica fragrans* seed essential oil preparations with and without aluminometasilicate.

2. Materials and Methods

2.1. Plant Material

The dried seeds of nutmeg (*Myristica fragrans*) were from Grenada. Seeds were identified by Jurga Bernatoniene, Medical Academy, Lithuania University of Health Sciences, Kaunas, Lithuania. A voucher specimen (I 18922) was placed for storage at the Herbarium of the Department of Drug Technology and Social Pharmacy. The seeds had a characteristic odor, a strong, bitter, and spicy flavour, and they were a brown-beige color. The seeds were ground into a powder (using laboratory mill), with particles smaller than 0.5 mm. All powder samples were kept in a dark and airtight container at 20 ± 2 °C.

2.2. Essential Oil and Hydrolat

The essential oil from nutmeg seeds was prepared by using hydrodistillation. The modified Clevenger type apparatus was used. Two samples of essential oil were prepared: one without excipient and the other with 1% of magnesium aluminometasilicate. Each sample was prepared with 15 g of nutmeg powder and 300 mL distilled water, and 1% magnesium aluminometasilicate was used as an excipient in one of the samples. Also, hydrolat of these two essential oils was used. It was collected from Clevenger apparatus. This material was collected first, followed by the essential oil. All samples were obtained and stored in airtight bottles in the refrigerator. The hydrodistillation took 4 h.

2.3. Antioxidant Activity by DPPH Radical Scavenging Assay

Antioxidant activity of nutmeg essential oil and hydrolat were evaluated using DPPH (Sigma Aldrich, St. Louis, MO, USA) [15]. First of all, 0.1 mM 96% DPPH solution in 96% ethanol was prepared. A total of 1 mL of DPPH solution was placed in a spectrophotometer cuvette and 100 µL of ethanolic essential oil solution at concentrations ranging from 0.2% to 20% was added. For an antioxidant activity evaluation of hydrolat, the absolute hydrolat was used. 1 mL DPPH solution and nutmeg hydrolat from 0.1 mL to 1 mL were mixed in a cuvette. All samples were incubated in the dark for 20 min and absorbance was taken at 515 nm. The antioxidant activity was performed on a UV Spectrophotometer UV-1800 (Shimadzu, Kyoto, Japan). The quantity of DPPH radical scavenging activity was calculated by using this formula:

$$DPPH\ scavenging\ effect\ \% = \frac{A_{control} - A_{sample}}{A_{control}} \times 100, \quad (1)$$

where $A_{control}$ and A_{sample} are the absorbance of the control sample (0.1 mM DPPH solution, solvent is 96% ethanol) and the experiment sample.

2.4. Antimicrobial Activity

The method used for antimicrobial activity was serial dilutions in liquid medium [23]. The broth liquid medium was dispensed into test tubes to give a final volume of 10 mL (with a sample of essential oil). The medium was sterilized. The physiological solution was dispensed into 5 mL individual tubes and used for preparation of suspension of the following bacteria: *Klebsiella pneumoniae*; *Salmonella enterica* 24 SPn06; *Pseudomonas aeruginosa* 17–331; *Acinetobacter baumanni* 17–380, *Proteus mirabilis*; 6MRSA M87fox; *Enterococcus faecalis* 86; *Enterococcus faecium* 103; *Bacillus cereus* 18 01; *Streptococcus mutans* (referent); *Enterobacter cloacae*; *Citrobacter freundii*; *Staphylococcus epidermidis*; *Staphylococcus haemolyticus*; *Pasteurella multocida* strains. All bacteria were isolated from clinical material. For each bacterial culture, three tubes of Mueller Hinton broth were used (9.94 mL, 9.97 mL, and 9.98 mL each). The tubes were inoculated with 10 µL of bacterial suspension with the essential oil at concentration 0.1%, 0.2%, and 0.5%. After 48 h of incubation, each tube was inoculated with 10 µL of suspension on soy-tryptone agar (Thermo Fisher, Hampshire, UK). MIC of essential oil was evaluated based on the presence of bacterial growth (bacterial colonies growing (+)/non growing (−).

2.5. Cell Culture and Treatments

Human fibroblasts (BJ-5ta, hTERT, LGC Standards Ltd. Middlesex, UK) were grown in 75 cm^2 flasks in Dulbecco's Modified Eagle Medium (DMEM) with Glutamax (Thermo Fisher Scientific, Waltham, MA, USA), 10% fetal bovine serum and 100 IU/mL Penicilin/Streptomycin according to standard supplier protocol. At 70–90% confluence, the cells were detached by 0.025% Trypsin/EDTA and plated in 96 well plates at a density of 2×10^5 cells/well. A total of 24 h after plating, the cells were treated with 1 µg/mL Poly I:C to simulate viral dsRNR-induced inflammatory response. For cell culture treatments, the essential oils were dissolved in 96% ethanol at a concentration of 5% (v/v). For determination of cell viability and LD$_{50}$, the solutions of essential oils and absolute hydrolats were used as a range of dilutions in cell culture medium starting from essential oil preparation to a medium v/v ratio of 1:1000 and finishing with 1:5. For the control, the same dilutions with solvent (ethanol) were performed. For anti-inflammatory activity evaluation, the solutions of essential oils at v/v dilutions of 1:100 or 1:200, or absolute hydrolats at v/v dilutions of 1:40, 1:100, and 1:200 were applied simultaneously with Poly I:C treatment.

2.6. Determination of Cell Viability and Determination on LD$_{50}$

Cell viability was assessed by using double nuclear fluorescent staining with Hoechst 33342 (10 µg/mL) and propidium iodide (PI, 5 µg/mL) according to standard supplier protocol for 5 min at 37 °C. PI-positive nuclei indicating lost nuclear membrane integrity were considered to be necrotic. Cells were visualized under fluorescent microscope OLYMPUS IX71S1F-3, counted in fluorescent micrographs and expressed as percentage of total cell number per image. The data is presented as averages ± standard deviation. LD$_{50}$ was calculated by SigmaPlot v.13 (Systat Software Inc., San Jose, CA, USA) using the equation selected by a dynamic curve fitting tool.

2.7. Assessment of Interleukin-6 Concentration

Medium collected after cell culture treatments was used to measure the concentration of pro-inflammatory cytokine interleukin-6 (IL-6) by ELISA kit (Thermo Fisher Scientific, Waltham, MA, USA) following the standard supplier protocol. The spectrophotometric readings were performed in a plate reader Infinite 200 Pro M Nano Plex (Tecan, Mannedorf, Svizzera).

2.8. Statistical Analysis

The results are presented as means of 3–7 replicates ± standard deviation. The statistical data analysis was performed by applying ANOVA with Tukey HSD post hoc test. Differences were considered statistically significant when $p < 0.05$. The data were processed using Microsoft Office Excel 2010 (Microsoft, Redmond, WA, USA) software.

3. Results

First, the antioxidant activity in essential oils was compared. The two samples of each category were analyzed: the essential oil without excipient, or pure essential oil (EO1), and essential oil with 1% of magnesium aluminometasilicate (EO2). The results are presented in Table 1. The range of essential oil concentration in this examination was from 0.2 to 20%.

Table 1. Antioxidant activity of nutmeg essential oils applied at different concentrations.

Sample	Essential Oil Concentration (%)						
	0.2	0.5	1	2	5	10	20
EO1	12.63 ± 0.53	16.34 ± 1.23	26.35 ± 0.88	30.58 ± 1.39	44.53 ± 0.84	61.01 ± 0.26	84.01 ± 0.78
EO2	12.65 ± 2.05	19.12 ± 2.24	27.03 ± 0.98	37.15 ± 0.80 *	44.92 ± 0.63	62.11 ± 0.43	72.71 ± 0.79 *

*—significant difference compared to EO1, $p < 0.05$, $n = 3$.

Both essential oils demonstrated similar antioxidant activity increasing in a concentration-dependent manner, except for some small fluctuations at 2% (EO2 had slightly higher antioxidant activity compared to EO1) and 20% (EO2 antioxidant activity was slightly lower than EO1). Both essential oil preparations at 10% concentration had higher than 50% antioxidant activity (more than half of DPPH radicals were bound).

Next in the study, the antioxidant activity of essential oil hydrolats was tested by using DPPH radical scavenging method. Hydrolat from EO1 was named EOH1, and from hydrolat from EO2 was named EOH2. The data are provided in Table 2.

Table 2. Antioxidant activity of nutmeg essential oil hydrolats.

Sample	Hydrolat Quantity (mL)				
	0.1	0.2	0.3	0.5	1
EOH1	12.97 ± 1.25	31.43 ± 1.55	36.21 ± 3.20	48.09 ± 3.96	56.42 ± 3.23
EOH2	15.22 ± 5.14	27.24 ± 1.63	33.52 ± 2.11	36.55 ± 0.68 *	44.19 ± 1.09 *

*—significant difference compared to EOH1, $p < 0.05$, $n = 3$.

At small quantities of up to 0.3 mL, the antioxidant activities of both hydrolat preparations were similar, but at 0.5 mL and 1 mL, the EOH1 antioxidant activity was significantly higher compared to that of EOH2. EOH1 at 1 mL had an antioxidant activity greater than 50%, and this activity level was similar to 5%–10% of EO1 activity. The results show that free radical scavenging activity of 0.2 mL of hydrolat is higher than that of 1% or less concentrated essential oil.

Summarizing the antioxidant activity results, magnesium aluminometasilicate did not improve, and even slightly decreased the antioxidant activity of nutmeg essential oil and its hydrolat.

Next in the study, antibacterial properties of nutmeg essential oil and hydrolats were investigated on 15 pathogenic clinical isolate strains by using a dilution range assay. The results are presented in Table 3.

Table 3. Antimicrobial study results of nutmeg essential oil and its hydrolats.

Sample	Microorganisms														
	1	2	3	4	5	6	7	8	9	10	11	12	13	14	15
EO1	+	+	+	+	+	+	+	+	+	+	+	+	+	+	0.2
EO2	+	+	+	+	+	+	0.5	+	+	≤0.1	+	+	+	+	0.2
EOH1	+	+	+	+	+	+	+	+	+	+	+	+	+	+	+
EOH2	+	+	+	+	+	+	+	+	+	0.5	+	+	+	+	+

+ means the pathogens growth. 1. Klebsiella pneumoniae, 2. Salmonella enterica 24 SPn06, 3. Pseudomonas aeruginosa 17-331, 4. Acinetobacter baumanni 17-380, 5. Proteus mirabilis, 6. 6MRSA M87fox, 7. Enterococcus faecalis 86, 8. Enterococcus faecium 103, 9. Bacillus cereus 18 01, 10. Streptococcus mutans (referent), 11. Enterobacter cloacae, 12. Citrobacter freundii, 13. Staphylococcus epidermidis, 14. Staphylococcus haemolyticus, 15. Pasteurella multocida. Where the growth of bacteria were inhibited, minimal inhibitory concentrations were provided in %.

EO1 only suppressed *Pasteurella multocida* growth, with the minimal concentration to achieve this effect being 0.2%. However, nutmeg essential oil with 1% of magnesium aluminometasilicate (EO2) had a broader effect. Next to *P. multocida*, it inhibited *E. faecalis* and *S. mutans*, and the efficient concentrations were rather low. A mere 0.5% was enough to completely suppress *E. faecalis*, and for the *S. mutans* strain even less than 0.1% was effective. In the case of EOH1 and EOH2, only the hydrolat with aluminometasilicate suppressed the growth of *S. mutans*. Thus, the results indicate that the excipient magnesium aluminometasilicate broadens the spectrum of antimicrobial activity of nutmeg essential oil.

One of the most important pharmacological activities of plant essential oils is related to anti-inflammatory properties. Nutmeg essential oil is also known for inflammation reducing activity [24]. Therefore, next in the study, we have assessed nutmeg seed essential oil and hydrolat preparations in a virus mimetic Poly I:C-induced inflammation in vitro model by using human fibroblast cell culture. Before starting the treatments, the general toxicity test of the oils and hydrolats was performed and LD_{50} doses as well as safe concentrations were established.

As indicated in Figure 1a, there were no significant difference in cell viability detected after treatment with both EO1 and EO2 essential oil solutions up to the dilution ratio 1:100. Further increases in concentration up to the dilution ratio 1:40 significantly decreased viability of the fibroblasts. After treatment with essential oil solutions at 1:40, the viability dropped from 97 ± 2% in control to 70 ± 12% in the case of EO1, and to 42 ± 10% in the case of EO2. At the dilution ratio of 1:5, the percentage of viable cells in the cultures was lower than 10% in the case of both essential oil preparations. The dilution ratios corresponding to LD_{50} calculated for EO1, EO2, and 96% ethanol were 0.047, 0.022, and 0.055, respectively. Thus, EO2 was significantly more toxic for the cells compared to EO1, and also to ethanol. In contrast, the toxicity pattern of EO1 was very close to that of ethanol, indicating there were no or very little toxic compounds in this essential oil preparation.

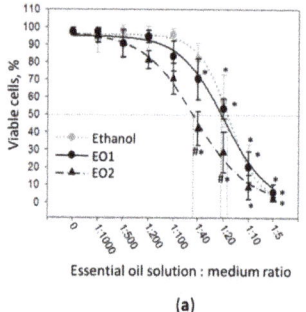 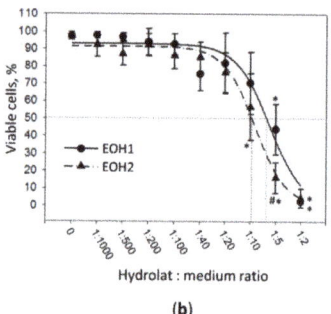

(a)   (b)

Figure 1. The effect of nutmeg essential oil ethanol solutions (**a**) and nutmeg essential oil hydrolats (**b**) on viability of cultured human fibroblasts. EO1—essential oil without excipient solution, EO2—essential oil with 1% magnesium aluminometasilicate solution, EOH1—hydrolat from EO1, and EOH2—hydrolat from EO2. In addition, 96% ethanol was assessed as solvent control for the essential oil. Punctured lines indicate the dilution ratios corresponding to LD_{50}. *—statistically significant difference compared to untreated control, #—compared to EO1 in (**a**) or EOH1 in (**b**), respectively, when $p < 0.05$.

Evaluation of cell viability after 24 h treatment with nutmeg seed essential oil hydrolats revealed that both EOH1 and EOH2 were not toxic up to a dilution of 1:20 (Figure 1b). After cell incubation with 1:10 EOH2, the percentage of viable cells in the cultures decreased to 57 ± 19%, making a significant difference compared with the untreated control. A significant viability drop in EOH1 treatment series was achieved when the dilution ratio 1:5 was applied. The level of viable cells in this treated cultures was 44 ± 14%. After treatment with EOH1 and EOH2 at the ratio 1:2, nearly all cells in the cultures were found to be necrotic. The dilution ratios corresponding to LD_{50} calculated for EOH1 and EOH2 were 0.160 and 0.105, respectively. Toxicity evaluation of the hydrolats indicated that EOH2 is slightly more toxic compared to EOH1.

The next task in this work was to evaluate the efficiency of nutmeg seed essential oil and hydrolat preparations to reduce toxicity and signaling in viral inflammation in vitro model. To stimulate inflammatory response, human fibroblast cell culture was treated with 1 µg/mL virus double stranded RNR mimetic polyinosinic: polycytidylic acid (Poly I:C) for 24 h, with or without nutmeg seed essential oil solutions or hydrolats. After toxicity assessment, the dilution ratios selected for anti-inflammatory property testing were 1:200 and 1:100 for the essential oil solutions, and 1:100 and 1:40 for the hydrolats. Anti-inflammatory assessment results are presented in Figure 2.

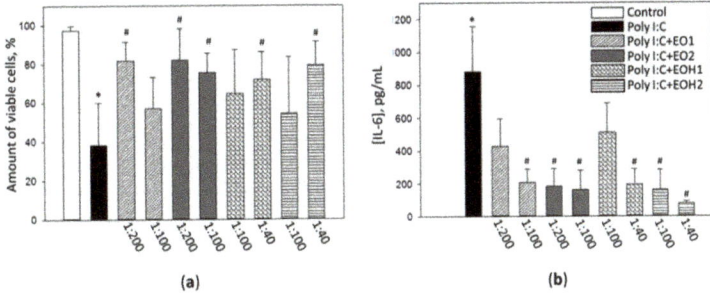

Figure 2. The effect of nutmeg seed essential oil ethanol solutions and the essential oil hydrolats on Poly I:C-treated human fibroblast cell viability (**a**) and cytokine IL-6 release from the cells (**b**). EO1—essential oil without the excipient solution, EO2—essential oil with 1% magnesium aluminometasilicate solution, EOH1—hydrolat from EO1, and EOH2—hydrolat from EO2. *—statistically significant difference compared to untreated control, #—compared to Poly I:C-only treatment, when $p < 0.05$; n = 5–7.

After fibroblast cell culture treatment with 1 mg/mL Poly I:C, the amount of viable cells decreased by 59% (Figure 1a). Addition of nutmeg essential oil preparations to the cell culture medium increased cell viability in the Poly I:C-affected cultures. Statistically significant differences compared with Poly I:C samples were found after treatment with 1:200 EO1, 1:200 and 1:100 EO2, as well as 1:40 EOH1 and 1:40 EOH2. The percentages of viable nuclei in these samples were 81 ± 9%, 82 ± 16%, 76 ± 10%, 72 ± 14%, and 79 ± 12%, respectively. Thus, EO2 has demonstrated the highest cytoprotective capacity in a virus mimetic inflammation model.

Evaluation of the release of IL-6 to the incubation medium revealed that after 24 h of Poly I:C treatment, the level of this pro-inflammatory cytokine jumped from nearly a "zero" value to 883 ± 273 pg/mL (Figure 2b). Nutmeg essential oil preparations applied together with Poly I:C significantly reduced the concentration of IL-6 in the medium. The significant drop in the IL-6 level was in the samples incubated with 1:100 EO1, 1:200 and 1:100 EO2, 1:40 EOH1, and 1:100 and 1:40 EOH2. IL-6 concentration in these samples was found in the range between 162 ± 123 pg/mL (with 1:40 EOH2) and 206 ± 83 pg/mL (with 1:100 EO1). The assessment of IL-6 release indicates that both the solution of nutmeg seed essential oil with magnesium aluminometasilicate and the hydrolat from this essential oil are most efficient against Poly I:C-induced release of this inflammatory cytokine.

4. Discussion

The purpose of our study was to compare the biological activity of nutmeg seed essential oil and hydrolat without excipient and using magnesium aluminometasilicate as the excipient. To our knowledge, it is the first application of aluminometasilicate as an excipient in essential oil studies. Essential oils have strong antioxidant activity, and some of them are used as preservation agents protecting food or cosmetics from oxidation-induced spoilage [25,26]. Antioxidant activity studies help to elucidate essential oil capacity to protect food from free radical damage [27]. The DPPH radical scavenging method is widely used for this purpose because it is simple and cost-efficient, and gives reliable results. Therefore, it was selected to evaluate the essential oil and hydrolat preparations in our study. Our previous study [22] showed that magnesium aluminometasilicate had influence not only on the yield of essential oil, but also on its chemical composition. Magnesium aluminometasilicate significantly increased the quantity of sabinene, α-pinene, and limonene. Dai et al. (2013) study with *Wedella Prostrata* essential oil (containing 11.38% limonene and 10.74% α-pinene) had a lower antioxidant activity than 100 μg/mL limonene but a higher antioxidant activity than the pure α-pinene [28]. Such a result suggests that limonene is a more prominent antioxidant compared to α-pinene. Our nutmeg seed essential oil (EO2) had 11.66 ± 3.39% α-pinene and 4.91 ± 0.71% limonene [22]. Based on the results of the study where the presence of limonenen together with α-pinene resulted in higher antioxidant activity [28], we can predict that our EO2 has higher antioxidant activity than the pure α-pinene sample.

The *Juniperus scopulorum* 10% essential oil had a 54.7% antioxidant activity (composition: sabinene 50.7%, α-pinene 3.23%, limonene 2.22%, cis sabinene hydrate 0.58%) [29]. Our study shows that 10% essential oils EO1 and EO2 have 61.01 ± 0.26% and 62.11 ± 0.43% antioxidant activity, respectively. Predominant compounds in EO2 were sabinene 61.42%, cis sabinene hydrate 0.3%, limonene 5.62%, and α-pinene 15.05%. The EO1 essential oil had more β-pinene [22], which could increase its antioxidant activity. Other authors have demonstrated nutmeg essential oil with higher antioxidant activity besides α-pinene had also β-pinene [27]. Misharina et al. (2009) have also studied antioxidant properties of nutmeg essential oil and found that 16.5% concentrated solution had approximately 50% antioxidant activity [30]. Such differences may be due to the distinct technique of the research and the variation in nutmeg seed material.

Hydrolats analyzed in our study had lower antioxidant activity compared to essential oils, most likely because of the lower concentrations of volatile compounds. We have searched the literature data, but could not find any studies about the antioxidant activity of nutmeg seed hydrolats so far. Hydrolats prepared from other plant sources had different antioxidant properties. For example, *Salvia officinalis* 0.1 g/mL had about 30% radical scavenging activity. The same concentration of *Rosmarinus officinalis* hydrolat had about 50% activity [31]. Our hydrolats (0.1 g/mL) had 31.43 ± 1.55% (EO1) and 27.24 ± 1.63% (EO2) antioxidant activity—the same as *Salvia officinalis*. Nutmeg seeds hydrolats at the highest concentration tested (0.5 g/mL) had 56.42% and 44.19% antioxidant activity (EOH1 and EOH2, respectively).

Essential oils are known for bioactive compounds with antibacterial activity, therefore they are used as antimicrobial agents in medicine, pharmacy, cosmetology, and other fields [18]. However, different essential oils affect microorganisms in distinct ways—some suppress gram-positive effects, others suppress gram-negative effects [19]. Also, the effective concentration of essential oils vary. There are various methods for determining antibacterial activity (the agar disk-diffusion method, antimicrobial gradient method, dilution methods, and other methods) [32]. Dilution methods are the simplest methods used to determine whether the essential oil suppresses the growth of bacteria or not [23]. There are many techniques and methods used for antimicrobial activity evaluation, and therefore, it is difficult to compare the results obtained from the different studies. In our study, the EO1 essential oil (0.2%) only suppressed *Pasteurella multocida*. The essential oil EO2 with a higher quantity of sabinene, α-pinene, and limonene [22] had antimicrobial activity against three pathogens. Next to *P. multocida*, it also prevented growth *E. faecalis* of and *S. mutans*. The increased efficiency of EO2 against pathogenic strains can be explained by the higher quantity of volatile compounds. Nurjanah et al.'s (2017) study showed (an in vitro disc diffusion antimicrobial activity method) that *Myristica fragrans* essential oil (60% concentration was used) from Central Java inhibited the largest areas [33]. The inhibition areas were from 12.96 mm to 16.79 mm, with the control at 0 mm (*S. aureus*, *S. dysenteriae*, *S. typhi*, and *S. epidermidis*). In the essential oil used for the above-mentioned study, sabinene, α-pinene, and β-pinene quantities were the highest out of all of the chemical compounds (the concentrations were 18.82%, 16.54%, and 13.82, respectively). The essential oil EO2 investigated in this study has a similar composition, meaning it could also be efficient against these pathogens at higher concentrations. In another study, the nutmeg essential oil with similar quantity of volatile compounds had a significant effect on the inhibition of the growth of *E. coli* and *S. aureus* [34]. In this study, we found that essential oil EA2 (0.5%) inhibited *E. faecalis*. This bacteria resides in infected canals of teeth and is often found in the oral cavity after tooth canal repair [35]. Repeated oral care products with chlorhexidine promotes the development of *E. faecalis* resistance [36]. Since EA2 showed activity against *E. faecalis*, nutmeg essential oil could be recommended as a safe protective component for oral care products in the future.

The investigation of human fibroblast cell culture affected by virus mimetic Poly I:C showed that nutmeg essential oils and hydrolats have an anti-inflammatory effect protecting cell viability and significantly reducing the release of cytokine IL-6. EO2 had a higher effect on preventing Poly I:C-induced necrosis and both EO2 and EOH2 more efficiently protected against IL-6 release compared to preparations without aluminometasilicate EO1 and EOH1. This is most likely due to the increased amount and content of active substances (sabinene, α-pinene, and limonene) in the preparations that is a result of the use of the excipient. α-Pinene significantly decreases the LPS-induced production of IL-6, TNF-α and nitric oxide in bacterial lipopolysaccharide (LPS)-treated macrophages [37]. Sabinene from *Oenanthe crocata* essential oil significantly inhibits nitric oxide production in LPS and IFNγ-treated macrophages [38]. Limonene has a significantly decreased manifestation of inflammatory signals in rat models of ulcerative colitis via regulation of iNOS, cyclooxygenase-2 (COX-2), PGE2, and ERK [39]. However, there are not many studies about the anti-inflammatory effect of nutmeg essential oil preparations. Zhang et al. have demonstrated the anti-inflammatory activity of nutmeg oil in complete Freund's adjuvant-injected rats [16]. Their study shows that nutmeg oil is effective in inflammatory pain relief via inhibition of the COX-2 pathway and substance P release. Another in vivo study exploring carrageenan-induced paw edema in rats have also confirmed the anti-inflammatory properties of nutmeg oil [24]. To the best of our knowledge, there are no in vitro studies on virus-induced anti-inflammatory activity of nutmeg oil. Dewi et al. have found that *M. fragrans* seed ethanolic extract and pure quercetin extract from *M. fragrans* inhibited NO production and the release of inflammatory cytokines, such as TNF-α, IL-6, and IL-1β from bacterial LPS-stimulated murine macrophages (RAW 264.7) in a dose-dependent manner. The essential oil of *Monodora myristica* was found to inhibit inflammation-related lipoxygenase [40]. Because of the high content of bioactive volatile compounds, which have been widely studied and characterized by gas chromatographic techniques [41] the essential oils might be good candidates for inhalation treatment of respiratory tract infections. Most common respiratory infections are induced by respiratory viruses, such as influenza or respiratory syncytial virus [42]. As a result, we examined the anti-inflammatory efficiency of nutmeg essential oil preparations in a virus mimetic Poly I:C mediated inflammation. Fibroblasts are multifunctional cells that are responsible for support of other, more tissue-specific cell types, regeneration, wound healing, extracellular matrix production, and inflammatory response [43]. They significantly contribute to the response to infection by secreting cytokines for monocyte/macrophage attraction and their conversion to inflammatory phenotype [44]. The release of IL-6 is one of the key inflammatory signals causing activation of matrix metalloproteinases, macrophages, neutrophil production, and is also involved in autoimmune responses in the condition such as chronic arthritis, osteoporosis, and psoriasis [41,45–48].

The results of our study indicate that nutmeg essential oil preparations have anti-inflammatory properties that might be exploited further for treatment or prevention of viral inflammation-related pathologies, taking the recent emerging nanotechnological and nutraceutical approaches in the field into account [49–51]. However, to increase the applicability of these substances, more studies have to be performed analyzing the mechanism of action of the essential compounds contained in the preparations.

5. Conclusions

Nutmeg essential oil prepared with and without magnesium aluminometasilicate as an excipient has similar antioxidant activity. Nutmeg essential oil hydrolat prepared without excipient has a higher antioxidant activity compared to that with magnesium aluminometasilicate as an excipient.

Nutmeg essential oil with aluminometasilicate has extended antibacterial properties compared to the pure oil without additions. Both preparations prevent growth of *P. multocida* strain, but the oil with aluminometasilicate also inhibits *E. faecalis* and *S. mutans* (referent).

Nutmeg essential oil preparations with aluminometasilicate have stronger anti-inflammatory activity in Poly I:C-affected fibrolast cell culture. The oil with the excipient has a higher degree of cytoprotection from Poly I:C-induced necrosis, and both the oil and hydrolats with excipient more efficiently prevent IL-6 release compared to the preparations without aluminometasilicate.

The results show that the application of magnesium aluminometasilicate as an excipient in hydrodistillation could help to increase the biological activity of essential oil and hydrolats.

Author Contributions: Conceptualization, I.M. and J.B.; methodology, I.M., A.J., D.M.K., M.R., A.S. and E.B.; validation, V.S., P.Z., J.B., A.S. and A.J.; investigation, I.M., L.J., P.Z., M.R. and V.S.; data curation, J.B.; writing-original draft preparation, I.M., A.J., P.Z., D.M.K.; writing-review and editing, J.B. and I.M.; visualization, L.J., I.M., G.L., A.S., V.S.; supervision, J.B. All authors have read and agreed to the published version of the manuscript.

Funding: This research received no external funding.

Acknowledgments: The authors would like to thank Open Access Centre for the Advanced Pharmaceutical and Health Technologies (Lithuanian University of Health Sciences) and for the opportunity to use modern infrastructure and perform this research.

Conflicts of Interest: The authors declare no conflict of interest.

References

1. Muchtaridi; Subarnas, A.; Apriyantono, A.; Mustarichie, R. Identification of Compounds in the Essential Oil of Nutmeg Seeds (*Myristica Fragrans* Houtt.) That Inhibit Locomotor Activity in Mice. *Int. J. Mol. Sci.* **2010**, *11*, 4771–4781. [CrossRef]
2. Barceloux, D.G. Nutmeg (*Myristica Fragrans* Houtt.). *Disease-A-Month* **2009**, *55*, 373–379. [CrossRef] [PubMed]
3. Baser, K.H.; Bunchbauer, G. *Handbook of Essential Oils: Science, Technology, and Applications*; CRC Press NW: Boca Raton, FL, USA, 2010.
4. Djilani, A.; Dicko, A. The Therapeutic Benefits of Essential Oils. In *Nutrition, Well-Being and Health*; IntechOpen Ltd.: London, UK, 2012; pp. 154–178. [CrossRef]
5. Gupta, A.D.; Bansal, V.K.; Babu, V.; Maithil, N. Chemistry, Antioxidant and Antimicrobial Potential of Nutmeg (*Myristica Fragrans* Houtt). *J. Genet. Eng. Biotechnol.* **2013**, *11*, 25–31. [CrossRef]
6. Lanari, D.; Marcotullio, M.; Neri, A.A. Design of Experiment Approach for Ionic Liquid-Based Extraction of Toxic Components-Minimized Essential Oil from *Myristica Fragrans* Houtt. Fruits. *Molecules* **2018**, *23*, 2817. [CrossRef] [PubMed]
7. Morsy, S.; Nashwa, F. A Comparative Study of Nutmeg (*Myristica Fragrans* Houtt.) Oleoresins Obtained by Conventional and Green Extraction Techniques. *J. Food Sci. Technol.* **2016**, *53*, 3770–3777. [CrossRef] [PubMed]
8. Chatterjee, S.; Gupta, S.; Variyar, S. Comparison of Essential Oils Obtained from Different Extraction Techniques as an Aid in Identifying Aroma Significant Compounds of Nutmeg (*Myristica Fragrans*). *Nat. Prod. Commun.* **2015**, *10*, 1443–1446. [CrossRef] [PubMed]
9. Ehrenpreis, J.E.; Deslauriers, C.; Lank, P. Nutmeg Poisonings: A Retrospective Review of 10 Years Experience from the Illinois Poison Center. *J. Med. Toxicol.* **2014**, *10*, 148–151. [CrossRef]
10. Abourashed, E.A.; El-Alfy, A.T. Chemical Diversity and Pharmacological Significance of the Secondary Metabolites of Nutmeg (*Myristica Fragrans* Houtt). *Phytochem. Rev.* **2016**, *15*, 1035–1056. [CrossRef]
11. Shafiei, Z.; Shuhairi, N.N.; Fazly, N.; Yap, S.; Sibungkil, C.H.; Latip, J. Antibacterial Activity of Myristica Fragrans against Oral Pathogens. *Evid. Based Complement. Altern. Med.* **2012**. [CrossRef]
12. Sanghai-vaijwade, D.N.; Kulkarni, S.R.; Sanghai, N.N. Nutmeg: A promising antibacterial agent for stability of sweets. *Int. J. Res. Pharm. Chem.* **2011**, *1*, 403–407.
13. Takikawa, A.; Abe, K.; Yamamoto, M.; Ishimaru, S.; Yasui, M.; Okubo, Y.; Yokoigawa, K. Antimicrobial Activity of Nutmeg against Escherichia Coli O157. *J. Biosci. Bioeng.* **2002**, *94*, 315–320. [CrossRef]
14. Firouzi, R.; Shekarforoush, S.S.; Nazer, A.H.; Borumand, Z.; Jooyandeh, A.R. Effects of Essential Oils of Oregano and Nutmeg on Growth and Survival of Yersinia Enterocolitica and Listeria Monocytogenes in Barbecued Chicken. *J. Food Prot.* **2007**, *70*, 2626–2630. [CrossRef] [PubMed]
15. Piaru, S.P.; Mahmud, R.; Abdul Majid, A.M.; Ismail, S.; Man, C.N. Chemical Composition, Antioxidant and Cytotoxicity Activities of the Essential Oils of Myristica Fragrans and Morinda Citrifolia. *J. Sci. Food Agric.* **2012**, *92*, 593–597. [CrossRef] [PubMed]
16. Zhang, W.K.; Tao, S.S.; Li, T.T.; Li, Y.S.; Li, X.J.; Tang, H.B.; Cong, R.H.; Ma, F.L.; Wan, C.J. Nutmeg Oil Alleviates Chronic Inflammatory Pain through Inhibition of COX-2 Expression and Substance P Release in Vivo. *Food Nutr. Res.* **2016**, *60*, 1–10. [CrossRef] [PubMed]
17. Perez-Roses, R.; Risco, E.; Vila, R.; Penalver, P.; Canigueral, S. Biological and Nonbiological Antioxidant Activity of Some Essential Oils. *J. Agric. Food Chem.* **2016**, *64*, 4716–4724. [CrossRef]

18. Perez-Roses, R.; Risco, E.; Vila, R.; Penalver, P.; Canigueral, S. Effect of Some Essential Oils on Phagocytosis and Complement System Activity. *J. Agric. Food Chem.* **2015**, *63*, 1496–1504. [CrossRef]
19. Filly, A.; Fabiano-Tixier, A.S.; Louis, C.; Fernandez, X.; Chemat, F. Water as a Green Solvent Combined with Different Techniques for Extraction of Essential Oil from Lavender Flowers. *Comptes Rendus Chim.* **2016**, *19*, 707–717. [CrossRef]
20. Kara, N.; Erbaş, S.; Baydar, H. The Effect of Seawater Used for Hydrodistillation on Essential Oil Yield and Composition of Oil-Bearing Rose (Rosa Damascena Mill.). *Int. J. Second. Metab.* **2017**, *4*, 482–487. [CrossRef]
21. Charchari, S.; Abdelli, M. Enhanced Extraction by Hydrodistillation of Sage (*Salvia Officinalis* L.) Essential Oil Using Water Solutions of Non-Ionic Surfactants. *J. Essent. Oil-Bear. Plants* **2014**, *17*, 1094–1099. [CrossRef]
22. Matulyte, I.; Marksa, M.; Ivanauskas, L.; Kalveniene, Z.; Lazauskas, R.; Bernatoniene, J. GC-MS Analysis of the Composition of the Extracts and Essential Oil from Myristica Fragrans Seeds Using Magnesium Aluminometasilicate as Excipient. *Molecules* **2019**, *24*, 1062. [CrossRef]
23. Canillac, N.; Mourey, A. Antibacterial Activity of the Essential Oil of Picea Excelsa on Listeria, Staphylococcus Aureus and Coliform Bacteria. *Food Microbiol.* **2001**, *18*, 261–268. [CrossRef]
24. Olajide, O.A.; Ajayi, F.F.; Ekhelar, A.I.; Awe, S.O.; Makinde, J.M.; Alada, A.R.A. Biological Effects of Myristica Fragrans (Nutmeg) Extract. *Phytother. Res.* **1999**, *345*, 344–345. [CrossRef]
25. Emami, S.A.; Abedindo, B.F.; Hassanzadeh-Khayyat, M. Antioxidant Activity of the Essential Oils of Different Parts of Juniperus excelsa M. Bieb. subsp. excelsa and J. excelsa M. Bieb. subsp. polycarpos (K. Koch) Takhtajan (Cupressaceae). *Iran. J. Pharm. Res.* **2011**, *10*, 799–810. [PubMed]
26. Kong, B.; Zhang, H.; Xiong, Y.L. Antioxidant Activity of Spice Extracts in a Liposome System and in Cooked Pork Patties and the Possible Mode of Action. *Meat Sci.* **2010**, *85*, 772–778. [CrossRef] [PubMed]
27. Nishad, J.; Koley, T.K.; Varghese, E.; Kaur, C. Synergistic Effects of Nutmeg and Citrus Peel Extracts in Imparting Oxidative Stability in Meat Balls. *Food Res. Int.* **2018**. [CrossRef] [PubMed]
28. Dai, J.; Liang, Z.; Yang, L.; Qui, J. Chemical Composition, Antioxidant and Antimicrobial Activities of Essential Oil from Wedelia Prostrata. *EXCLI J.* **2013**, *12*, 479–490. [CrossRef] [PubMed]
29. Zheljazkov, V.D.; Astatkie, T.; Jeliazkova, E.A.; Adrienne, O.; Schlegel, V.; Zheljazkov, V.D.; Astatkie, T.; Jeliazkova, E.A. Distillation Time Alters Essential Oil Yield, Composition and Antioxidant Activity of Female Juniperus Scopulorum Trees. *J. Essent. Oil Res.* **2013**, *25*, 62–69. [CrossRef]
30. Misharina, T.A.; Terenina, M.B.; Krikunova, N.I. Antioxidant Properties of Essential Oils. *Prikl. Biokhim. Mikrobiol.* **2009**, *45*, 710–716. [CrossRef]
31. Aazza, S.; Lyoussi, B.; Miguel, M.G. Antioxidant Activity of Some Morrocan Hydrosols. *J. Med. Plants Res.* **2011**, *5*, 6688–6696. [CrossRef]
32. Balouiri, M.; Sadiki, M.; Ibnsouda, S.K. Methods for in Vitro Evaluating Antimicrobial Activity: A Review. *J. Pharm. Anal.* **2016**, *6*, 71–79. [CrossRef]
33. Nurjanah, S.; Putri, I.L.; Sugiarti, D.P. Antibacterial Activity of Nutmeg Oil. *KnE Life Sci.* **2017**, *2*, 563. [CrossRef]
34. Cui, H.; Zhang, X.; Zhou, H.; Zhao, C.; Xiao, Z.; Lin, L.; Changzhu, L. Antibacterial Properties of Nutmeg Oil in Pork and Its Possible Mechanism. *J. Food Saf.* **2015**, *35*, 370–377. [CrossRef]
35. Funk, B.; Kirmayer, D.; Sahar-heft, S.; Gati, I.; Friedman, M.; Steinberg, D. Efficacy and Potential Use of Novel Sustained Release Fillers as Intracanal Medicaments against Enterococcus Faecalis Biofilm in Vitro. *BMC Oral Health* **2019**, *19*, 1–9. [CrossRef] [PubMed]
36. Kitagawa, H.; Izutani, N.; Kitagawa, R.; Maezono, H.; Yamaguchi, M.; Imazato, S. Evolution of Resistance to Cationic Biocides in Streptococcus Mutans and Enterococcus Faecalis. *J. Dent.* **2016**, *47*, 18–22. [CrossRef]
37. Kim, D.S.; Lee, H.J.; Jeon, Y.D.; Han, Y.H.; Kee, J.Y.; Kim, H.J.; Shin, H.J.; Kang, J.; Lee, B.S.; Kim, S.H.; et al. Alpha-Pinene Exhibits Anti-Inflammatory Activity Through the Suppression of MAPKs and the NF-KB Pathway in Mouse Peritoneal Macrophages. *Am. J. Chin. Med.* **2015**, *43*, 731–742. [CrossRef]
38. Valente, J.; Zuzarte, M.; Gonçalves, M.J.; Lopes, M.C.; Cavaleiro, C.; Salgueiro, L.; Cruz, M.T. Antifungal, Antioxidant and Anti-Inflammatory Activities of Oenanthe Crocata L. Essential Oil. *Food Chem. Toxicol.* **2013**, *62*, 349–354. [CrossRef]
39. Yu, L.; Yan, J.; Sun, Z. D-Limonene Exhibits Anti-Inflammatory and Antioxidant Properties in an Ulcerative Colitis Rat Model via Regulation of INOS, COX-2, PGE2 and ERK Signaling Pathways. *Mol. Med. Rep.* **2017**, *15*, 2339–2346. [CrossRef]

40. Akinwunmi, K.F.; Oyedapo, O.O. In Vitro Anti-Inflammatory Evaluation of African Nutmeg (Monodora Myristica) Seeds. *Eur. J. Med. Plants* **2015**, *8*, 167–174. [CrossRef]
41. Romano, R.; Giordano, A.; Le Grottaglie, L.; Manzo, N.; Paduano, A.; Sacchi, R.; Santini, A. Volatile compounds in intermittent frying by gas chromatography and nuclear magnetic resonance. *Eur. J. Lipid Sci. Technol.* **2013**, *115*, 764–773. [CrossRef]
42. Crowe, J.E., Jr. Common Viral Respiratory Infections. In *Harrison's Principles of Internal Medicine*, 20th ed.; Jameson, J.L., Fauci, A.S., Kasper, D.L., Hauser, S.L., Longo, D.L., Loscalzo, J., Eds.; McGraw-Hill Education: New York, NY, USA, 2018; Chapter 194.
43. Cells, S.; Smith, R.S.; Smith, T.J.; Blieden, T.M.; Phipps, R.P. Commentary Synthesis of Chemokines and Regulation of Inflammation. *Am. J. Pathol.* **1997**, *151*, 317–322.
44. Richards, C.D. Innate Immune Cytokines, Fibroblast Phenotypes, and Regulation of Extracellular Matrix in Lung. *J. Interf. Cytokine Res.* **2017**, *37*. [CrossRef] [PubMed]
45. Naik, S.P.; Mahesh, P.A.; Jayaraj, B.S.; Madhunapantula, S.V.; Jahromi, S.R.; Yadav, M.K. Evaluation of Inflammatory Markers Interleukin-6 (IL-6) and Matrix Metalloproteinase-9 (MMP-9). *J. Asthma* **2017**, *54*, 584–593. [CrossRef] [PubMed]
46. Ishihara, K.; Hirano, T. IL-6 in Autoimmune Disease and Chronic Inflammatory Proliferative Disease. *Cytokine Growth Factor Rev.* **2002**, *13*, 357–368. [CrossRef]
47. Sundararaj, K.P.; Samuvel, D.J.; Li, Y.; Sanders, J.J.; Lopes-Virella, M.F.; Huang, Y. Interleukin-6 Released from Fibroblasts Is Essential for Up-Regulation of Matrix Metalloproteinase-1 Expression by U937 Macrophages in Coculture: cross-talikng between fibroblast and U937 macrofages exposed high glucose. *J. Biol. Chem.* **2009**, *284*, 13714–13724. [CrossRef] [PubMed]
48. Fernandes, A.R.; Martins-Gomes, C.; Santini, A.; Silva, A.M.; Souto, E.B. Psoriasis vulgaris—Pathophysiology of the disease and its classical treatment versus new drug delivery systems. In *Design of Nanostructures for Versatile Therapeutics Applications*; Pharmaceutical Nanotechnology; Grumezescu, A.M., Ed.; Elsevier: Oxford, UK, 2018; Chapter 9; pp. 379–406. ISBN 978-0-12-813667-6.
49. Daliu, P.; Santini, A.; Novellino, E. A decade of nutraceutical patents: where are we now in 2018? *Expert Opin. Ther. Pat.* **2018**, *28*, 875–882. [CrossRef] [PubMed]
50. Campos, J.R.; Severino, P.; Ferreira, C.S.; Zielinska, A.; Santini, A.; Souto, S.B.; Souto, E.B. Linseed Essential Oil—Source of Lipids as Active Ingredients for Pharmaceuticals and Nutraceuticals. *Curr. Med. Chem.* **2019**, *26*, 1–22. [CrossRef] [PubMed]
51. Severino, P.; Resende Diniz, F.; Cardoso Cordeiro, J.; do Céu Teixeira, M.; Santini, A.; Kovačević, A.B.; Souto, E.B. Essential oils with antimicrobial properties formulated in lipid carriers—Review of the state of the art. In *Essential Oils and Nanotechnology for the Cure of Microbial Diseases*; Rai, M., Derita, M., Zacchino, S., Eds.; CRC Press: Boca Raton, FL, USA, 2017; Chapter 1; pp. 1–13. ISBN 978-1-1386-3072-7.

© 2020 by the authors. Licensee MDPI, Basel, Switzerland. This article is an open access article distributed under the terms and conditions of the Creative Commons Attribution (CC BY) license (http://creativecommons.org/licenses/by/4.0/).

Article

Elderberry (*Sambucus nigra* L.) Fruit Extract Alleviates Oxidative Stress, Insulin Resistance, and Inflammation in Hypertrophied 3T3-L1 Adipocytes and Activated RAW 264.7 Macrophages

Joanna Zielińska-Wasielica [1], Anna Olejnik [1,*], Katarzyna Kowalska [1], Mariola Olkowicz [2] and Radosław Dembczyński [1]

1 Department of Biotechnology and Food Microbiology, Poznan University of Life Sciences, Wojska Polskiego 48, 60-627 Poznan, Poland
2 Department of Chemistry, University of Waterloo, 200 University Avenue West, Waterloo, ON N2L 3G1, Canada
* Correspondence: anna.olejnik@up.poznan.pl; Tel.: +48-61-846-60-08

Received: 11 July 2019; Accepted: 4 August 2019; Published: 8 August 2019

Abstract: Oxidative stress and inflammation in hypertrophied adipose tissue with excessive fat accumulation play a crucial role in the development of obesity and accompanying metabolic dysfunctions. This study demonstrated the capacity of elderberry fruit (EDB) extract to decrease the elevated production of reactive oxygen species in hypertrophied 3T3-L1 adipocytes. Treatment with the EDB extract resulted in modulation of mRNA expression and protein secretion of key adipokines in hypertrophied adipocytes. Expression of leptin and adiponectin was, respectively, down- and up-regulated. Moreover, glucose uptake stimulation was noticed in mature adipocytes, both sensitive to insulin and insulin resistant. This may suggest a positive effect of EDB extract on insulin resistance status. The extract was also found to alleviate the inflammatory response in activated RAW 264.7 macrophages by down-regulating the expression of proinflammatory genes (*TNF-α*, *IL-6*, *COX-2*, *iNOS*) and suppressing the enhanced production of inflammatory mediators (TNF-α, IL-6, PGE$_2$, NO). *In vitro* experiments showed that the EDB extract could inhibit digestive enzymes, including α-amylase, α-glucosidase, and pancreatic lipase, leading to reduced intestinal absorption of dietary lipids and carbohydrates. Further *in vivo* studies could be postulated to support EDB as a functional food component for the prevention and treatment of obesity and metabolic-immune comorbidities.

Keywords: elderberry polyphenols; functional food; obesity; digestive enzymes; fat cells; intracellular reactive oxygen species; adipokines; glucose uptake; immune-metabolic effects

---

1. Introduction

Obesity is associated with excessive adipose tissue growth, which occurs through two possible mechanisms: hypertrophy (expansion of existing adipocytes) and hyperplasia (recruitment of new adipocytes). Hypertrophic adipose tissue growth is mainly considered to be related to insulin resistance and other obesity metabolic comorbidities [1]. Abnormal expansion of adipose tissue is accompanied by local hypoxia, adipocyte death, enhanced cytokine and chemokine secretion, dysfunctional fatty acid metabolism and accumulation, and immune cell infiltration. Dysregulation of lipid metabolism in adipose tissue leads to enhanced release of free fatty acids, which initiates inflammatory signaling cascades in the infiltrating cell population. Chronic low-grade inflammation, found in abnormal fat tissue, negatively affects the insulin signal transduction pathway, and promotes insulin resistance [2,3].

Recent scientific preclinical studies have shown that bioactive dietary compounds may specifically influence hypertrophic adipose cells and mitigate the effects of extensive adipose tissue growth by affecting various adverse phenomena, including oxidative stress, inflammation, disturbances in adipokine secretion, fatty acid release, and others. Berry fruits have been recognized as capable of counteracting obesity and obesity-related metabolic disorders, through the inhibition of adipocyte differentiation, a decrease in lipogenesis, an increase in lipolysis, or mitigation of inflammatory and insulin resistance status [4].

A promising candidate capable of attenuating obesity and complications related to excessive fat tissue growth might be *Sambucus nigra* L. (European elderberry) fruit as a valuable source of polyphenolic compounds, primarily flavonols, flavanols, phenolic acids, proanthocyanidins, and anthocyanins [5]. The unique polyphenol composition is responsible for the high biological potential of elderberry fruit (EDB), including antiviral and antimicrobial activity, as well as chemopreventive, neuroprotective, and anti-inflammatory effects that have been documented in several scientific reports [6–10]. Also, it has been suggested that EDB may be an effective remedy for diabetes, obesity, and metabolic dysfunctions [9]. Animal studies have shown the ability of *Sambucus nigra* preparations to improve glucose and lipid metabolism and diabetic osteoporosis status [11–14].

Anthocyanin-rich EDB extract has been proved to attenuate systemic inflammation and insulin resistance in high-fat diet-induced obese mice. Pro-inflammatory markers of low-grade chronic inflammation, including serum monocyte chemoattractant protein-1 (MCP-1) and tumor necrosis factor-α (TNF-α), were significantly reduced in EDB-fed mice. Also, the high-fat diet supplemented with EDB extract mitigated some metabolic disturbances by lowering serum triglycerides and improving insulin sensitivity [12]. Lowered insulin resistance was found in diabetic rats fed with a high-fat diet supplemented with EDB extracts rich in triterpenic acids or polyphenol compounds. The extracts modulated glucose metabolism by correcting hyperglycemia or reducing insulin secretion, respectively [13]. The anthocyanin-rich EDB extract protected against inflammation-related impairments in high-density lipoprotein (HDL) function in a mouse model of hyperlipidemia and HDL dysfunction. The decrease in total cholesterol content of the aorta in EDB-fed mice suggested limiting atherosclerosis progression [14]. Scientific reports indicate that EDB extracts possess the unique potential to modulate the immune response depending on the immune stimuli and inflammatory disorders. The EDB bioactives have evoked different immune effects by controlling pro- and anti-inflammatory cytokines and mediators (Reactive oxygen species, NO, IL-6, TNF-α, MCP-1, IL-1, IL-8, IL-10, PGE$_2$, COX-2, iNOS, INF-γ), that play a crucial role in acute and chronic low-grade inflammatory diseases associated with obesity, diabetes, dyslipidemia, cardiovascular disturbances, and neurodegenerative diseases [7,8,10–16].

Over the last decade, significant advances in knowledge about the health-beneficial potential of EDB fruit have been achieved through extensive preclinical studies. However, the results obtained only in the few clinical trials have not enabled to express an unambiguous opinion and, so far, have not provided strong evidence of the therapeutic effects of *Sambucus nigra* fruit in obesity and metabolic disorders [9]. Recently, the scientific community has stated the need for further research on the health-promoting properties of this valuable plant as a natural constituent of food products and beneficial component of a healthy diet [6,9].

This study aimed to evaluate the capacity of *Sambucus nigra* fruit extract to mitigate obesity-related metabolic complications through the carbohydrate and lipid metabolism regulation, glucose uptake improvement, and insulin sensitivity controlling. Also, the goal of the study was the assessment of the ability of the extract to alleviate the inflammatory response in activated macrophages, which are recruited into excessively growing fat tissue and may be a primary source of locally produced pro-inflammatory mediators.

2. Materials and Methods

2.1. Preparation of Elderberry Fruit Extract

The fruits of elderberry (*Sambucus nigra* L.) cultivar Sampo, obtained from Bio Berry Poland (Warsaw, Poland), were homogenized to fruit pulp, which was subsequently frozen at −80 °C and subjected to freeze-drying at a vacuum pressure of 0.1 mbar and temperature of 20 °C for 23 h and post-drying at 23 °C for 3 h using a freeze dryer (LMC-1, Martin Christ Gefriertrocknungsanlagen GmbH, Germany). The lyophilized EDB were finely ground and packaged under nitrogen atmosphere. The EDB extract was obtained by dissolving the EDB powder in complete culture medium with the pH adjustment to 7.4. The EDB suspension was then centrifuged (3000 g, 5 min) and filtered through a 0.22 µm membrane (Merck, Germany).

2.2. Determination of Individual Phenolic Compounds Using HPLC-DAD-MS^n Analysis

Analyses of phenolic compounds were performed on an Agilent 1200 series HPLC system (Agilent Technologies, Inc., Santa Clara, CA, USA) that was equipped with a G1315D photodiode array detector and coupled online with an Agilent 6224 time-of-flight MS system. Phenolic compounds were identified using a mass spectrometer fitted with an electrospray ionization (ESI) source that was operated in positive-ion or negative-ion mode. Analyses were carried out using full MS scan mode, and full mass spectra were recorded in the range of 100 to 1700 *m/z*. Technical specification of apparatus and major HPLC/MS parameters and analysis conditions were described in detail in our previous work [17].

For quantification purposes, all anthocyanins conjugates were expressed as cyanidin-3-glucoside equivalents; all flavan-3-ols and their polymers as catechin equivalents; hydroxybenzoic acid glucoside and hydrolysable tannins as gallic acid equivalents; phenolic acids derivatives as chlorogenic acid equivalents; and flavonol glycosides as quercetin equivalents.

2.3. T3-L1 Cell Culture, Differentiation, and Treatment

The mouse embryo 3T3-L1 cell line was purchased from the American Type Culture Collection (ATCC, CL-173). The 3T3-L1 preadipocytes were grown, passaged, and differentiated into adipocytes as described previously [18]. The 3T3-L1 cells were grown in Dulbecco's Modified Eagle's Medium (DMEM) with 10% calf serum supplementation (Sigma-Aldrich, Merck Group, Darmstadt, Germany). Cell differentiation was induced in post-confluent cell cultures by a differentiation mixture consisting of 1 µM insulin, 0.25 µM dexamethasone (DEX), and 0.5 mM 3-isobutyl-1-methylxanthine (IBMX) in DMEM with 10% fetal bovine serum (FBS) (Gibco, Thermo Fisher Scientific Polska, Warsaw, Poland).

Fully differentiated 3T3-L1 cells were exposed to the EDB extract at concentrations of 5, 10, and 20 mg/mL for 24 h. The levels of intracellular ROS generation and lipid accumulation in mature adipocytes were determined. Also, the viability and metabolic activity of the mature adipocytes were analyzed after the treatment.

After completion of the differentiation process, insulin resistance was induced in 3T3-L1 adipocytes by 10 ng/mL murine TNF-α (Sigma-Aldrich) for 5 days, with medium/TNF-α replacement every 2 days. Glucose uptake measurement was performed in insulin-resistant and insulin-sensitive adipocytes subjected to the EDB treatment.

2.4. Macrophage Cell Culture and Anti-Inflammatory Experiment Procedure

RAW 264.7 murine macrophage line was obtained from the European Collection of Authenticated Cell Cultures (ECACC, 91062702) and supplied by Sigma-Aldrich. Cells were grown in DMEM supplemented with 10% heat-inactivated FBS at 37 °C in a humidified, 5% CO_2, 95% air atmosphere. The 24-h cultures of RAW 264.7 macrophages, seeded at a density of 5×10^5 cells/cm^2, were treated with EDB extract prepared in DMEM at the concentrations of 0.1, 1, and 10 µg/mL and incubated for 2 h in standard culture conditions. Controls were treated with DMEM only. Subsequently, macrophages were stimulated with 5 ng/mL of lipopolysaccharide (LPS) from *Escherichia coli* O-127 (Sigma-Aldrich).

After 3-h macrophage activation, the culture media and cells were harvested to analyze the protein secretion and gene expression of pro-inflammatory mediators.

2.5. Cell Viability Assay

The viability and metabolic activity of differentiated 3T3-L1 adipocytes and LPS-stimulated RAW 264.7 macrophages were analyzed using the MTT (3-(4,5-dimethylthiazol-2-yl)-2,5-diphenyltetrazolium bromide) test (Sigma–Aldrich) following the protocol described previously [8].

2.6. Measurement of Reactive Oxygen Species in Adipocytes

The intracellular ROS generation was determined using nitro blue tetrazolium (NBT) according to the procedure described by Choi et al. [19]. The cells were incubated in 0.2% NBT solution for 90 min, washed with phosphate-buffered saline (PBS), fixed with methanol, and then air-dried. The formazan extraction was performed using KOH and DMSO for dissolving. The absorbance was measured at 620 nm using a Tecan M200 Infinite microplate reader (Tecan Group Ltd., Männedorf, Switzerland).

2.7. Measurement of Intracellular Triglyceride Content in Adipocytes

Total concentrations of triglycerides (TG) in differentiated 3T3-L1 adipocytes were determined using Adipogenesis Assay Kit (Sigma-Aldrich) according to the manufacturer's protocol. Intracellular TG content was measured by a coupled enzyme assay, which resulted in a fluorometric product detected at λ_{ex} = 535 nm and λ_{em} = 587 nm (Tecan M200 Infinite), which was proportional to the TG present. The TG concentration was calculated based on the curve plotted for TG standards.

2.8. Glucose Uptake Measurement in Adipocytes

Glucose uptake assay was performed according to the modified method of Alonso-Castro and Salazar-Olivo [20]. Mature 3T3-L1 adipocytes, cultured on 24-well plates for fluorescence-based assays, were starved in serum-free medium (MEM containing BSA 0.5%) overnight. Subsequently, the medium was replaced with Krebs Ringer phosphate HEPES (KRPH) buffer containing 0.2% BSA (KRPH/BSA) and incubated for 60 min. The cells were then exposed for 60 min to EDB extract suspended in KRPH/BSA buffer supplemented with 80 µM 2-NBDG (2-N-7-(nitrobenz-2-oxa-1,3-diazol-4-yl) amino-2-deoxy-D-glucose) (Sigma-Aldrich) used as fluorescent glucose analogue. The control cultures were treated with 100 nM insulin or 10 µM rosiglitazone (Sigma-Aldrich). After incubation, cultures were immediately washed three times with ice-cold PBS. The fluorescence intensity of 2-NBDG was measured at λ_{ex} = 485 nm and λ_{em} = 535 nm (Tecan M200 Infinite).

2.9. Determination of Adipokine Production in 3T3-L1 Adipocytes

The leptin and adiponectin concentrations were measured using ELISA kits (Sigma-Aldrich, Merck Group) following the manufacturer's instructions. The adipokine concentrations were expressed in ng/mL of culture medium, which was equivalent to the amount of protein per 1×10^6 cells.

2.10. Determination of IL-6, TNF-α, and PGE$_2$ Production in RAW 264.7 Macrophages

The secretion of IL-6 and TNF-α cytokines as well as generation of PGE$_2$ by LPS-stimulated RAW 264.7 macrophages were determined with ELISA kits (R&D Systems, Inc, Minneapolis, MN, USA) according to the manufacturer's instructions. Protein concentrations were expressed in pg/mL of culture supernatant, which was equivalent to the amount of protein per 1×10^6 cells.

2.11. Determination of NO Production in RAW 264.7 Macrophages

Griess method was applied to determine nitrite as an indicator of NO production. Equal volumes of the Griess reagent (Sigma-Aldrich) and RAW 264.7 culture supernatant were mixed and incubated at room temperature for 15 min. The absorbance was measured at 540 nm (Tecan M200 Infinite). The standard curve plotted for sodium nitrite was used to calculate NO concentration.

2.12. Quantification of Gene Expression Using Real-Time PCR

The analysis of gene expression was carried out in accordance with the detailed protocol presented in the previous work [17]. The TRI reagent (Sigma-Aldrich) was used to isolate total RNA, Synthesis cDNA Transcriptor First-Strand kit (Roche Diagnostics GmbH, Mannheim, Germany) for first-strand cDNA synthesis, and SYBR1 Select Master Mix (Life Technologies, Carlsbad, CA, USA) for real-time PCR. The primers used for the amplification of cDNAs are listed in Table 1.

Table 1. The primers sequence used for real-time PCR.

Gene	Accession	No. Sequence (5′–3′)	Amplicon (bp)
Mm *LEP*	NM-008493	F: GGA TCA GGT TTT GTG GTG CT R: TTG TGG CCC ATA AAG TCC TC	187
Mm *GLUT-4*	NM-001359114.1	F: TGC TGG GCA CAG CTA CCC R: CGG TCA GGC GCT TTA GAC	162
Mm *ADIPOQ*	NM-009605	F: CTG GCC ACT TTC TCC TCA TT TC R: GGC ATG ACT GGG CAG GAT TA	120
Mm *IL-6*	NM-031168.1	F: TCT GAA GGA CTC TGG CTT TG R: GAT GGA TGC TAC CAA ACT GGA	142
Mm *NOS-2*	NM-010927.3	F: TGA AGA AAA CCC CTT GTG CT R: TTC TGT GCT GTC CCA GTG AG	100
Mm *PTGS2*	NM-011198.3	F: GGC GCA GTT TAT GTT GTC TGT R: CAA GAC AGA TCA TAA GCG AGG A	107
Mm *TNF-α*	NM-001278601.1	F: AGG GTC TGG GCC ATA GAA CT R: CCA CCA CGC TCT TCT GTC TAC	103
Mm *NOX-4*	NM-015760.5	F: GAT CAC AGA AGG TCC CTA GCA G R: GTT GAG GGC ATT CAC CAA GT	134
Mm *SOD2*	NM-013671.3	F: CGT GTC TGT GGG AGT CCA AGG TTC AG R: GTC AAT CCC CAG CAG CGG AAT AAG	139
Mm *CATALASE*	NM-009804.2	F: CCT CCT CGT TCA GGA TGT GGT T R: CGA GGG TCA CGA ACT GTG TCA G	243
Mm *GPx*	NM-008160.6	F: GGG CAA GGT GCT GCT CAT TG R: AGA GCG GGT GAG CCT TCT CA	269
Mm *ACTB*	NM-007393	F: CCA CAG CTG AGA GGG AAA TC R: AAG GAA GGC TGG AAA AGA GC	193

The relative expression of each gene was calculated using the $2^{-\Delta\Delta CT}$ method. The mRNA levels in the control cells were designated as 1, and the relative levels of the gene transcripts in the samples were expressed as the fold change.

2.13. Digestive Enzyme Inhibition Assays

2.13.1. Measurement of Pancreatic Lipase Inhibition

The EDB inhibitory activity against pancreatic lipase (EC 3.1.1.3) was evaluated according to the method of Boath et al. with minor modification [21]. The p-nitrophenyl laurate (pNP laurate) was used as a substrate. The pNP laurate was dissolved to 0.08% in 5 mM sodium acetate (pH 5.2) containing 1% Triton X-100 and 0.05% Arabic gum. The reaction mixture consisting of 350 µL of assay buffer (100 mM Tris, pH 8.2), 50 µL of EDB extract, 150 µL of pancreatic lipase type II from porcine pancreas (10 mg/mL), and 450 µL of substrate solution was incubated at 37 °C for 2 h. Orlistat, a known porcine pancreatic lipase inhibitor, was applied as a positive control. After incubation, the sample was centrifuged at 13,000 rpm for 3 min and read at 400 nm of wavelength (Tecan M200).

2.13.2. Measurement of α-Amylase Inhibition

The inhibition of α-amylase (EC 3.2.1.1) activity was determined using the method of Tan et al. with slight modification [22]. The reaction mixture consisting of 200 μL of distilled water, 50 μL of EDB extract, 250 μL of α-amylase from porcine pancreas (30 mg/mL), and 500 μL of 0.5% starch was incubated at 37 °C for 10 min. Acarbose, a known pancreatic α-amylase inhibitor, was applied as a positive control. Enzymatically released reducing sugars were determined by DNS reagent solution (96 mM 3,5-dinitrosalicylic acid, 5.31 M sodium potassium tartrate in 2 M NaOH) after heating at 95 °C for 10 min. Then, the mixture was diluted with distilled water and the absorbance was measured at 540 nm (Tecan M200 Infinite).

2.13.3. Measurement of α-Glucosidase Inhibition

The inhibition assay of α-glucosidase (EC 3.2.1.20) was adopted from Tan et al. [22]. The p-nitrophenyl-α-D-glucuronide (pNPG) dissolved to 4 mM in 0.1 M HEPES (pH 6.8) was used as a substrate. The reaction mixture consisting of 350 μL of HEPES (pH 6.8), 50 μL of EDB extract, 150 μL of α-glucosidase (20 mg/mL), and 450 μL of substrate solution was incubated at 37 °C for 2 h. The release of p-nitrophenol from the pNPG substrate was measured at 410 nm (Tecan M200 Infinite). As a positive control, the glucosidase inhibitor, acarbose, was used.

All reagents used in digestive enzyme inhibition assays were provided by Sigma-Aldrich.

2.13.4. Data Analysis

Enzyme activity in the presence of inhibitor (EDB extract or reference inhibitor) was expressed as a percentage of the non-inhibited enzyme activity and plotted versus inhibitor concentration. Based on the dose-response curve, the inhibitor concentration required for 10% and 50% inhibition of enzyme activity (IC_{10} and IC_{50}) was determined as a measure of inhibitory potency. The percentage of the non-inhibited enzyme activity was calculated by following equation:

$$\% \text{ non-inhibited enzyme activity} = [(A_{Inhibitor} - A_{Inhibitor\ blank})/(A_{Control} - A_{Control\ blank})] \times 100\%$$

where $A_{Control}$ is the absorbance of the sample without EDB extract/reference inhibitor; $A_{Inhibitor}$ is the absorbance of the sample containing EDB extract/reference inhibitor; $A_{Inhibitor\ blank}$ is the absorbance of the sample with EDB extract/reference inhibitor, but without enzyme addition; $A_{Control\ blank}$ is the absorbance of the sample without EDB extract/reference inhibitor and enzyme addition.

2.14. Statistical Analysis

All data are expressed as the means ± SD from three independent experiments. Statistical analysis was performed using the STATISTICA version 13.3 software (Statsoft, Inc., Tulsa, OK, USA). One-way analysis of variance (ANOVA) followed by Tukey's post hoc test was used to determine the differences between the mean values of multiple groups. The T-student's test was applied to determine the significant difference between two independent groups. The equality of variances assumption was verified with the Levene's test.

3. Results

3.1. Polyphenol Composition in the Elderberry Fruit Extract

HPLC-DAD-ESI-MS^n analysis of the EDB extract revealed the presence of 22 polyphenolic compounds, including anthocyanins (peaks 1–5), hydroxybenzoic acid derivative (peak 6), flavan-3-ols (peaks 7–8), polymers, tentatively identified as hydrolysable tannins (peaks 9–10), hydroxycinnamic acids (peaks 11–15), and flavonols (peaks 16–22). HPLC-DAD chromatograms and chromatographic characteristics with mass spectral data of polyphenols identified in the EDB extract are presented in Figure 1 and Table 2, respectively.

Table 2. HPLC-MS identification of phenolic compounds in elderberry fruit extract in positive (electrospray ionization (ESI) +) and negative (ESI −) ionization mode.

Peak No.	RT (min)	UV λ max (nm)	[M]+/[M + H]+ (m/z)	[M − H]− (m/z)	MS/MS (m/z)	Tentative Identification	Concentration (mg/g) *
1	14.01	280, 520	611.1651 743.2095	-	287.0583 287.0579	Cyanidin-3,5-O-diglucoside Cyanidin-3-O-sambubiosyl-5-O-glucoside (co-elution)	3.27 ± 0.25
2	16.90	280,520	449.1133		287.0632	Cyanidin-3-O-glucoside	Trace amounts
3	19.84	280,520	595.1734		287.0578	Cyanidin-3-O-rutinoside	Trace amounts
4	23.92	280,520	433.1187		271.0640	Pelargonidin-3-O-glucoside	0.31 ± 0.04
5	29.85	280,520	581.1635		287.0633	Cyanidin-3-O-sambubioside	9.76 ± 0.68
6	4.48	275		299.2506	—	4-Hydroxybenzoic acid glucoside	1.60 ± 0.12
7	9.05	280		289.1139	245.1203	(+)/(−)-Catechin	1.07 ± 0.06
8	9.99	280		289.1140	245.1210	(+)/(−)-Epicatechin	1.41 ± 0.08
9	13.96	268		597.4616	—	Hydrolysable tannin	3.45 ± 0.22
10	14.38	268		597.4628	—	Hydrolysable tannin	0.92 ± 0.06
11	9.99	299,325		353.2873	191.1737	Neochlorogenic acid	0.50 ± 0.03
12	11.00	299,325		353.2886	191.1749	Chlorogenic acid	0.59 ± 0.04
13	12.54	300,325		353.2865	191.1729	Cryptochlorogenic acid	0.26 ± 0.02
14	17.00	310,234		337.0917	173.0443	P-coumaroylquinic acid	0.78 ± 0.06
15	17.52	316,234		371.3016	163.0396	P-Coumaric acid hexoside	2.85 ± 0.18
16	22.83	268,354		463.0882	301.0354	Quercetin-3-O-glucoside	0.29 ± 0.01
17	24.39	255,355		609.1461	301.0349	Quercetin-3-O-rutinoside	2.27 ± 0.19
18	28.63	255,358		505.0872	301.0366	Quercetin 3-O-(6"-acetyl-glucoside)	0.09 ± 0.01
19	29.02	266,348		447.0935	285.0540	Kaempferol-3-O-glucoside	0.09 ± 0.01
20	30.34	255,352		593.1515	285.0542	Kaempferol-3-O-rutinoside	1.36 ± 0.12
21	33.15	255,370		301.0356	151.0031	Quercetin	0.09 ± 0.02
22	36.14	255,352		623.1044	315.0449	Isorhamnetin-3-O-rutinoside	0.07 ± 0.01

* mg/g of lyophilized elderberry powder, values were expressed as mean ± SEM for three independent experiments.

Anthocyanins accounted for 43% of all polyphenolics; cyanidin-based anthocyanin compounds (-3,5-O-diglucoside, -3-O-sambubiosyl-5-O-glucoside, -3-O-glucoside, -3-O-rutinoside, and -3-O-sambubioside) were the main group of anthocyanins, with a significant predominance of cyanidin-3-O-sambubioside ([M + H]$^+$ at m/z 287) constituting 73.2% of all anthocyanins. Peak 1 contained two compounds, identified as cyanidin-3,5-O-diglucoside ([M + H]$^+$ at m/z 611) and cyanidin-3-O-sambubiosyl-5-O-glucoside ([M + H]$^+$ at m/z 743). These anthocyanins represented 25.0% of total anthocyanin compounds, quantitatively determined in the EDB extract. In contrast, cyanidin-3-O-glucoside ([M + H]$^+$ at m/z 449) and cyanidin-3-O-rutinoside ([M + H]$^+$ at m/z 595) were only detected in trace amounts in the EDB extract (Table 2). The anthocyanin with a molecular ion of m/z 433, that yielded on MS^2 fragment at m/z 271, was identified as pelargonidin-3-O-glucoside and quantified in small amounts estimated at 2.3% of all anthocyanins.

Other groups of compounds: Flavan-3-ols, hydroxycinnamic acids, and flavonols amounted to 22.1%, 16.0%, and 13.7% of the total content of polyphenols, respectively. Moreover, the presence of 4-hydroxybenzoic acid glucoside (5.2%) was found in the extract. The group of non-anthocyanin compounds with the largest share in the polyphenol pool were flavan-3-ols, among which catechin and epicatechin (36%), and tannins (64%) were identified. A total of five compounds were detected within another group of hydroxycinnamic acid derivatives, including p-coumaric acid hexoside (57%), p-coumaroylquinic acid (16%), chlorogenic acid, and its isomers: Neochlorogenic and cryptochlorogenic acids (27%). Concerning flavonols, the results of the HPLC-DAD analysis revealed the presence of quercetin, kaempferol, and isorhamnetin derivatives, with quercetin-3-O-rutinoside (53%) and kaempferol-3-O-rutinoside (32%) quantified as the dominant compounds within this class.

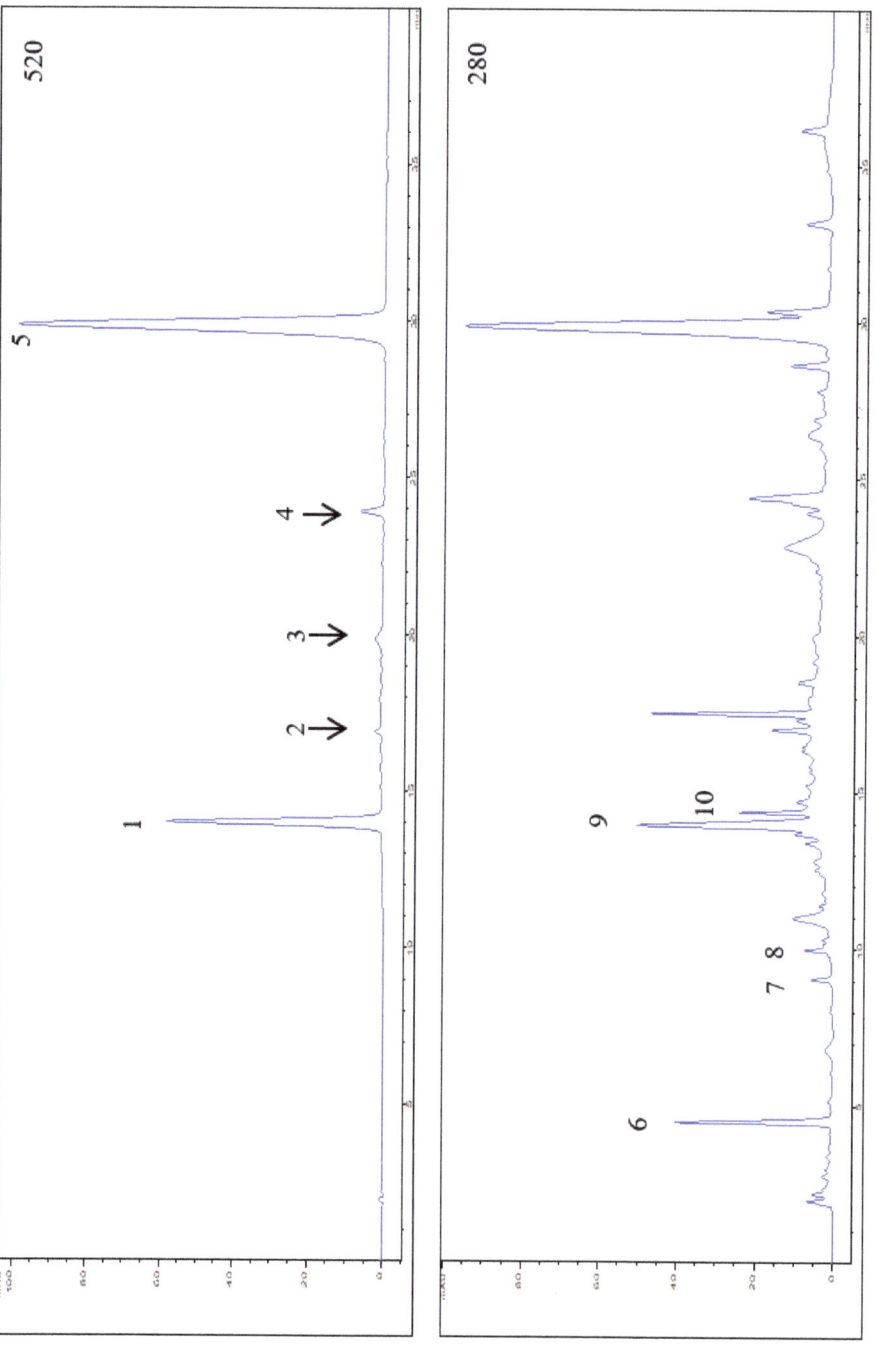

Figure 1. Cont.

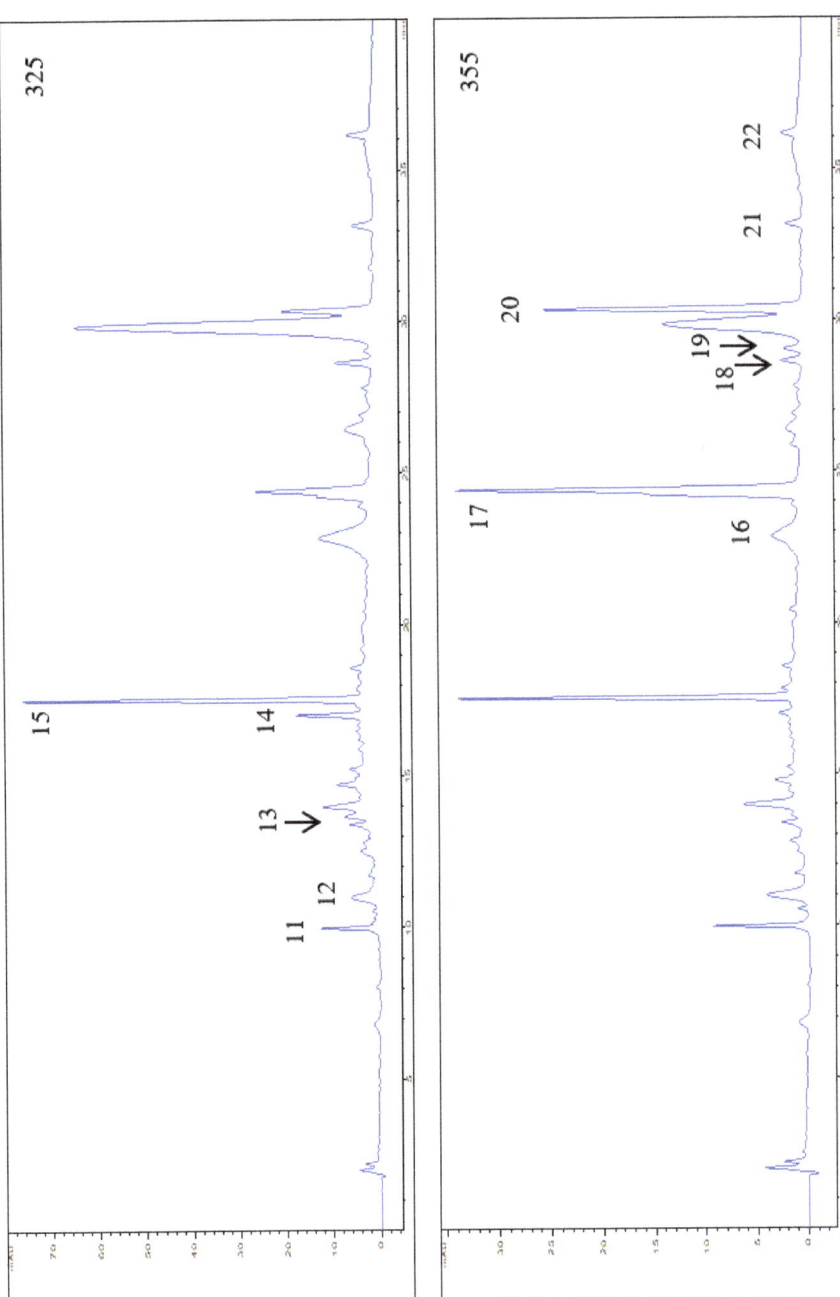

Figure 1. HPLC-DAD chromatograms of elderberry (*Sambucus nigra* L.) fruit extract recorded at 520, 280, 325, and 355 nm, respectively. Peak numbers and retention times refer to compounds indicated in Table 2.

The polyphenolic compounds, described above, have been previously identified in berries of *Sambucus nigra* [7,23,24]. However, the content of individual polyphenols in EDB fruits varies, depending on the EDB genotypes as well as specific growth conditions. The total polyphenol content in the extract analyzed was determined to 31.03 mg/g of EDB lyophilized powder, including 13.34 mg of anthocyanins, 6.85 mg of flavan-3-ols, 4.98 mg of hydroxycinnamic acid derivatives, 4.26 mg of flavonols, and 1.6 mg of hydroxybenzoic acid glucoside (Table 2).

3.2. Digestive Enzyme Activity Inhibition by Elderberry Fruit Extract

The EDB extract was evaluated for the ability to inhibit digestive enzymes, including α-glucosidase, α-amylase and lipase. The extract showed similar α-glucosidase and α-amylase inhibitions (Figure 2a,b), with the same IC_{10} value of 1.25 mg/mL and non-significantly different IC_{50} values of 6.38 mg/mL and 6.70 mg/mL, respectively (Table 3).

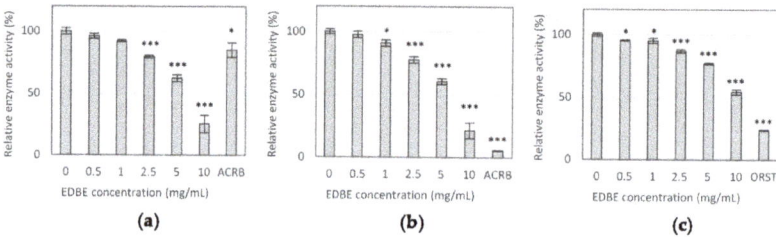

Figure 2. The effect of elderberry fruit extract (EDBE) on the activity of digestive enzymes: α-glucosidase (a), α-amylase (b), and lipase (c). Enzyme activity is expressed in relation to the negative control (without extract addition). As positive controls, reference enzyme inhibitors, including acarbose (ACRB, 100 µg/mL) and orlistat (ORST, 5 µg/mL) were used in the experiment. The values represent the means ($n = 3$) ± SD. * $p < 0.05$, *** $p < 0.001$ vs. control group.

Table 3. Inhibitory concentrations IC_{10} and IC_{50} (the inhibitor concentration required for 10% and 50% inhibition of enzyme activity) of the elderberry fruit extract (EDBE), acarbose (ACRB), and orlistat (ORST) as reference pharmacological inhibitors.

Enzyme Inhibitor	α-Glucosidase		α-Amylase		Lipase	
	IC_{10}	IC_{50}	IC_{10}	IC_{50}	IC_{10}	IC_{50}
EDBE (mg/mL)	1.25 ± 0.09	6.70 ± 0.56	1.25 ± 0.10	6.38 ± 0.44	2.09 ± 0.30	10.98 ± 0.47
ACRB (µg/mL)	85.12 ± 0.63	> 100	1.30 ± 0.06	8.23 ± 0.16	-	-
ORST (µg/mL)	-	-	-	-	0.2 ± 0.02	1.43 ± 0.15

Lower potency of the extract was observed in inhibiting pancreatic lipase activity. The extract concentrations reducing lipase activity by 10% and 50% were determined at 2.09 mg/mL and 10.98 mg/mL, respectively (Table 3). In inhibiting α-amylase and lipase activity, the drugs acarbose (100 µg/mL) and orlistat (5 µg/mL) were more potent than the EDB extract (Figure 2b,c). However, in α-glucosidase inhibition, the extract at concentrations of 5 mg/mL and 10 mg/mL evoked the stronger effects than acarbose at a dose of 100 µg/mL.

3.3. The Effect of Elderberry Fruit Extract on Hypertrophied Adipocytes

To determine whether EDB extract affects the condition of mature adipocytes, we examined its effect on the viability, lipid accumulation, and ROS production, as well as regulation of leptin and adiponectin expression in terminally differentiated 3T3-L1 cells. The obtained results indicated that EDB extract did not exert any significant effect on both adipocyte viability (Figure 3a) and intracellular lipid content (Figure 3b,g–i). Nevertheless, it dose-dependently inhibited the intracellular ROS generation in hypertrophied adipocytes. EDB extract caused 36%, 53%, and 58% decrease in ROS level when applied at 5, 10, and 20 mg/mL, respectively, in comparison to untreated cells ($p < 0.001$) (Figure 3c). Moreover, treatment with EDB extract led to down-regulation of *NADPH oxidase 4* (*NOX-4*) mRNA expression by approximately 49 ± 6%, independently of the extract dose. Treatment with the extract at the maximum concentration (20 mg/mL) resulted in approximately 2-fold increased mRNA expression of *superoxide dismutase* (*SOD*) and *glutathione peroxidase* (*GPx*). In contrast, the extract did not influence the mRNA expression level of *catalase* (*CAT*) (Figure 3d).

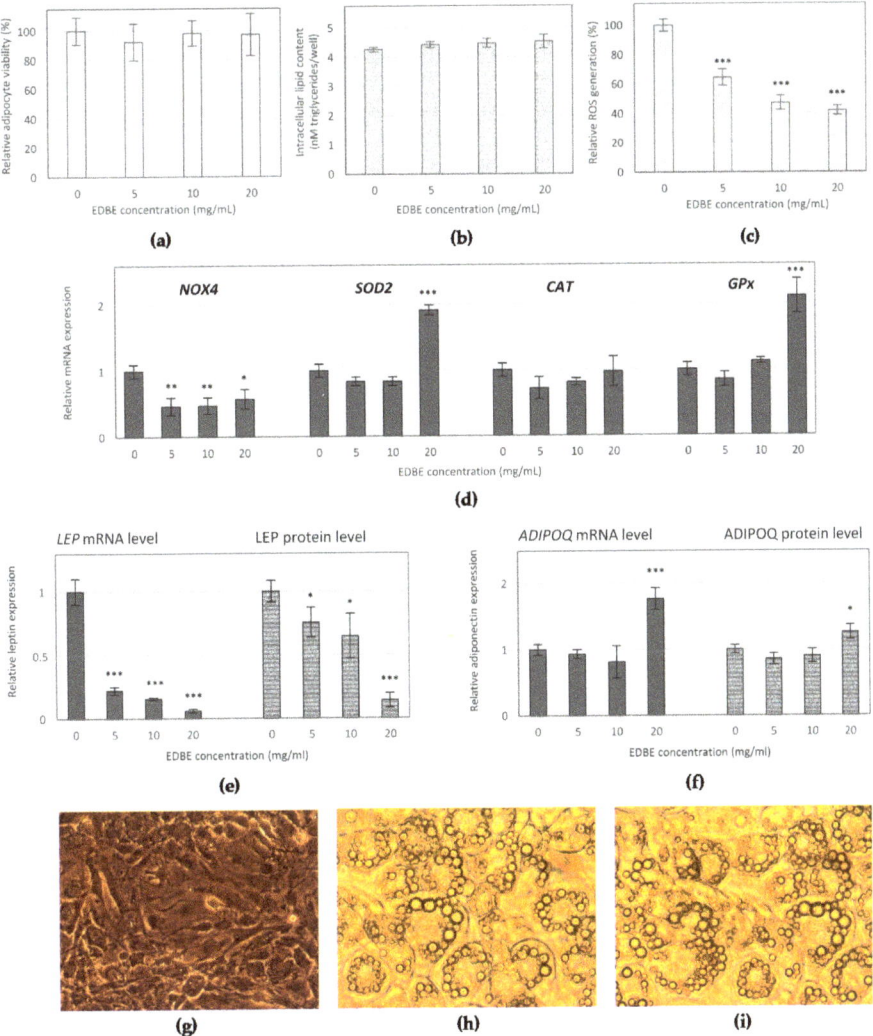

Figure 3. Changes in cell viability (**a**), intracellular triglyceride content (**b**), reactive oxygen species production (**c**), and mRNA expression of *NOX4*, *SOD2*, *catalase* (*CAT*), and *GPx* enzymes (**d**) as well as leptin (*LEP*) (**e**) and adiponectin (*ADIPOQ*) (**f**) expression upon the treatment of 3T3-L1 mature adipocytes with the elderberry fruit extract (EDBE). The photos present 3T3-L1 preadipocytes (**g**), fully differentiated 3T3-L1 adipocytes non-treated (**h**) and treated with EDBE at the concentration of 20 mg/mL (**i**). The cells were photographed at magnification of 100 ×. The results were expressed as the means ± SD ($n = 3$). * $p < 0.05$, ** $p < 0.01$, *** $p < 0.001$ vs. control group.

Supplementation of the fully differentiated adipocyte cultures with EDB extract considerably affected the expression of key adipokines in treated adipocytes. The decrease in *leptin (LEP)* mRNA levels between 78% and 94% was observed in mature adipocytes exposed to the extract at concentrations ranging from 5 mg/mL to 20 mg/mL ($p < 0.001$) (Figure 3e). In contrast to leptin, the expression of adiponectin was significantly up-regulated. The highest dose of EDB extract elevated *adiponectin (ADIPOQ)* mRNA level by 77% compared to the control ($p < 0.001$) (Figure 3f). The exposure of mature adipocytes to EDB extract resulted in a decrease of leptin secretion. The extract at concentrations of 5 mg/mL and 10 mg/mL significantly reduced leptin synthesis ($p < 0.05$). Whereas, the highest inhibitory effect with the reduction of leptin by 86% ($p < 0.001$) was observed in the cells treated with the extract at maximum concentration. In contrast, EDB extract at the highest dose of 20 mg/mL stimulated the adiponectin secretion in treated cells. The level of adiponectin was increased by 36% respect to the control ($p < 0.05$) (Figure 3f).

3.4. The Effect of Elderberry Fruit Extract on Glucose Uptake in Mature 3T3-L1 Adipocytes

The effect of EDB extract on the glucose analogue (2-NBDG) uptake in mature 3T3-L1 adipocytes was analyzed to determine whether the extract affects the glucose uptake by adipocytes. As shown in Figure 4a, EDB extract caused a significant increase in 2-NBDG uptake at all assayed concentrations ($p < 0.001$). The extract stimulated 2-NBDG uptake by 40%, 44%, and 62% tested at 5, 10, and 20 mg/mL, respectively, in comparison to the control system. Unexpectedly, the stimulatory effect of the extract on the glucose uptake in insulin-sensitive cells was more effective than that observed with rosiglitazone (Figure 4a). Subsequently, the expression of *glucose transporter type 4 (GLUT-4)* gene in mature 3T3-L1 cells exposed to EDB extract was investigated. The results indicated no significant effect of EDB extract on *GLUT-4* mRNA level (Figure 4d).

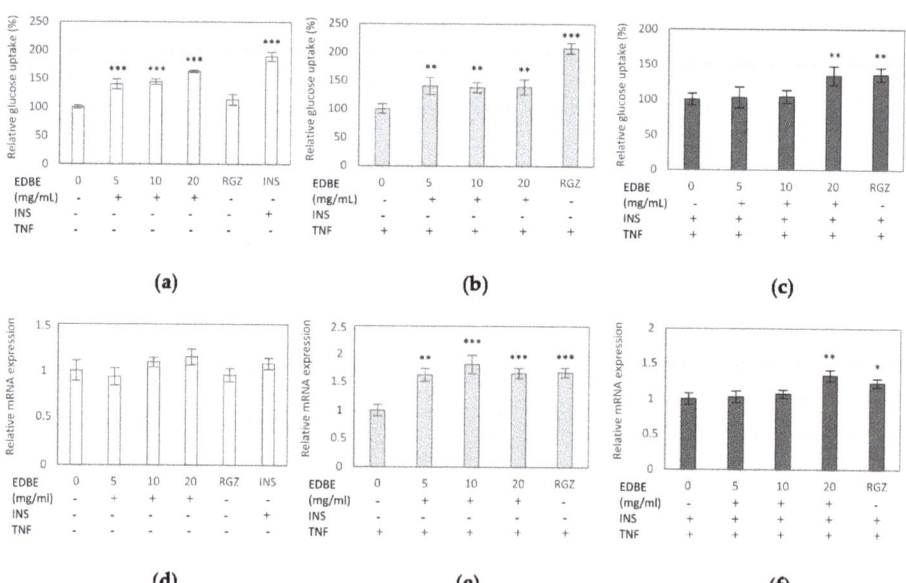

Figure 4. Effect of the elderberry fruit extract (EDBE) on the glucose uptake (**a–c**) and glucose transporter *GLUT-4* mRNA expression (**d–f**) in mature 3T3-L1 adipocytes non-induced (**a,d**) and induced by TNF-α (**b–f**) and non-treated (**b,e**) and treated (**c,f**) with insulin (INS). The cells were exposed to EDBE at concentrations of 5, 10, and 20 mg/mL, and to rosiglitazone (RGZ). Data are mean values ± SD ($n = 3$). The significance of the main effects of EDBE was determined by Tukey post hoc test; the control (INS, RGZ) significance was analyzed by T-student test; * $p < 0.05$, ** $p < 0.01$, *** $p < 0.001$.

In the parallel experiment, the influence of EDB extract on 2-NBDG incorporation into insulin resistant adipocytes treated with TNF-α was determined. The cells were incubated with the extract in the absence or presence of insulin at 100 nM. In the insulin-resistant adipocytes cultured without insulin supplementation, the EDB extract at concentrations of 5, 10, and 20 mg/mL enhanced 2-NBDG uptake by 40%, 38%, and 39%, respectively (Figure 4b). In this case, quantitative PCR analysis revealed that treatment with EDB extract at doses from 5 mg/mL to 20 mg/mL significantly up-regulated the mRNA expression of *GLUT-4* with an increase ranging from 63% to 82% (Figure 4e).

In the experiments on insulin resistant adipocytes exposed to insulin, the EDB extract at the highest concentration showed insulin-sensitizing properties, stimulating 2-NBDG incorporation by 34% ($p < 0.01$). Its efficacy was comparable to that of rosiglitazone, which enhanced the 2-NBDG uptake by 36% ($p < 0.01$) (Figure 4c). In this study, the extract at a dose of 20 mg/mL up-regulated the expression of *GLUT-4* by 33% ($p < 0.01$), compared to control adipocytes (Figure 4f).

3.5. Anti-Inflammatory Effects of Elderberry Fruit Extract

Anti-inflammatory effects of EDB extract were evaluated in activated RAW 264.7 macrophages, considering proinflammatory cytokines and mediators determined at both molecular and cellular levels, using low non-cytotoxic extract doses of 0.1, 1, and 10 µg/mL. The obtained results demonstrated the ability of EDB extract to alleviate the cellular inflammatory response induced by LPS. The extract at concentrations of 1 and 10 µg/mL suppressed mRNA expression of *IL-6* by 34% ($p < 0.01$) and 69% ($p < 0.001$), respectively, compared to control macrophages. Moreover, a 28% decrease in *TNF-α* mRNA level ($p < 0.05$) was observed following treatment with EDB extract at a dose of 10 µg/mL (Figure 5a). The down-regulation of *IL-6* and *TNF-α* expression was consistent with the inhibited secretion of these cytokines. Namely, EDB extract assayed at 1 µg/mL reduced production of IL-6 and TNF-α by 44% and 26%, respectively ($p < 0.05$). At the extract concentration of 10 µg/mL, the 60% decrease in IL-6 secretion ($p < 0.01$) was observed while synthesis of TNF-α was reduced by 52% ($p < 0.001$) (Figure 5b).

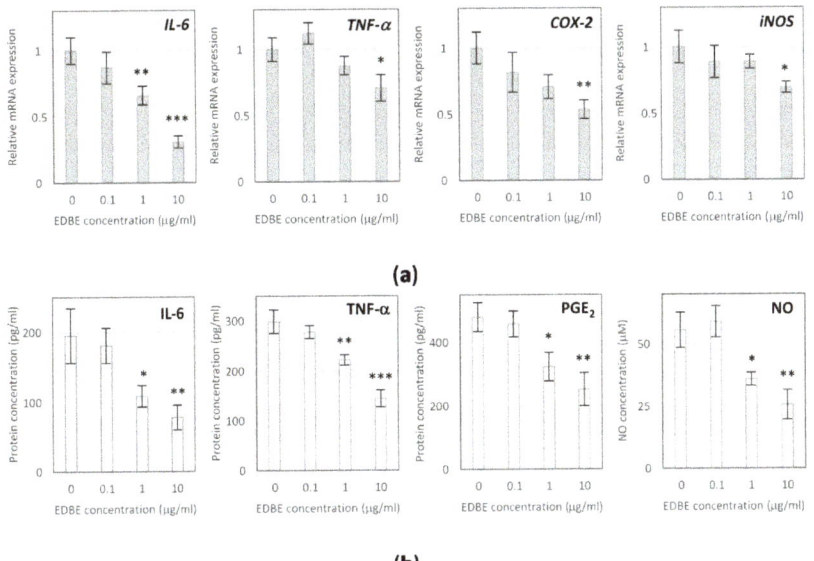

Figure 5. Effect of the elderberry fruit extract (EDBE) on mRNA expression of *IL-6*, *TNF-α*, *COX-2*, and *iNOS* (**a**) and on the production of IL-6, TNF-α, PGE$_2$ protein, and NO (**b**) in the lipopolysaccharide (LPS)-activated RAW 264.7 macrophages. Data are mean values ± SD ($n = 3$). * $p < 0.05$, ** $p < 0.01$, *** $p < 0.001$ when compared with control.

The extract at a dose of 10 µg/mL also affected *COX-2* expression, causing a 46% reduction of *COX-2* transcripts level in activated macrophages ($p < 0.01$) (Figure 5a). As a consequence, the decrease in PGE$_2$ production was detected after cell exposure to the extract. Namely, PGE$_2$ level lowered by 33% and 47%, respectively, for a dose of 1 µg/mL and 10 µg/mL (Figure 5b).

Furthermore, the obtained results demonstrated the reduction in NO production as well as down-regulation of *inducible NO synthase (iNOS)* expression in LPS-stimulated RAW 264.7 macrophages in response to the EDB extract treatment. The extract reduced the NO synthesis by 35% ($p < 0.05$) when assayed at 1 µg/mL and by 54% ($p < 0.01$) when applied at 10 µg/mL (Figure 5b). A significant inhibitory effect of EDB extract on *iNOS* expression was noted only at the highest tested dose, which decreased *iNOS* mRNA level by 30% ($p < 0.01$) (Figure 5a).

4. Discussion

Excessive fat accumulation in hypertrophic adipose tissue associated with obesity is responsible for oxidative stress, chronic inflammation, and dysregulated adipokine secretion [25]. It is believed that the therapeutic potential of natural dietary compounds against obesity and obesity-related disorders should focus on improving the fat function in pathogenic hypertrophic adipocytes by reducing oxidative stress, alleviating inflammation, and regulating underproduction or overproduction of clinically relevant adipocyte factors. However, most bioactive compounds or extracts strongly affect preadipocytes, their viability, proliferation, and differentiation into mature fat cells, without any significant effects on the pathological status of hypertrophic adipocytes. Therefore, in this work, the influence of the *Sambucus nigra* fruit extract on mature fully differentiated insulin-resistant 3T3-L1 adipocytes was investigated.

In our study, we found no reduction in cell viability and lipid content in hypertrophic 3T3-L1 adipocytes after exposure to EDB extract. However, as a result of the treatment, the intracellular ROS generation was significantly down-regulated and probably, oxidative stress accompanying excessive fat accumulation was also importantly reduced. Oxidative stress induced by enhanced lipid content is reported to be involved in the pathogenesis of obesity-related comorbidities including insulin resistance and diabetes, cardiovascular complications, and cancer [26]. It was found that ROS are intensively generated in visceral adipose tissue by adipocytes during the metabolism of excess nutrients and also by macrophages, which accumulate in adipose tissue in obesity state. The increased release of fatty acids from overproduced fat accumulated in adipose tissue, activate NADPH oxidases (NOX) and induce or aggravate ROS production. Other factors that also contribute oxidative stress to obesity include hyperleptinemia, low antioxidant defense, or chronic inflammation [27]. Our results showed that EDB extract could reduce ROS generation by lowering the expression of NOX4, the major NOX isoform in adipocytes. Treatment of hypertrophied 3T3-L1 adipocytes with EDB extract caused a significant decreasing in *NOX4* mRNA expression. Furthermore, up-regulation of mRNA expression of antioxidant enzymes, like SOD and GPx, could also contribute to enhancing adipocyte antioxidant defense efficiency. Numerous studies have shown the high antioxidant capacity of *Sambucus nigra* fruit [6,10]. However, the antioxidant effects of EDB on adipocytes have not yet been reported in the literature. In the present study, we demonstrated that introduction of EDB extract to the culture of hypertrophic adipocytes resulted in decreased ROS generation in cells. The antioxidant action of EDB extract in adipocytes may be a potential protective mechanism against obesity-associated pathological risk factors, including insulin resistance and chronic inflammation.

Additionally, EDB extract treatment modulated the leptin and adiponectin gene expression and protein secretion in hypertrophic 3T3-L1 adipocytes. Leptin and adiponectin are adipocytokines, which influence energy homeostasis, glucose and lipid metabolism, cardiovascular function, and immune response [28]. Leptin is primarily secreted by fully differentiated adipocytes, and its crucial role is to regulate energy intake and expenditure through controlling appetite and glucose metabolism. Reflecting the increased amount of adipose tissue, obese individuals often have elevated leptin concentration and the simultaneous apparent loss of efficacy of leptin, which is a result of leptin resistance, the state that leads to uncontrolled food intake, pro-inflammatory state, diabetes mellitus, and other obesity-related complications [29]. In contrast to leptin, adiponectin is down-regulated in obesity, and the circulating adiponectin levels are inversely correlated with body fat amount. Adiponectin enhances energy metabolism and fatty acid oxidation, promotes insulin sensitivity, improves glucose tolerance, and exerts anti-inflammatory effects [28]. Low serum adiponectin and high serum leptin levels are considered as risk factors for developing type 2 diabetes (T2DM), obesity, dyslipidemia, hypertension, and cardiovascular diseases. In this study, a remarkable decrease in leptin expression and secretion was observed in response to EDB extract treatment of hypertrophied 3T3-L1 adipocytes, which may help counteract the leptin resistance state. Whereas, adiponectin mRNA expression and protein secretion in treated adipocytes were significantly increased. The effect of EDB extract on adiponectin production may indicate anti-inflammatory potential and insulin-sensitizing activity of *Sambucus nigra* fruit.

The association of visceral obesity with T2DM is a long-recognized phenomenon. The primary determinant of this correlation is the fact that central obesity is the critical factor in the emergence of insulin resistance. The insulin-resistant state results in defective insulin-stimulated glucose uptake and consequently in hyperglycemia, elevated circulating free fatty acids level, abnormal fat accumulation, and dysregulation of hepatic glucose production, that, in combination with a paucity of insulin secretion by pancreatic β-cells, leads to T2DM [30]. These metabolic abnormalities may arise from impairment in insulin signaling pathways and subsequent defect in translocation of insulin-responsive glucose transporter protein (GLUT-4) and in adipose tissue, also from down-regulation of *GLUT-4* gene [31].

The effects of EDB extract on glucose uptake and *GLUT-4* expression were evaluated in this study. Experiments were performed both with mature 3T3-L1 adipocytes sensitive to insulin and adipocytes treated with TNF-α to induce an inflammatory status and insulin resistance. Analysis revealed that EDB extract stimulated the 2-NBDG uptake in both types of adipocytes and up-regulated mRNA expression of *GLUT-4* in insulin-resistant cells, suggesting insulin-like and insulin-sensitizing activities of the extract. The signaling pathways involved in the development of these activities will be further examined in future studies. This is the first study assessing the effects of EDB extract on glucose uptake in 3T3-L1 cells. Although several recent reports have suggested the anti-diabetic and hypoglycemic properties of elderberry, it has been found that EDB methanolic extracts markedly stimulate glucose uptake in liver HepG$_2$ cells and also exert inhibitory effect towards carbohydrate hydrolyzing enzyme [32]. Furthermore, EDB extracts, EDB anthocyanins, mainly cyanidin-3-glucoside and cyanidin-3-sambubioside, procyanidins, and their metabolites were found to enhance glucose uptake in human skeletal muscle cells [33]. Whereas, EDB lipophilic and polar extracts were reported to modulate glucose metabolism or lower insulin secretion contributing to the mitigation of insulin resistance in T2DM rats [13].

Anti-obesity and anti-diabetic activity of EDB extract could be related to the inhibition of dietary fat and sugar absorption from the intestinal tract. There is some evidence that polyphenols from berry fruits, such as strawberry, raspberry, blueberry, bilberry, black and red currant, lingonberry, red and green gooseberry, cranberry, and chokeberry, contribute to the inhibition of digestive enzymes involved in the hydrolysis of dietary lipids and carbohydrates [34]. Based on our research, the *Sambucus nigra* fruit may be included in the class of berries considered as effective inhibitors of α-amylase, α-glucosidase, and pancreatic lipase activity.

Obesity is known to be accompanied by metaflammation—low-grade chronic inflammation condition triggered by excess nutrients in metabolic cells [35]. An attribute of obesity-related inflammation is enhanced infiltration of macrophages into expanding adipose tissue, activation of specialized immune cells, and secretion of proinflammatory cytokines such as TNF-α, IL-6, and MCP-1 leading to an unresolved inflammatory response, which affects normal metabolism and insulin action [35]. Inhibition of obesity-induced inflammation could, thus, be a therapeutic intervention against adipose tissue dysfunction and related co-morbidities. In recent years, the use of anti-inflammatory nutrients provided through diet as a potential approach against obesity has been extensively studied [36,37].

In the present study, we evaluated anti-inflammatory effects of EDB extract in LPS-stimulated RAW 264.7 macrophages. Activated macrophages produce cytokines such as TNF-α, IL-1β, and IL-6 as well as pro-inflammatory mediators, such as NO and PGE$_2$ [38]. IL-6 and TNF-α are potent proinflammatory cytokines, which play a central role in inflammatory response and are characterized by a broad spectrum of functions with various effects in adipose tissue. TNF-α substantially influences lipid metabolism and adipocytes apoptosis. It can disrupt insulin signaling pathway promoting insulin resistance and adipocytes dysfunction [39]. TNF-α has, thus, been believed to be the crucial mediator in the detrimental paracrine loop between adipocytes and macrophages [40]. IL-6 has a pivotal role in acute phase reactions. It also influences hormonal balance and energy homeostasis and may affect the increase of free fatty acids level. Circulating levels of IL-6 and TNF-α are elevated in obese individuals and patients with insulin resistance [41]. In general, the regulation of TNF-α and IL-6 secretion is considered to be a potent treatment strategy for inflammation-associated diseases [42].

The research presented in this work suggests that EDB extract dose-dependently down-regulates mRNA expression and protein production of TNF-α and IL-6 in activated RAW 264.7 macrophages and therefore alleviates the cellular inflammatory response induced by LPS. In addition to TNF-α and IL-6, EDB extract significantly reduced the production of inflammatory mediators—PGE$_2$ and NO. Increased level of PGE$_2$ is observed in obese adipose tissue due to remarkable up-regulation of COX-2—the key enzyme in eicosanoid metabolism, of which expression is induced in inflammation state [43]. It has been suggested that COX-2-mediated inflammation in visceral fat is responsible for insulin resistance and fatty liver development in high-fat-induced obese rats [44]. The same study revealed that COX-2 inhibition significantly reversed adipocyte hypertrophy, macrophage infiltration, and decreased markers of adipocyte differentiation. Nitric oxide formed by iNOS is a short-lived vasodilator that acts as an important regulator of physical homeostasis, while its overproduction has been closely correlated with the pathological conditions including septic shock, osteoporosis and rheumatoid arthritis, insulin resistance, and inflammation [45]. In the present study, EDB extract was found to suppress PGE$_2$ and NO production via down-regulation of *COX-2* and *iNOS* expression. These findings indicate that inhibition of PGE$_2$ and NO generation is one of the anti-inflammatory mechanisms of the extract. Several recent studies have shown the anti-inflammatory potential of EDB fruit preparations. In our previous study, we demonstrated the anti-inflammatory potential of gastrointestinally digested EDB extract following intestinal absorption in a co-culture model of intestinal epithelial Caco-2 cells and LPS-stimulated RAW 264.7 macrophages [7]. The analyzed extract down-regulated the expression of genes (IL-1β, IL-6, TNF-α, COX-2) involving in the inflammatory pathway in a range comparable to that of budesonide. This study demonstrated adequate bioavailability and intestinal permeability of EDB compounds that are probably sufficient to evoke systemic anti-inflammatory effects [7]. Moreover, there is increasing evidence that the EDB bioactives can penetrate the blood–brain barrier and modulate the immune response induced in different types of brain injuries, including ischemic stroke. It has been found that EDB extract and its phenolic components significantly inhibit activation of microglia, considered to be resident macrophages responsible for the initial immune response to brain injuries. Treatment of activated microglial bv-2 cells with EBD extract led to diminishing ROS and NO generation, and as a consequence, attenuating the neuroinflammatory process [16].

Results of the study, as discussed above, indicate that *Sambucus nigra* fruit extract may offer substantial preventive and therapeutic potential for the treatment of obesity and obesity-related disorders, accompanied by oxidative stress, inflammationm and insulin resistance. Moreover, the extract can inhibit digestive enzyme activity, and consequently, significantly reduce the intestinal absorption of dietary lipids and carbohydrates, which is an effective strategy for the prevention and treatment of obesity and metabolic comorbidities.

Considering the findings of *in vitro* studies, we can postulate a nutraceutical application of the *Sambucus nigra* fruit extract. The scientific community focuses great attention on introducing nutraceuticals into the daily diet to prevent the occurrence of the pathological conditions, to delay or avoid the need for drug treatment and to support pharmacological therapy. Nutraceuticals as pharmafoods should be evaluated in the clinical aspects regarding safety, side effects, bioavailability, beneficial health effects, mechanisms of action and efficacy, and any possible interactions between food and drugs assumed together with them [46,47]. Thus, the developing of clinical studies will be of significant importance for clinically justified promotion of the *Sambucus nigra* fruit extract as a safe nutraceutical with the capacity of prevention or treatment of obesity and obesity-related immune-metabolic disorders.

Author Contributions: Conceptualization, J.Z.-W. and A.O.; methodology, J.Z.-W., A.O., M.O. and K.K.; formal analysis, A.O. and R.D.; investigation, J.Z.-W. and M.O.; resources, R.D.; writing—original draft preparation, J.Z.-W.; writing—review and editing, A.O.; project administration, A.O.; funding acquisition, A.O.

Funding: This research was funded by THE NATIONAL SCIENCE CENTRE, POLAND, grant number 2015/19/B/NZ9/01054.

Conflicts of Interest: The authors declare no conflict of interest.

References

1. Jo, J.; Gavrilova, O.; Pack, S.; Jou, W.; Mullen, S.; Sumner, A.E.; Cushman, S.W.; Periwal, V. Hypertrophy and/or Hyperplasia: Dynamics of Adipose Tissue Growth. *PLoS Comput. Biol.* **2009**, e1000324. [CrossRef] [PubMed]
2. Schuster, D.P. Obesity and the development of type 2 diabetes: The effects of fatty tissue inflammation. *Diabetes Metab. Syndr. Obes.* **2010**, *3*, 253–262. [CrossRef] [PubMed]
3. Tateya, S.; Kim, F.; Tamori, Y. Recent advances in obesity-induced inflammation and insulin resistance. *Front. Endocrinol.* **2013**, *4*, 93. [CrossRef] [PubMed]
4. Kowalska, K.; Olejnik, A. Current evidence on the health-beneficial effects of berry fruits in the prevention and treatment of metabolic syndrome. *Curr. Opin. Clin. Nutr. Metab. Care* **2016**, *19*, 446–452. [CrossRef] [PubMed]
5. Veberic, R.; Jakopic, J.; Stampar, F.; Schmitzer, V. European elderberry (*Sambucus nigra* L.) rich in sugars, organic acids, anthocyanins and selected polyphenols. *Food Chem.* **2009**, *114*, 511–515. [CrossRef]
6. Sidor, A.; Gramza-Michałowska, A. Advanced research on the antioxidant and health benefit of elderberry (*Sambucus nigra*) in food—A review. *J. Funct. Foods* **2015**, *18*, 941–958. [CrossRef]
7. Olejnik, A.; Kowalska, K.; Olkowicz, M.; Rychlik, J.; Juzwa, W.; Myszka, K.; Dembczyński, R.; Białas, W. Anti-inflammatory effects of gastrointestinal digested *Sambucus nigra* L. fruit extract analysed in co-cultured intestinal epithelial cells and lipopolysaccharide-stimulated macrophages. *J. Funct. Foods* **2015**, *19*, 649–660. [CrossRef]
8. Olejnik, A.; Olkowicz, M.; Kowalska, K.; Rychlik, J.; Dembczyński, R.; Myszka, K.; Juzwa, W.; Białas, W.; Moyer, M.P. Gastrointestinal digested *Sambucus nigra* L. fruit extract protects *in vitro* cultured human colon cells against oxidative stress. *Food Chem.* **2016**, *197*, 648–657. [CrossRef] [PubMed]
9. Młynarczyk, K.; Walkowiak-Tomczak, D.; Łysiak, G.P. Bioactive properties of *Sambucus nigra* L. as a functional ingredient for food and pharmaceutical industry. *J. Funct. Foods* **2018**, *40*, 377–390. [CrossRef]
10. Neves, D.; Valentao, P.; Bernardo, J.; Oliveira, M.C.; Ferreira, J.M.G.; Pereira, D.M.; Andrade, P.B.; Videira, R.A. A new insight on elderberry anthocyanins bioactivity: Modulation of mitochondrial redox chain functionality and cell redox state. *J. Funct. Foods* **2019**, *56*, 145–155. [CrossRef]
11. Badescu, M.; Badulescu, O.; Badescu, L.; Ciocoiu, M. Effects of *Sambucus nigra* and *Aronia melanocarpa* extracts on immune system disorders within diabetes mellitus. *Pharm. Biol.* **2015**, *53*, 533–539. [CrossRef] [PubMed]
12. Farrell, N.J.; Norris, G.H.; Ryan, J.; Porter, C.M.; Jiang, C.; Blesso, C.N. Black elderberry extract attenuates inflammation and metabolic dysfunction in diet-induced obese mice. *Br. J. Nutr.* **2015**, *114*, 1123–1131. [CrossRef] [PubMed]
13. Salvador, Â.C.; Król, E.; Lemos, V.C.; Santos, S.A.O.; Bento, F.P.M.S.; Costa, C.P.; Almeida, A.; Szczepankiewicz, D.; Kulczyński, B.; Krejpcio, Z.; et al. Effect of Elderberry (*Sambucus nigra* L.) Extract Supplementation in STZ-Induced Diabetic Rats Fed with a High-Fat Diet. *Int. J. Mol. Sci.* **2017**, *18*, 13. [CrossRef] [PubMed]

14. Farrell, N.; Norris, G.; Lee, S.G.; Porter, C.M.; Chun, O.K.; Blesso, C.N. Anthocyanin-rich black elderberry extract improves markers of HDL function and reduces aortic cholesterol in hyperlipidemic mice. *Food Funct.* **2015**, *6*, 1278–1287. [CrossRef] [PubMed]
15. Ho, G.T.; Wangensteen, H.; Barsett, H. Elderberry and elderflower extracts, phenolic compounds, and metabolites and their effect on complement, RAW 264.7 macrophages and dendritic cells. *Int. J. Mol. Sci.* **2017**, *18*, 584. [CrossRef] [PubMed]
16. Simonyi, A.; Chen, Z.; Jiang, J.; Zong, Y.; Chuang, D.Y.; Gu, Z.; Lu, C.H.; Fritsche, K.L.; Greenlief, C.M.; Rottinghaus, G.E.; et al. Inhibition of microglial activation by elderberry extracts and its phenolic components. *Life Sci.* **2015**, *128*, 30–38. [CrossRef]
17. Kowalska, K.; Olejnik, A.; Zielińska-Wasielica, J.; Olkowicz, M. Inhibitory effects of lingonberry (*Vaccinium vitis-idaea* L.) fruit extract on obesity-induced inflammation in 3T3-L1 adipocytes and RAW 264.7 macrophages. *J. Funct. Foods* **2019**, *54*, 371–380. [CrossRef]
18. Kowalska, K.; Olejnik, A.; Szwajgier, D.; Olkowicz, M. Inhibitory activity of chokeberry, bilberry, raspberry and cranberry polyphenol-rich extract towards adipogenesis and oxidative stress in differentiated 3T3-L1 adipose cells. *PLoS ONE* **2017**, *12*, e0188583. [CrossRef]
19. Choi, H.S.; Kim, J.W.; Cha, Y.-N.; Kim, C. A quantitative nitroblue tetrazolium assay for determining intracellular superoxide anion production in phagocytic cells. *J. Immunoass. Immunoch.* **2006**, *27*, 31–44. [CrossRef]
20. Alonso-Castro, A.J.; Salazar-Olivo, L.A. The anti-diabetic properties of *Guazuma ulmifolia* Lam are mediated by the stimulation of glucose uptake in normal and diabetic adipocytes without inducing adipogenesis. *J. Ethnopharmacol.* **2008**, *118*, 252–256. [CrossRef]
21. Boath, A.S.; Grussu, D.; Stewart, D.; McDougall, G.J. Berry Polyphenols Inhibit Digestive Enzymes: A Source of Potential Health Benefits? *Food Dig.* **2012**, *3*, 1–7. [CrossRef]
22. Tan, Y.; Chang, S.K.C.; Zhang, Y. Comparison of α-amylase, α-glucosidase and lipase inhibitory activity of the phenolic substances in two black legumes of different genera. *Food Chem.* **2017**, *214*, 259–268. [CrossRef] [PubMed]
23. Mikulic-Petkovsek, M.; Schmitzer, V.; Slatnar, A.; Stampar, F.; Veberic, R. Composition of sugars, organic acids, and total phenolics in 25 wild or cultivated berry species. *J. Food Sci.* **2012**, *77*, C1064–C1070. [CrossRef] [PubMed]
24. Mikulic-Petkovsek, M.; Schmitzer, V.; Slatnar, A.; Todorovic, B.; Veberic, R.; Stampar, F.; Ivancic, A. Investigation of anthocyanin profile of four elderberry species and interspecific hybrids. *J. Agric. Food Chem.* **2014**, *62*, 5573–5580. [CrossRef] [PubMed]
25. Bays, H.E.; González-Campoy, J.M.; Bray, G.A.; Kitabchi, A.E.; Bergman, D.A.; Schorr, A.B.; Rodbard, H.W.; Henry, R.R. Pathogenic potential of adipose tissue and metabolic consequences of adipocyte hypertrophy and increased visceral adiposity. *Expert Rev. Cardiovasc. Ther.* **2008**, *6*, 343–368. [CrossRef] [PubMed]
26. Manna, P.; Jain, S.K. Obesity, Oxidative Stress, Adipose Tissue Dysfunction, and the Associated Health Risks: Causes and Therapeutic Strategies. *Metab. Syndr. Relat. Disord.* **2015**, *13*, 423–444. [CrossRef] [PubMed]
27. Matsuda, M.; Shimomura, I. Increased oxidative stress in obesity: Implications for metabolic syndrome, diabetes, hypertension, dyslipidemia, atherosclerosis, and cancer. *Obes. Res. Clin. Pract.* **2013**, *7*, e330–e341. [CrossRef] [PubMed]
28. Meier, U.; Gressner, A.M. Endocrine regulation of energy metabolism: Review of pathobiochemical and clinical chemical aspects of leptin, ghrelin, adiponectin, and resistin. *Clin. Chem.* **2004**, *50*, 1511–1525. [CrossRef] [PubMed]
29. Bravo, P.E.; Morse, S.; Borne, D.M.; Aguilar, E.A.; Reisin, E. Leptin and hypertension in obesity. *Vasc. Health Risk Manag.* **2006**, *2*, 163–169. [CrossRef]
30. Zeyda, M.; Stulnig, T.M. Obesity, inflammation, and insulin resistance—A mini-review. *Gerontology* **2009**, *55*, 379–386. [CrossRef]
31. Kahn, B.B.; Flier, J.S. Obesity and insulin resistance. *J. Clin. Investig.* **2000**, *106*, 473–481. [CrossRef] [PubMed]
32. Ho, G.T.T.; Nguyen, T.K.Y.; Kase, E.T.; Tadesse, M.; Barsett, H.; Wangensteen, H. Enhanced Glucose Uptake in Human Liver Cells and Inhibition of Carbohydrate Hydrolyzing Enzymes by Nordic Berry Extracts. *Molecules* **2017**, *22*, 1806. [CrossRef] [PubMed]

33. Ho, G.T.T.; Kase, E.T.; Wangensteen, H.; Barsett, H. Phenolic Elderberry Extracts, Anthocyanins, Procyanidins, and Metabolites Influence Glucose and Fatty Acid Uptake in Human Skeletal Muscle Cells. *J. Agric. Food Chem.* **2017**, *65*, 2677–2685. [CrossRef] [PubMed]
34. Podsędek, A.; Majewska, I.; Redzynia, M.; Sosnowska, D.; Koziołkiewicz, M. In vitro inhibitory effect on digestive enzymes and antioxidant potential of commonly consumed fruits. *J. Agric. Food Chem.* **2014**, *62*, 4610–4617. [CrossRef] [PubMed]
35. Hotamisligil, G.S. Inflammation and metabolic disorders. *Nature* **2006**, *444*, 860–867. [CrossRef] [PubMed]
36. Gregor, M.F.; Hotamisligil, G.S. Inflammatory mechanisms in obesity. *Annu. Rev. Immunol.* **2011**, *29*, 415–445. [CrossRef]
37. Jayarathne, S.; Koboziev, I.; Park, O.-H.; Oldewage-Theron, W.; Shen, C.-L.; Moustaid-Moussa, N. Anti-Inflammatory and Anti-Obesity Properties of Food Bioactive Components: Effects on Adipose Tissue. *Prev. Nutr. Food Sci.* **2017**, *22*, 251–262. [CrossRef]
38. Aderem, A.; Ulevitch, R.J. Toll-like receptors in the induction of the innate immune response. *Nature* **2000**, *406*, 782–787. [CrossRef]
39. Rodríguez-Hernández, H.; Simental-Mendía, L.E.; Rodríguez-Ramírez, G.; Reyes-Romero, M.A. Obesity and Inflammation: Epidemiology, Risk Factors, and Markers of Inflammation. *Int. J. Endocrinol.* **2013**, *2013*, 678159. [CrossRef]
40. Suganami, T.; Nishida, J.; Ogawa, Y. A paracrine loop between adipocytes and macrophages aggravates inflammatory changes: Role of free fatty acids and tumor necrosis factor alpha. *Arterioscler. Thromb. Vasc. Biol.* **2005**, *25*, 2062–2068. [CrossRef]
41. Popko, K.; Gorska, E.; Stelmaszczyk-Emmel, A.; Plywaczewski, R.; Stoklosa, A.; Gorecka, D.; Pyrzak, B.; Demkow, U. Proinflammatory cytokines Il-6 and TNF-α and the development of inflammation in obese subjects. *Eur. J. Med. Res.* **2010**, *15*, 120–122.
42. Chen, L.; Deng, H.; Cui, H.; Fang, J.; Zuo, Z.; Deng, J.; Li, Y.; Wang, X.; Zhao, L. Inflammatory responses and inflammation-associated diseases in organs. *Oncotarget* **2018**, *9*, 7204–7218. [CrossRef]
43. García-Alonso, V.; Titos, E.; Alcaraz-Quiles, J.; Rius, B.; Lopategi, A.; López-Vicario, C.; Jakobsson, P.J.; Delgado, S.; Lozano, J.; Clària, J. Prostaglandin E2 Exerts Multiple Regulatory Actions on Human Obese Adipose Tissue Remodeling, Inflammation, Adaptive Thermogenesis and Lipolysis. *PLoS ONE* **2016**, *11*, e0153751. [CrossRef]
44. Hsieh, P.-S.; Jin, J.-S.; Chiang, C.-F.; Chan, P.-C.; Chen, C.-H.; Shih, K.-C. COX-2-mediated inflammation in fat is crucial for obesity-linked insulin resistance and fatty liver. *Obesity* **2009**, *17*, 1150–1157. [CrossRef]
45. Lin, H.-Y.; Juan, S.-H.; Shen, S.-C.; Hsu, F.-L.; Chen, Y.-C. Inhibition of lipopolysaccharide-induced nitric oxide production by flavonoids in RAW264.7 macrophages involves heme oxygenase-1. *Biochem. Pharmacol.* **2003**, *66*, 1821–1832. [CrossRef]
46. Santini, A.; Novellino, E. Nutraceuticals: Shedding light on the grey area between pharmaceuticals and food. *Expert Rev. Clin. Pharmacol.* **2018**, *11*, 545–547. [CrossRef]
47. Daliu, P.; Santini, A.; Novellino, E. From pharmaceuticals to nutraceuticals: Bridging disease prevention and management. *Expert Rev. Clin. Pharmacol.* **2019**, *12*, 1–7. [CrossRef]

© 2019 by the authors. Licensee MDPI, Basel, Switzerland. This article is an open access article distributed under the terms and conditions of the Creative Commons Attribution (CC BY) license (http://creativecommons.org/licenses/by/4.0/).

Article

Fermented Sea Tangle (*Laminaria japonica* Aresch) Suppresses RANKL-Induced Osteoclastogenesis by Scavenging ROS in RAW 264.7 Cells

Jin-Woo Jeong [1], Seon Yeong Ji [2,3], Hyesook Lee [2,3], Su Hyun Hong [2,3], Gi-Young Kim [4], Cheol Park [5], Bae-Jin Lee [6], Eui Kyun Park [7], Jin Won Hyun [8], You-Jin Jeon [4] and Yung Hyun Choi [2,3,*]

1. Nakdonggang National Institute of Biological Resources, Sangju 37242, Korea
2. Department of Biochemistry, Dong-eui University College of Korean Medicine, Busan 47227, Korea
3. Anti-Aging Research Center, Dong-eui University, Busan 47227, Korea
4. Department of Marine Life Sciences, Jeju National University, Jeju 63243, Korea
5. Department of Molecular Biology, College of Natural Sciences, Dong-eui University, Busan 47340, Korea
6. Marine Bioprocess Co. Ltd., Busan 46048, Korea
7. Department of Oral Pathology and Regenerative Medicine, School of Dentistry, Kyungpook National University, Daegu 41940, Korea
8. Department of Biochemistry, School of Medicine, Jeju National University, Jeju 63243, Korea
* Correspondence: choiyh@deu.ac.kr; Tel.: +82-51-850-7413

Received: 2 July 2019; Accepted: 24 July 2019; Published: 26 July 2019

Abstract: Sea tangle (*Laminaria japonica* Aresch), a brown alga, has been used for many years as a functional food ingredient in the Asia-Pacific region. In the present study, we investigated the effects of fermented sea tangle extract (FST) on receptor activator of nuclear factor-κB (NF-κB) ligand (RANKL)-stimulated osteoclast differentiation, using RAW 264.7 mouse macrophage cells. FST was found to inhibit the RANKL-stimulated activation of tartrate-resistance acid phosphatase (TRAP) and F-actin ring structure formation. FST also down-regulated the expression of osteoclast marker genes like TRAP, matrix metalloproteinase-9, cathepsin K and osteoclast-associated receptor by blocking RANKL-induced activation of NF-κB and expression of nuclear factor of activated T cells c1 (NFATc1), a master transcription factor. In addition, FST significantly abolished RANKL-induced generation of reactive oxygen species (ROS) by activation of nuclear factor-erythroid 2-related factor 2 (Nrf2) and its transcriptional targets. Hence, it seems likely that FST may have anti-osteoclastogenic potential as a result of its ability to inactivate the NF-κB-mediated NFATc1 signaling pathway and by reducing ROS production through activation of the Nrf2 pathway. Although further studies are needed to inquire its efficacy in vivo, FST appears to have potential use as an adjunctive or as a prophylactic treatment for osteoclastic bone disease.

Keywords: fermented sea tangle; osteoclast differentiation; receptor of activator of nuclear factor kappa-B ligand (RANKL); nuclear factor-κB (NF-κB); reactive oxygen species (ROS)

1. Introduction

Bone remodeling is an active physiological process involving bone deposition and bone resorption by osteoblast and osteoclast, respectively. Imbalance of these processes in favor of resorption may lead to the formation of osteolytic lesions and an increase in bone disease-related disorders and morbidity [1–3]. Receptor activator of nuclear factor-κB (NF-κB) ligand (RANKL) and macrophage colony-stimulating factor (M-CSF) are cytokines that play important roles in osteoclast differentiation and maturation. RANKL belongs to the tumor necrosis factor (TNF) superfamily and is regarded as the key promoter of osteoclastogenesis. M-CSF by contrast, is involved in the maintenance of mature osteoclast survival and mobility [4,5]. The binding of RANKL to its receptor RANK results in the activation of various signaling pathways, including the NF-κB pathway [6,7], which then enhances the activation of nuclear factor of activated T cell c1 (NFATc1), which then in turn promotes osteoclast formation by up-regulating the expression of osteoclast-specific genes [8,9]. In addition, a number of previous studies have shown that reactive oxygen species (ROS) are also critical messengers for osteoclast differentiation [10,11] and increased activity of the Nrf2 signaling system can block this activation [12–14]. These findings suggest that suppression of ROS production in combination with increasing activity of Nrf2 may provide a means to block osteoclast activity. Although various drugs have been used clinically to inhibit bone resorption, all have severe side effects when used long-term [15] and as a result, research into the prevention and treatment of osteolytic diseases using natural products has greatly increased in recent years.

Many marine algae extract or components of these extracts have been shown to exhibit potential for preventing and treating bone resorption related diseases [16,17] and fermented marine algal extracts have attracted the attention of the food and medical care industries [17,18]. The sea tangle, *Laminaria japonica* Aresch, is one of the most well-known edible brown seaweeds and has long been used as an important food supplement in Pacific and Asian countries [19]. This seaweed is rich in polysaccharides, dietary fiber, minerals, carbohydrates, polyphenols and proteins [20,21] and has been reported to protect against obesity, inflammation and cancer [22–25]. Interestingly, Lee et al. [26] developed a fermented form of sea tangle using *Lactobacillus brevis* with high antioxidant activity and showed that a fermented sea tangle extract (FST) protected against liver damage better than a non-fermented sea tangle extract [27,28]. They speculated that glutamate in the sea tangle which converted to gamma-aminobutyric acid through the fermentation process, was the reason behind the increased antioxidant capacity. It has been reported that FST supplementation reduce obesity and improve stress management [29]. Furthermore, previous studies have shown that FST can protect against age-associated short-term memory loss and reduced physical functioning [30–32]. However, the effect of FST on bone has not previously been investigated and therefore we decided to investigate whether FST had any inhibitory effect on RANKL-stimulated osteoclast differentiation using RAW 264.7 mouse macrophage cells.

2. Materials and Methods

2.1. Reagents and Antibodies

Dulbecco's modified Eagle's medium (DMEM), fetal bovine serum (FBS) and other reagents for cell culture were purchased from WelGENE Inc. (Daegu, Republic of Korea). RANKL and osteoprotegerin (OPG) were obtained from Abcam (Cambridge, MA, USA) and Peprotech (Rocky Hill, NJ, USA), respectively. 3-(4,5-dimethylthiazol-2-yl)-2,5-diphenyltetrazolium bromide (MTT), tartrate-resistant acid phosphatase (TRAP) assay kit, bovine serum albumin (BSA), 4′,6-diamidino-2-phenylindole (DAPI), 5,6-carboxy-2′,7′-dichlorofluorescein diacetate (DCF-DA) and N-acetyl cysteine (NAC) were purchased from Sigma-Aldrich Chemical Co. (St. Louis, MO, USA). NE-PER™ nuclear and cytoplasmic extraction reagents kit and polyvinylidene difluoride (PVDF) membranes were obtained from Pierce Biotechnology (Rockford, IL, USA) and Schleicher & Schuell (Keene, NH, USA), respectively. Fluorescein isothiocyanate (FITC)-phalloidin solution was purchased from Thermo Fisher Scientific (Waltham, MA, USA). Primary and secondary antibodies were obtained from Santa Cruz Biotechnology Inc. (Santa Cruz, CA, USA), Cell Signaling Technology Inc. (Beverly, MA, USA), Abcam, Novus (Novus Biologicals, LLC., Littleton, CO, USA), Thermo Fisher Scientific and R&D system. Appropriate horseradish-peroxidase (HRP)-linked secondary antibodies and enhanced chemiluminescence (ECL) detection solution were purchased from Amersham Corp. (Arlington Heights, IL, USA). All reagents not specifically identified were purchased from Sigma-Aldrich Chemical Co. (St. Louis, MO, USA).

2.2. Preparation of FST

FST received from Marine Bioprocess Co. Ltd. (Busan, Korea) was extracted as previously described [30]. In brief, yeast extract and glucose were added to water at a ratio of 1:15 (w/v) and sea tangle (*L. japonica* Aresch) was then added and sterilized in an autoclave at 121 °C for 30 min. After autoclaving, culture broth of *L. brevis* BJ20 (accession no. KCTC 11377BP) was added to the mix at a concentration of 1.2% (v/v) and the mixture was incubated at 37 °C for 2 days. The fermented product was obtained by filtration and lyophilized. The dried extract (FST) so obtained was dissolved in Milli-Q Water to produce a 10 mg/mL stock solution.

2.3. Cell Culture and Viability Analysis

RAW 264.7 cell line was purchased from the American Type Culture Collection (Manassas, VA, USA). The cells were cultured in DMEM containing 10% heat inactivated FBS, penicillin (100 units/mL) and streptomycin (100 g/mL) at 37 °C in a humidified 5% CO_2 atmosphere and subcultured every 3 days. The viability of the cells was assessed by MTT assay as previously described [14]. Briefly, the cells were treated with the desired concentrations of FST with or without 100 ng/mL RANKL for 72 h and then incubated with 50 µg/mL MTT solution for 3 h. Formazan crystals were dissolved in DMSO and the absorbance was measured using an enzyme-linked immunosorbent assay (ELISA) microplate reader (Dynatech Laboratories, Chantilly, VA, USA) at 540 nm.

2.4. Osteoclast Differentiation and TRAP Assay

Osteoclast formation was measured by quantifying cells positively stained by TRAP. Briefly, the cells were fixed in 4% paraformaldehyde (pH 7.4) at room temperature for 10 min and then stained with commercial TRAP staining kit according to the manufacturer's instructions. Osteoclasts were defined as TRAP-positive multinuclear cells containing 3 or more nuclei, under a phase-contrast microscope (Carl Zeiss, Oberkochen, Germany). TRAP activity was determined in culture media using a TRAP assay kit, in accordance with the manufacturer's instructions. TRAP activities were expressed as percentages of control activities.

2.5. F-Actin Ring Staining

As described previously, evaluation of actin ring formation was performed [14]. Briefly, the cells were fixed with 4% paraformaldehyde, permeabilized with 0.1% Triton X-100 in PBS for 5 min and then stained with an anti-actin antibody at 4 °C overnight. After washing with PBS, the cells were incubated with FITC-conjugated phalloidin for 30 min at 37 °C and then counterstained with 2.5 µg/mL DAPI for 20 min. F-actin rings were analyzed by fluorescence microscopy (Carl Zeiss, Oberkochen, Germany).

2.6. Western Blot Analysis

As described previously, total protein was extracted from the cells using the Bradford Protein assay kit [14]. Nuclear and cytosolic proteins were prepared using a NE-PER nuclear and cytoplasmic extraction reagents kit according to the manufacturer's instructions. Equal amounts of protein from samples were loaded and separated by sodium dodecyl sulfate (SDS)-polyacrylamide gel electrophoresis and transferred onto PVDF membranes. The membranes were blocked with 5% non-fat skim milk in trisbuffered saline containing 0.1% Triton X-100 (TBST) for 1 h and probed with specific primary antibodies at 4 °C overnight (Table 1). After washing three times with TBST, the membranes were incubated with the appropriate HRP-conjugated secondary antibodies for 2 h. Protein expression was detected by an ECL kit and visualized by Fusion FX Image system (Vilber Lourmat, Torcy, France).

Table 1. Information of primary and secondary antibodies.

Antibody	Manufacturer	Item No.
β-actin	Santa Cruz	sc-1615
CTSK	Santa Cruz	sc-48353
HO-1	Millipore	374090
IkBα	Santa Cruz	sc-371
Lamin B	Santa Cruz	sc-6216
MMP-9	Abcam	38898
NFATc1	Santa Cruz	sc-7294
NF-κB p65	Santa Cruz	sc-109
Phospho- NF-κB p65	Cell signaling	3033
Nrf2	Santa Cruz	sc-13032
phospho-Nrf2	Abcam	76026
NQO-1	Novus	NB200-209
OSCAR	R&D system	MAB1633
TRAP	Thermo Fisher Scientific	PA5-42729
Goat anti-mouse IgG-HRP	Santa Cruz	sc-2005
Goat anti-rabbit IgG-HRP	Santa Cruz	sc-2004
Bovine anti-goat IgG-HRP	Santa Cruz	sc-2350
Mouse anti-rabbit igG-TR	Santa Cruz	Sc-3917

CTSK: cathepsin K; HO-1: heme oxygenase-1; IκBα: inhibitory proteins of kappa B, alpha; MMP-9: matrix metalloproteinase-9; NFATc1: nuclear factor of activated T cells c1; NF-κB: nuclear factor-kappa B; Nrf2: nuclear factor-erythroid 2-related factor 2; NQO-1: NAD(P)H quinone oxidoreductase 1; OSCAR: osteoclast-associated receptor; TRAP: tartrate-resistance acid phosphatase; HRP: horseradish-peroxidase.

2.7. Immunofluorescence Staining for NF-κB

RAW 264.7 cells were seeded on gelatin-coated glass coverslips. After it was cultured for 24 h, cells were treated with RANKL in the presence or absence of various concentrations of FST for 24 h, fixed in 4% paraformaldehyde for 15 min, permeabilized with 0.2% Triton X-100 in PBS for 15 min and blocked with PBS containing 5% BSA. Cells were stained with primary antibody against phosphoNF-κB p65 at 4 °C overnight and incubated with a fluorescein-conjugated anti-rat IgG in the dark at 37 °C for 1 h. Cells were mounted on slides and then analyzed by fluorescence microscope.

2.8. Measurement of Intracellular ROS Levels

The production of intracellular ROS was measured by a flow cytometer with DCF-DA as described previously [14]. Briefly, the cells were treated with FST in the presence or absence of 100 ng/mL RANKL. In the last 20 min of treatment, 10 µM DCF-DA was added to the incubated cells in the dark. Following incubation, the cells were washed twice with PBS and 10,000 cells were analyzed for intracellular ROS content by BD Accuri C6 software in a flow cytometer (BD Biosciences) at 480/520 nm. To observe ROS generation by fluorescence microscopy, cells were stimulated with RANKL in the presence or absence of FST for 1 h. Cells were then stained with DCF-DA and then fixed with 4% paraformaldehyde for 2.

2.9. Statistical Analysis

All experiments were performed at least three times. Data were analyzed using GraphPad Prism software (version 5.03; GraphPad Software, Inc., La Jolla, CA, USA) and expressed as the mean ± standard deviation (SD). Differences between groups were assessed using analysis of variance followed by ANOVA-Tukey's post hoc test and $p < 0.05$ was considered to indicate a statistically significant difference.

3. Results

3.1. Effect of FST on Cell Viability in RAW 264.7 Cells

RAW 264.7 cells were treated with various concentrations of FST for 72 h and then 3-(4,5-dimethylthiazol-2-yl)-2,5-diphenyltetrazolium bromide (MTT) assay was performed. Figure 1A shows that FST had no cytotoxicity on the cells at concentrations up to 800 μg/mL but relatively cytotoxic effect was observed in the 1000 μg/mL treatment group as compared with untreated controls. In the presence of 100 ng/mL RANKL or 100 ng/mL osteoprotegerin (OPG), a decoy receptor for RANKL that inhibits osteoclastogenesis [33,34], cell viability was not significantly reduced by FST at concentrations up to 800 ng/mL compared to that of control groups (Figure 1B). Hence, non-toxic concentrations (<800 μg/mL) were used to investigate the effect of FST on RANKL-induced osteoclast differentiation.

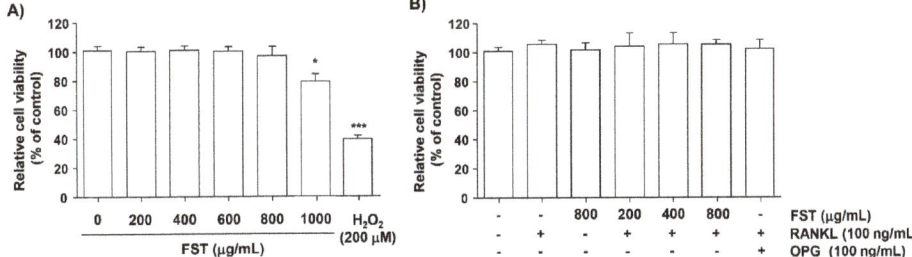

Figure 1. Effects of fermented sea tangle extract (FST) and receptor of activator of nuclear factor kappa-B ligand (RANKL) on the viability of RAW 264.7 mouse macrophage-like cells. Cells were treated with desired concentrations of FST in the absence (**A**) or presence (**B**) of 100 ng/mL RANKL and/or 100 ng/mL OPG for 72 h. H_2O_2 was used as a positive control. Cell viabilities were measured by 3-(4,5-dimethylthiazol-2-yl)-2,5-diphenyltetrazolium bromide (MTT) assay. Relative cell viability is expressed as percentages compared to treatment of naïve control cells. Results are presented as means ± SD of three independent experiments. * $p < 0.05$ and *** $p < 0.005$ indicates significant difference compared to the untreated control cells. OPG: osteoprotegerin; +: cells treated the reagent; -: cells untreated the reagent.

3.2. FST Suppresses RANKL-Induced Osteoclastogenesis in RAW 264.7 Cells

In order to examine the effect of FST on RANKL-induced osteoclastogenesis, RAW 264.7 cells were treated with FST in the presence of different concentrations of RANKL. As shown in Figure 2A, FST treatment markedly inhibited RANKL-induced osteoclast-like morphological changes. TRAP staining demonstrated that FST suppressed cell fusion and the conversion of RAW 264.7 cells into osteoclasts (Figure 2B FST suppressed numbers of TRAP-positive osteoclasts as compared with RANKL treated cells, dose-dependent manner (Figure 2C). These reductions in TRAP-positive osteoclast number were paralleled by the inhibition of TRAP activity (Figure 2D). As expected, RANKL-induced osteoclast differentiation and TRAP activity were completely suppressed in the presence of OPG.

3.3. FST Disrupts RANKL-Induced Formation of F-Actin Rich Adhesive Structures in RAW 264.7 Mouse Macrophage-Like Cells

Formation of the F-actin rich adhesive structures by osteoclasts is an essential step in bone resorption [35,36]. Figure 3 indicated that staining with FITC-conjugated phalloidin showed RANKL (100 ng/mL) stimulation increased well-defined F-actin sealing rings with a higher intensity ring height. However, the size of rings formed by RANKL-treated cells was remarkably and concentration-dependently reduced in cells co-treated with FST. Furthermore, OPG treatment complementally inhibited the F-actin sealing ring formation in RANKL-stimulated cells.

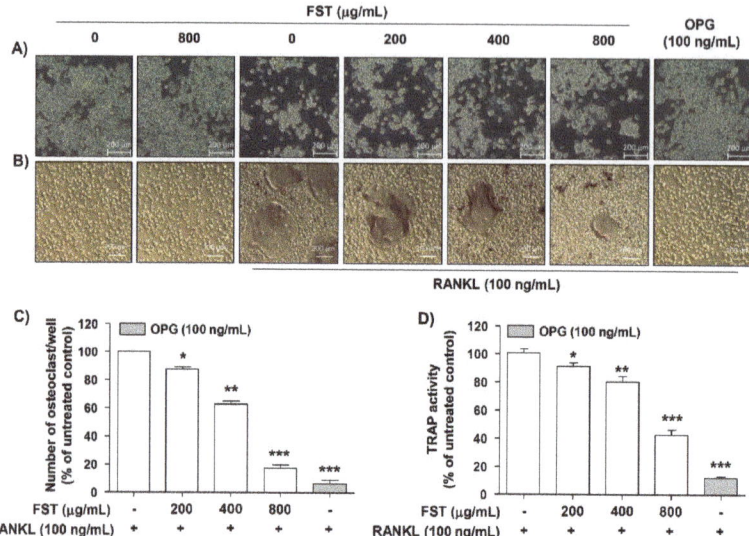

Figure 2. Inhibition of RANKL-stimulated osteoclast differentiation by FST in RAW 264.7 mouse macrophage-like cells. Cells were stimulated with 100 ng/mL RANKL in the presence or absence of FST or 100 ng/mL OPG for 5 days. (**A**) Representative photographs of the morphological changes are presented. (**B**) Cells were fixed and stained for TRAP and examined under an inverted microscope. (**C**) TRAP-positive multinucleated cells were counted to determine osteoclast numbers. (**D**) Supernatants were collected from cells grown under the same conditions and TRAP activities were measured using an ELISA reader. Results are presented as means ± SD of three independent experiments. * $p < 0.05$, ** $p < 0.01$ and *** $p < 0.001$ indicates significant difference compared to RANKL-treated cells. RANKL: receptor of activator of nuclear factor kappa-B ligand; FST: fermented sea tangle extract; OPG: osteoprotegerin; TRAP: tartrate-resistance acid phosphatase; +: cells treated the reagent; -: cells untreated the reagent.

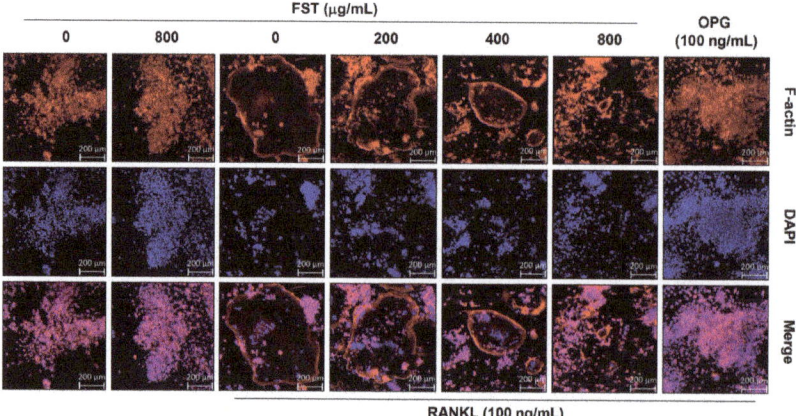

Figure 3. Suppression of F-actin ring formation by FST in RANKL-induced RAW 264.7 mouse macrophage-like cells. The cells were co-treated with 100 ng/mL RANKL in the presence or absence of FST or 100 ng/mL OPG for 5 days and stained for F-actin rich adhesive structures with fluorescein isothiocyanate (FITC)-phalloidin and 4′,6-diamidino-2-phenylindole (DAPI). The photographs are representative of the morphological changes observed under a fluorescence microscope. RANKL: receptor of activator of nuclear factor kappa-B ligand; FST: fermented sea tangle extract.

3.4. FST Inhibits the RANKL-Induced Nuclear Translocation of NF-κB and IκBα Degradation in RAW 264.7 Cells

Activation of NF-κB through nuclear translocation by RANKL is an essential step for initiation of osteoclast differentiation [6,7]. Therefore, we assessed whether FST affected the activation of NF-κB induced by RANKL. As shown in Figure 4A,B, our immunoblotting results reveal that the expression of NF-κB was markedly increased in the nuclei of RANKL treated cells but the expression of IκBα was reduced in the cytoplasm, which suggested that RANKL stimulated activation of NF-κB. However, FST suppressed the RANKL-mediated degradation of IκBα and the subsequent nuclear accumulation of NF-κB. Furthermore, immunofluorescence studies produced similar results. More specifically, phosphorylated NF-κB p65 was predominantly located in nuclei in RANKL-stimulated cells but not in FST and RANKL co-treated cells (Figure 4C).

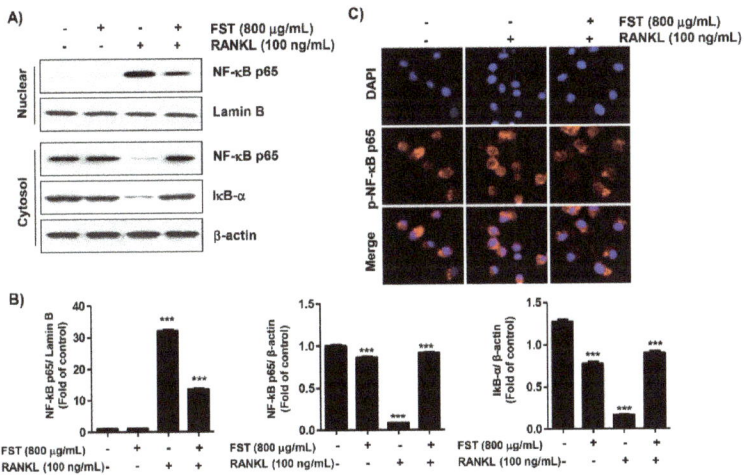

Figure 4. Effects of FST on the RANKL-induced activation of NF-κB in RAW 264.7 mouse macrophage-like cells. (**A**) After co-treating cells with 100 ng/mL RANKL in the presence or absence of FST for 1 h, nuclear and cytosolic proteins were isolated. The expression of NF-κB and IκB-α were determined by Western blotting. Lamin B and β-actin were used as internal controls for the nuclear and cytosolic fractions, respectively. (**B**) Densitometry quantifications of protein expressions were measured by ImageJ. Statistical analyses were conducted using analysis of variances between groups. *** $p < 0.0001$ when compared to control. (**C**) Cells grown on gelatin-coated glass coverslips were co-treated with 800 μg/mL FST with or without 100 ng/mL RANKL. Localization of phospho-NF-κB p65 was observed under a fluorescence microscope following staining with anti-phospho-NF-κB p65 antibody (red) and DAPI (nuclear stain; blue). Original magnification ×400. RANKL: receptor of activator of nuclear factor kappa-B ligand; FST: fermented sea tangle extract; IκBα: inhibitory proteins of kappa B, alpha; NF-κB: nuclear factor-kappa B; +: cells treated the reagent; -: cells untreated the reagent.

3.5. FST Down-Regulates RANKL-Induced Osteoclast-Associated Gene Expression in RAW 264.7 Cells

NFATc1 is considered to be the most important regulator of the transcriptional activation of osteoclast differentiation-associated genes by RANKL [8,9]. To examine in more detail the mechanism of FST-mediated inhibition of osteoclastogenesis, we assessed the expression of NFATc1 in RANKL-stimulated RAW 264.7 cells. Consistent with previous studies, the expression of NFATc1 was significantly increased by RANKL but was down-regulated in a concentration-dependent manner by FST (Figure 5). In addition, we investigated the effects of FST on the levels of specific marker for osteoclast such as TRAP, cathepsin (CTSK), matrix metallopeptidase-9 (MMP-9) and osteoclast-associated receptor

(OSCAR). Figure 5 showed that RANKL markedly up-regulated levels of these osteoclast-specific markers, which were effectively attenuated by the addition of FST. Co-treatment with OPG also completely prevented increases in these protein markers.

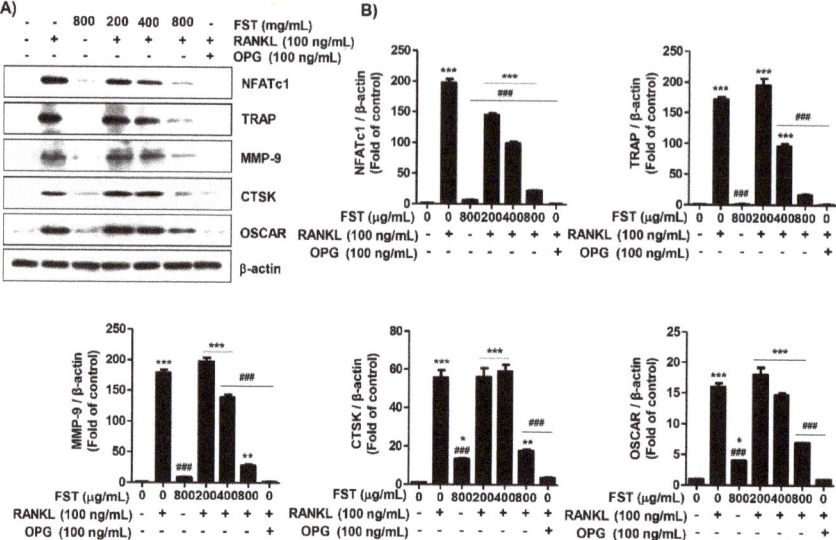

Figure 5. Inhibition of the RANKL-induced expressions of osteoclast-regulatory genes by FST in RAW 264.7 mouse macrophage-like cells. Cells were co-treated with various concentrations of FST or 100 ng/mL OPG in the presence or absence of 100 ng/mL RANKL for 5 days. (**A**) The expression levels of osteoclast-regulatory proteins were assessed by Western blot analysis. β-actin was used as the internal control. The results shown are representative of three independent experiments. (**B**) Densitometry quantifications of protein expression were measured by ImageJ. Statistical analyses were conducted using analysis of variances. * $p < 0.05$ and *** $p < 0.0001$ when compared to control. ### $p < 0.0001$ when compared to RANKL treatment. RANKL: receptor of activator of nuclear factor kappa-B ligand; FST: fermented sea tangle extract; OPG: osteoprotegerin; CTSK: cathepsin K; MMP-9: matrix metalloproteinase-9; NFATc1: nuclear factor of activated T cells c1; OSCAR: osteoclast-associated receptor; +: cells treated the reagent; -: cells untreated the reagent.

3.6. FST Attenuates RANKL-Induced Intracellular ROS Accumulation Associated with Activation of Nrf2 in RAW 264.7 Mouse Macrophage-Like Cells

Overproduction of intracellular ROS plays a critical step in RANKL-mediated osteoclastogenesis [12–14], thereby we examined whether FST inhibits the generation of ROS during RANKL-mediated osteoclastogenesis using DCF-DA, a cell permeant redox-sensitive dye. We demonstrated by flow cytometry that ROS levels were significantly increased by RANKL and that these up-regulation were abolished by FST (Figure 6A,D). Moreover, this effect of FST was supported by our fluorescence microscopic examination (Figure 6B) and further, co-treatment with N-acetyl cysteine (NAC), an intensive ROS scavenger, completely alleviated RANKL-induced ROS generation and F-actin ring formation (Figure 6C). In addition, Figure 6E,F shows that FST has the efficacy of equivalence and/or superiority compared with NAC and it was suggested that FST is a powerful anti-oxidant, thereby it has a suppressed RANKL-mediated ROS generation.

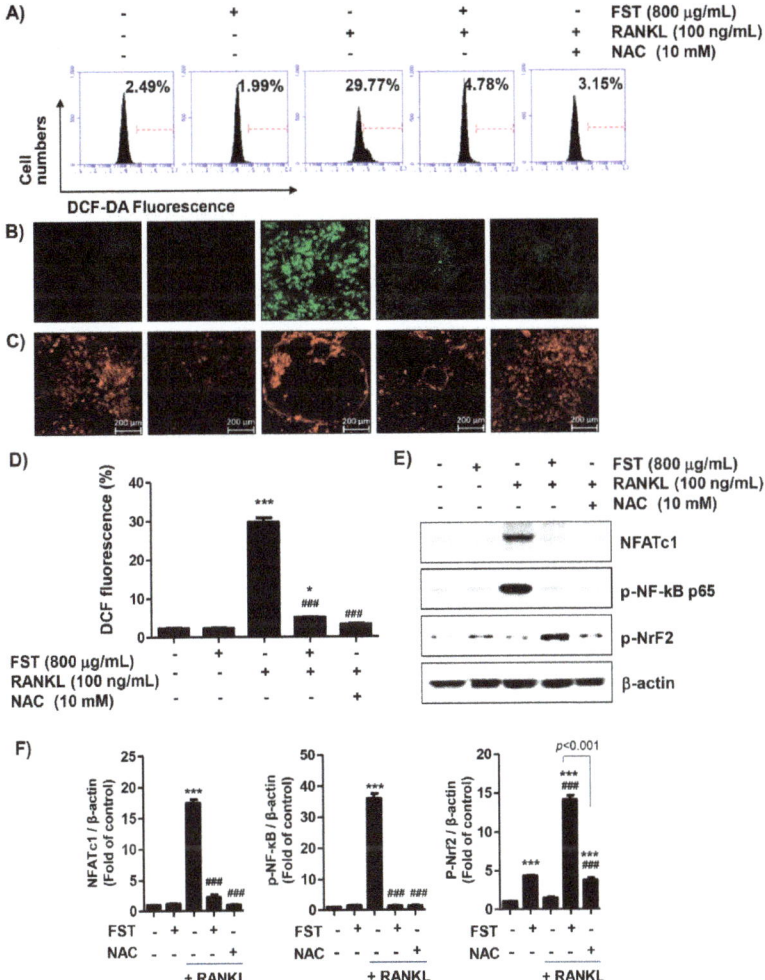

Figure 6. Effect of FST on RANKL-induced reactive oxygen species (ROS) generation in RAW 264.7 mouse macrophage-like cells. Cells were co-treated with 100 ng/mL RANKL for 1 h in the presence or absence of 800 μg/mL FST or 10 mM NAC. (**A,D**) Cells were stained with 5,6-carboxy-2′, 7′-dichlorofluorescein diacetate (DCF-DA) and DCF fluorescence was measured by flow cytometry. Results are means of two independent experiments. (**B**) After staining with DCF-DA, images were obtained using a fluorescence microscope. Images are representative of at least three independent experiments. (**C**) Cells cultured under the conditions used to induce osteoclast differentiation were fixed and stained for F-actin ring with FITC-phalloidin solution and imaged under a fluorescence microscope. Representative photographs of the morphological changes observed are presented. (**E**) Cellular proteins were isolated from cells and the expression of NFATc1, phospho-NF-κB and phosphor-Nrf2 by Western blot analysis. β-actin was used as the internal control. The results shown are representative of three independent experiments. (**F**) Statistical analyses were conducted using analysis of variances. * $p < 0.05$ and *** $p < 0.001$ when compared to control. ### $p < 0.0001$ when compared to RANKL treatment. RANKL: receptor of activator of nuclear factor kappa-B ligand; FST: fermented sea tangle extract; NAC: N-acetyl cysteine osteoprotegerin; NFATc1: nuclear factor of activated T cells c1; p- NF-κB p65: phosphorylated nuclear factor-kappa B p65; p-Nrf2: phosphorylated nuclear factor-erythroid 2-related factor 2; +: cells treated the reagent; -: cells untreated the reagent.

In addition, we show that FST increased the expression and phosphorylation of Nrf2 in RANKL-stimulated cells, which was associated with an increase in typical Nrf2-dependent cytoprotective enzymes such as heme oxygenase-1 (HO-1) and NAD(P)H: Quinone oxidoreductase 1 (NQO-1) (Figure 7A,B). Furthermore, we observed that Nrf2 translocation to the nucleus was promoted by FST treatment (Figure 7C,D).

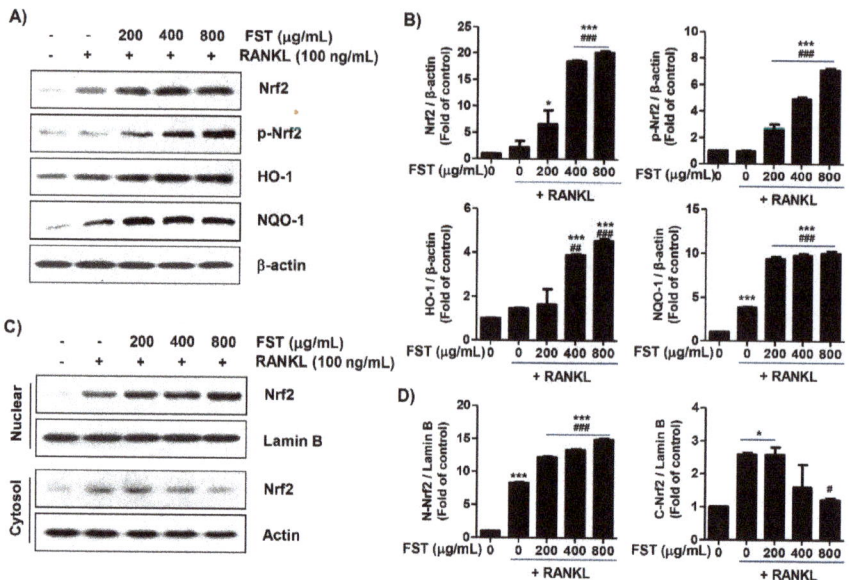

Figure 7. Activation of Nrf2 signaling pathway by FST in RAW 264.7 mouse macrophage-like cells. Cells were treated with FST with or without 100 ng/mL RANKL for 5 days. (**A**) Total cellular proteins were isolated from cells and the expression levels of Nrf2 and its regulatory proteins were assessed by Western blot analysis. β-actin was used as the internal control. (**C**) The expression of nuclear and cytosol Nrf2 were determined by Western blotting. Lamin B and β-actin were used as internal controls for the nuclear and cytosolic fractions, respectively. The results shown are representative of three independent experiments. (**B,D**) Statistical analyses were conducted using analysis of variances between groups. * $p < 0.05$ and *** $p < 0.0001$ when compared to control. # $p < 0.05$ and ### $p < 0.0001$ when compared to RANKL treatment. RANKL: receptor of activator of nuclear factor kappa-B ligand; FST: fermented sea tangle extract; Nrf2: nuclear factor-erythroid 2-related factor 2; p-Nrf2: phosphorylated nuclear factor-erythroid 2-related factor 2; HO-1: heme oxygenase-1; NQO-1: NAD(P)H quinone oxidoreductase 1; +: cells treated the reagent; -: cells untreated the reagent.

4. Discussion

Osteoclasts are multinucleated cells of hematopoietic origin which are derived from the monocyte/macrophage in their ability to resorb bone, whereas osteoblast are derived from pluripotent mesenchymal stem cells and are involved in bone formation [1,2]. Since excessive bone resorption by osteoclasts causes an imbalance in bone regeneration and induces osteolytic diseases, osteoclasts are considered prime targets for the management and treatment of bone diseases [2,3]. RANKL is a pro-osteoclastogenic cytokine and plays a crucial role in promoting osteoclastogenesis from osteoclast progenitor cells [4,5]. As has been well established in many earlier studies, RANKL binds to RANK expressed on the plasma membrane of osteoclast precursors and activates complex signaling cascades including NF-κB and NFATc1 for osteoclast differentiation [9,10]. Differentiation through activation of these signal transduction systems by RANKL is characterized by the formation of multinucleated giant cells [5,7]. This is a preliminary step in the maintenance, formation and function of the F-actin loop structure, which plays an important role in seal zone formation and resorption of bone mineral matrix in osteoclasts by activated TRAP [37,38]. According to the present findings, FST effectively inhibited RANKL-induced TRAP activation and F-actin ring formation without causing any significant cytotoxicity in RAW 264.7 cells, implying that FST suppressed osteoclast differentiation from osteoclast precursors at an early stage.

NF-κB, a transcription factor that plays a key role in inducing osteoclast differentiation, complexes with cytoplasmic IκB-α in the absence of osteoclastogenic induction signals and keeps it in an inactive form that tightly regulates its transcriptional activity for osteoclast differentiation [4,37]. However, the interaction of RANKL and RANK promotes the activation of the IκB kinase (IKK) complex, which phosphorylates IκB-α leading to ubiquitin-dependent degradation [6,7]. As a result, free NF-κB translocates to the nucleus and activates transcription of various genes involved in osteoclastogenesis [5]. Our results demonstrated that RANKL promoted the degradation of IκB-α in the cytoplasm and induced the translocation of NF-κB into the nucleus, both of which are essential for the activation of NF-κB but that these changes were completely inhibited by FST.

In the early stages of NF-κB activation and osteoclast differentiation, NFATc1 acts as a master regulator that enhances transcription of various osteoclast marker genes, which are highly expressed in the terminal differentiation stage to promote bone resorption [5,39]. In addition to blocking RANKL-induced NF-κB activation, FST inhibited NFATc1 expression in RANKL-treated cells. OPG treatment also completely blocked the expression of NFATc1. As further proof that FST was effectively inhibiting osteoclastogenesis, we showed that it attenuated RANKL-induced up-regulation of osteoclast marker genes such as TRAP, MMP-9, CTSK and OSCAR to levels seen in the control and OPG co-treated groups. Although further experiments are required to determine whether NFATc1 inhibition is the direct result of NF-κB inactivation, the present results indicate that inactivation of the NF-κB signaling pathway and inhibition of the expression of osteoclast marker genes associated with a decrease in NFATc1 expression are involved as important mechanisms in the anti-osteoclastogenic effect of FST.

A number of previous studies have shown that ROS, as specific secondary messengers, play a key role in the initiation of RANKL-stimulated osteoclast differentiation and bone resorption through similar pathways involving the activation of NF-κB and NFATc1 [10,12]. However, the accumulation of excessive ROS due to oxidative stress blocks osteoblast differentiation, suppresses osteoblast survival and acts to promote bone loss [12,13]. It has also been reported that a variety of natural products with antioxidant activity inhibit osteoclast differentiation by inhibiting ROS production [6,10,11]. Therefore, ROS can be considered a potential target for inhibition of osteoclast differentiation and prevention of bone loss. The present results showed that FST significantly suppressed ROS production by RANKL. Moreover, consistent with the results of previous studies [12,40], RANKL-induced osteoclast differentiation was completely inhibited when production of ROS was artificially blocked using NAC, indicating that FST blocks osteoclast differentiation by acting as a scavenger or inhibitor of ROS. In order to reduce the damage from oxidative stress in the face of excess production of ROS in cells, several transcription factors are known to be activated to increase the expression of downstream antioxidant enzymes [41,42]. One of these redox sensitive transcription factors, Nrf2 has recently been reported to attenuate osteoclast differentiation through the regulation of ROS production [42,43]. For example, Nrf2 deficiency improved RANKL-induced osteoclast differentiation [44], whereas local induction of nuclear Nrf2 weakened RANKL-mediated osteoclastogenesis [45]. Under normal conditions, Nrf2 is sequestered by Kelch-like ECH-associated protein 1 (Keap1) to the cytoplasm but becomes separated from Keap1 by oxidative or electrophilic stress and translocated into the nucleus. In the nucleus, Nrf2 binds to the antioxidant response elements to induce the transcription of target antioxidants and detoxifying enzymes including HO-1 and NQO-1 [42,43]. In this study, FST significantly increased expression of Nrf2 and its transcriptional targets, including HO-1 and NQO-1 in RANKL-treated RAW 264.7 cells. We also observed that FST increased phosphorylation and nuclear translocation of Nrf2 compared to the RANKL-alone stimulated group. The results presented, indicate that FST attenuates osteoclast differentiation by decreasing RANKL-induced oxidative stress in osteoclast precursor cells through the activation of Nrf2 and its downstream genes.

5. Conclusions

To assume the effect of FST on RANKL-mediated osteoclast differentiation, recombinant RANKL protein was used to differentiate murine monocyte/macrophage RAW 264.7 cells as osteoclast precursor cells into osteoclasts. Present results demonstrated that FST inhibited RANKL-induced osteoclastogenesis and reduced the expression of several key osteoclast-regulatory genes through the inactivation of NF-κB. In addition, FST blocked RANKL-induced oxidative stress, which was associated with the activation of Nrf2 signaling pathway. Although the present study provides new insights into the inhibition of osteoclastogenesis by FST, further investigation of the molecular mechanisms underlying this process as well as identification of the bioactive constituents of FST are needed.

Author Contributions: Y.H.C., Y.-J.J., C.P. and B.-J.L. conceived and designed the experiments; J.-W.J., H.L., S.H.H. and C.P. performed the experiments; S.Y.J., H.L., G.-Y.K., E.K.P. and J.W.H. analyzed the data; J.-W.J. wrote the paper and Y.H.C. edited the paper.

Funding: This research was a part of the project titled 'Development of functional food products with natural materials derived from marine resources (20170285)', funded by the Ministry of Oceans and Fisheries.

Conflicts of Interest: The authors declare no conflict of interest.

References

1. Kikuta, J.; Ishii, M. Osteoclast migration, differentiation and function: Novel therapeutic targets for rheumatic diseases. *Rheumatology* **2013**, *52*, 226–234. [CrossRef] [PubMed]
2. Teitelbaum, S.L.; Ross, F.P. Genetic regulation of osteoclast development and function. *Nat. Rev. Genet.* **2003**, *4*, 638–649. [CrossRef] [PubMed]
3. Galson, D.L.; Roodman, G.D. Pathobiology of Paget's disease of bone. *J. Bone Metab.* **2014**, *21*, 85–98. [CrossRef] [PubMed]
4. Ono, T.; Nakashima, T. Recent advances in osteoclast biology. *Histochem. Cell Biol.* **2018**, *149*, 325–341. [CrossRef] [PubMed]
5. Park, J.H.; Lee, N.K.; Lee, S.Y. Current understanding of RANK signaling in osteoclast differentiation and maturation. *Mol. Cells* **2017**, *40*, 706–713. [PubMed]
6. Bi, H.; Chen, X.; Gao, S.; Yu, X.; Xiao, J.; Zhang, B.; Liu, X.; Dai, M. Key triggers of osteoclast-related diseases and available strategies for targeted therapies: A review. *Front. Med.* **2017**, *4*, 234. [CrossRef] [PubMed]
7. Kuroda, Y.; Matsuo, K. Molecular mechanisms of triggering, amplifying and targeting RANK signaling in osteoclasts. *World J. Orthop.* **2012**, *3*, 167–174. [CrossRef]
8. Sundaram, K.; Nishimura, R.; Senn, J.; Youssef, R.F.; London, S.D.; Reddy, S.V. RANK ligand signaling modulates the matrix metalloproteinase-9 gene expression during osteoclast differentiation. *Exp. Cell Res.* **2007**, *313*, 168–178. [CrossRef]
9. Asagiri, M.; Sato, K.; Usami, T.; Ochi, S.; Nishina, H.; Yoshida, H.; Morita, I.; Wagner, E.F.; Mak, T.W.; Serfling, E.; et al. Autoamplification of NFATc1 expression determines its essential role in bone homeostasis. *J. Exp. Med.* **2005**, *202*, 1261–1269. [CrossRef]
10. Lee, S.H.; Jang, H.D. Scoparone attenuates RANKL-induced osteoclastic differentiation through controlling reactive oxygen species production and scavenging. *Exp. Cell Res.* **2015**, *331*, 267–277. [CrossRef]
11. Domazetovic, V.; Marcucci, G.; Iantomasi, T.; Brandi, M.L.; Vincenzini, M.T. Oxidative stress in bone remodeling: Role of antioxidants. *Clin. Cases Miner. Bone Metab.* **2017**, *14*, 209–216. [CrossRef] [PubMed]
12. Li, Z.; Chen, C.; Zhu, X.; Li, Y.; Yu, R.; Xu, W. Glycyrrhizin suppresses RANKL-Induced osteoclastogenesis and oxidative stress through inhibiting NF-κB and MAPK and activating AMPK/Nrf2. *Calcif. Tissue Int.* **2018**, *103*, 324–337. [CrossRef] [PubMed]
13. Thummuri, D.; Naidu, V.G.M.; Chaudhari, P. Carnosic acid attenuates RANKL-induced oxidative stress and osteoclastogenesis via induction of Nrf2 and suppression of NF-κB and MAPK signalling. *J. Mol. Med.* **2017**, *95*, 1065–1076. [CrossRef] [PubMed]

14. Kim, H.J.; Park, C.; Kim, G.Y.; Park, E.K.; Jeon, Y.J.; Kim, S.; Hwang, H.J.; Choi, Y.H. Sargassum serratifolium attenuates RANKL-induced osteoclast differentiation and oxidative stress through inhibition of NF-κB and activation of the Nrf2/HO-1 signaling pathway. *Biosci. Trends* **2018**, *12*, 257–265. [CrossRef] [PubMed]
15. Khan, M.; Cheung, A.M.; Khan, A.A. Drug-related adverse events of osteoporosis therapy. *Endocrinol. Metab. Clin.* **2017**, *46*, 181–192. [CrossRef] [PubMed]
16. Koyama, T. Extracts of marine algae show inhibitory activity against osteoclast differentiation. *Adv. Food Nutr. Res.* **2011**, *64*, 443–454. [PubMed]
17. Venkatesan, J.; Kim, S.K. Osteoporosis treatment: Marine algal compounds. *Adv. Food Nutr. Res.* **2011**, *64*, 417–427. [PubMed]
18. De Jesus Raposo, M.F.; De Morais, A.M.; De Morais, R.M. Emergent sources of prebiotics: Seaweeds and microalgae. *Mar. Drugs* **2016**, *14*, E27. [CrossRef] [PubMed]
19. Shirosaki, M.; Koyama, T. Laminaria japonica as a food for the prevention of obesity and diabetes. *Adv. Food Nutr. Res.* **2011**, *64*, 199–212.
20. Gao, J.; Lin, L.; Sun, B.; Zhao, M. A comparison study on polysaccharides extracted from Laminaria japonica using different methods: Structural characterization and bile acid-binding capacity. *Food Funct.* **2017**, *8*, 3043–3052. [CrossRef]
21. Machu, L.; Misurcova, L.; Ambrozova, J.V.; Orsavova, J.; Mlcek, J.; Sochor, J.; Jurikova, T. Phenolic content and antioxidant capacity in algal food products. *Molecules* **2015**, *20*, 1118–1133. [CrossRef] [PubMed]
22. Kim, Y.M.; Jang, M.S. Anti-obesity effects of Laminaria japonica fermentation on 3T3-L1 adipocytes are mediated by the inhibition of C/EBP-α/β and PPAR-γ. *Cell. Mol. Biol.* **2018**, *64*, 71–77. [CrossRef] [PubMed]
23. Jang, E.J.; Kim, S.C.; Lee, J.H.; Lee, J.R.; Kim, I.K.; Baek, S.Y.; Kim, Y.W. Fucoxanthin, the constituent of Laminaria japonica, triggers AMPK-mediated cytoprotection and autophagy in hepatocytes under oxidative stress. *BMC Complement. Altern. Med.* **2018**, *18*, 97. [CrossRef] [PubMed]
24. Je, J.Y.; Park, S.Y.; Ahn, C.B. Antioxidant and cytoprotective activities of enzymatic extracts from Rhizoid of Laminaria japonica. *Prev. Nutr. Food Sci.* **2017**, *22*, 312–319. [CrossRef] [PubMed]
25. Zeng, M.; Wu, X.; Li, F.; She, W.; Zhou, L.; Pi, B.; Xu, Z.; Huang, X. Laminaria Japonica polysaccharides effectively inhibited the growth of nasopharyngeal carcinoma cells in vivo and in vitro study. *Exp. Toxicol. Pathol.* **2017**, *69*, 527–532. [CrossRef] [PubMed]
26. Lee, B.J.; Senevirathne, M.; Kim, J.S.; Kim, Y.M.; Lee, M.S.; Jeong, M.H.; Kang, Y.M.; Kim, J.I.; Nam, B.H.; Ahn, C.B.; et al. Protective effect of fermented sea tangle against ethanol and carbon tetrachloride-induced hepatic damage in Sprague-Dawley rats. *Food Chem. Toxicol.* **2010**, *48*, 1123–1128. [CrossRef] [PubMed]
27. Kang, Y.M.; Lee, B.J.; Kim, J.I.; Nam, B.H.; Cha, J.Y.; Kim, Y.M.; Ahn, C.B.; Choi, J.S.; Choi, I.S.; Je, J.Y. Antioxidant effects of fermented sea tangle (Laminaria japonica) by Lactobacillus brevis BJ20 in individuals with high level of γ-GT: A randomized, double-blind, and placebo-controlled clinical study. *Food Chem. Toxicol.* **2012**, *50*, 1166–1169. [CrossRef]
28. Cha, J.Y.; Jeong, J.J.; Yang, H.J.; Lee, B.J.; Cho, Y.S. Effect of fermented sea tangle on the alcohol dehydrogenase and acetaldehyde dehydrogenase in Saccharomyces cerevisiae. *J. Microbiol. Biotechnol.* **2011**, *21*, 791–795. [CrossRef]
29. You, J.S.; Sung, M.J.; Chang, K.J. Evaluation of 8-week body weight control program including sea tangle (Laminaria japonica) supplementation in Korean female college students. *Nutr. Res. Pract.* **2009**, *3*, 307–314. [CrossRef]
30. Choi, W.C.; Reid, S.N.S.; Ryu, J.K.; Kim, Y.; Jo, Y.H.; Jeon, B.H. Effects of γ-aminobutyric acid-enriched fermented sea tangle (Laminaria japonica) on brain derived neurotrophic factor-related muscle growth and lipolysis in middle aged women. *Algae* **2016**, *31*, 175–187. [CrossRef]
31. Reid, S.N.S.; Ryu, J.K.; Kim, Y.; Jeon, B.H. GABA-enriched fermented Laminaria japonica improves cognitive impairment and neuroplasticity in scopolamine- and ethanol-induced dementia model mice. *Nutr. Res. Pract.* **2018**, *12*, 199–207. [CrossRef] [PubMed]
32. Reid, S.N.S.; Ryu, J.K.; Kim, Y.; Jeon, B.H. The effects of fermented Laminaria japonica on short-term working memory and physical fitness in the elderly. *Evid. Based Complement. Altern. Med.* **2018**, *2018*, 8109621.
33. Khosla, S. Minireview: The OPG/RANKL/RANK system. *Endocrinology* **2001**, *142*, 5050–5055. [CrossRef] [PubMed]
34. Hofbauer, L.C. Osteoprotegerin ligand and osteoprotegerin: Novel implications for osteoclast biology and bone metabolism. *Eur. J. Endocrinol.* **1999**, *141*, 195–210. [CrossRef] [PubMed]

35. Kong, X.; Yang, Y.; Wu, W.; Wan, H.; Li, X.; Zhong, M.; Su, X.; Jia, S.; Lin, N. Triterpenoid saponin W3 from Anemone flaccida suppresses osteoclast differentiation through inhibiting activation of MAPKs and NF-κB pathways. *Int. J. Biol. Sci.* **2015**, *11*, 1204–1214. [CrossRef]
36. Hong, S.; Huh, J.E.; Lee, S.Y.; Shim, J.K.; Rhee, S.G.; Jeong, W. TRP14 inhibits osteoclast differentiation via its catalytic activity. *Mol. Cell. Biol.* **2014**, *34*, 3515–3524. [CrossRef] [PubMed]
37. Soysa, N.S.; Alles, N. Osteoclast function and bone-resorbing activity: An overview. *Biochem. Biophys. Res. Commun.* **2016**, *476*, 115–120. [CrossRef]
38. Hayman, A.R. Tartrate-resistant acid phosphatase (TRAP) and the osteoclast/immune cell dichotomy. *Autoimmunity* **2008**, *41*, 218–223. [CrossRef]
39. Kim, Y.W.; Baek, S.H.; Lee, S.H.; Kim, T.H.; Kim, S.Y. Fucoidan, a sulfated polysaccharide, inhibits osteoclast differentiation and function by modulating RANKL signaling. *Int. J. Mol. Sci.* **2014**, *15*, 18840–18855. [CrossRef]
40. Zhang, Y.; Zhang, Y.; Kou, J.; Wang, C.; Wang, K. Role of reactive oxygen species in angiotensin II: Induced receptor activator of nuclear factor-κB ligand expression in mouse osteoblastic cells. *Mol. Cell. Biochem.* **2014**, *396*, 249–255. [CrossRef]
41. Lee, N.K.; Choi, Y.G.; Baik, J.Y.; Han, S.Y.; Jeong, D.W.; Bae, Y.S.; Kim, N.; Lee, S.Y. A crucial role for reactive oxygen species in RANKL-induced osteoclast differentiation. *Blood* **2005**, *106*, 852–859. [CrossRef] [PubMed]
42. Kang, K.A.; Hyun, J.W. Oxidative stress, Nrf2, and epigenetic modification contribute to anticancer drug resistance. *Toxicol. Res.* **2017**, *33*, 1–5. [CrossRef] [PubMed]
43. Sun, Y.X.; Xu, A.H.; Yang, Y.; Li, J. Role of Nrf2 in bone metabolism. *J. Biomed. Sci.* **2015**, *22*, 101. [CrossRef] [PubMed]
44. Hyeon, S.; Lee, H.; Yang, Y.; Jeong, W. Nrf2 deficiency induces oxidative stress and promotes RANKL-induced osteoclast differentiation. *Free Radic. Biol. Med.* **2013**, *65*, 789–799. [CrossRef] [PubMed]
45. Kanzaki, H.; Shinohara, F.; Kajiya, M.; Fukaya, S.; Miyamoto, Y.; Nakamura, Y. Nuclear Nrf2 induction by protein transduction attenuates osteoclastogenesis. *Free Radic. Biol. Med.* **2014**, *77*, 239–248. [CrossRef] [PubMed]

 © 2019 by the authors. Licensee MDPI, Basel, Switzerland. This article is an open access article distributed under the terms and conditions of the Creative Commons Attribution (CC BY) license (http://creativecommons.org/licenses/by/4.0/).

Article

Antioxidant Activities of *Solanum nigrum* L. Leaf Extracts Determined in In Vitro Cellular Models

Agata Campisi [1], Rosaria Acquaviva [1], Giuseppina Raciti [1], Anna Duro [2], Milena Rizzo [1,*] and Natale Alfredo Santagati [1]

[1] Department of Drug Science, University of Catania, Viale Andrea Doria 6, 95125 Catania, Italy; agcampisi@gmail.com (A.C.); racquavi@unict.it (R.A.); racitigi@unict.it (G.R.); santagat@unict.it (N.A.S.)
[2] Department of Biological, Geological and Environmental Sciences, University of Catania, Via A. Longo 19, 95125 Catania, Italy; annaduro@unict.it
* Correspondence: milena.rizzo@unict.it

Received: 31 December 2018; Accepted: 30 January 2019; Published: 8 February 2019

Abstract: Several medicinal foods abound in traditional medicine with antioxidant potentials that could be of importance for the management of several diseases but with little or no scientific justification to substantiate their use. Thus, the objective of this study was the assessment of the antioxidant effect of two leave extracts of *Solanum nigrum* L. (SN), which is a medicinal plant member of the *Solanaceae* family, mainly used for soup preparation in different parts of the world. Then methanolic/water (80:20) (SN1) and water (SN2) leaves extracts were prepared. The total polyphenolic content and the concentration of phenolic acids and flavones compounds were determined. In order to verify whether examined extracts were able to restore the oxidative status, modified by glutamate in primary cultures of astrocytes, the study evaluated the glutathione levels, the intracellular oxidative stress, and the cytotoxicity of SN1 and SN2 extracts. Both extracts were able to quench the radical in an in vitro free cellular system and restore the oxidative status in in vitro primary cultures of rat astroglial cells exposed to glutamate. These extracts prevented the increase in glutamate uptake and inhibited glutamate excitotoxicity, which leads to cell damage and shows a notable antioxidant property.

Keywords: *Solanum nigrum* L. leave extracts; natural products; antioxidant activity; functional food

1. Introduction

Several medicinal foods abound in traditional medicine with antioxidant activities. This could be of important for the management of several diseases but has little or no scientific justification to substantiate their use. In the tropics, as in Asia and in sub-Saharan Africa, green leafy vegetables are used as one of the major components of local dishes.

Solanum nigrum L. (SN) belongs to the *Solanaceae* family to Europe, Asia, and North America and was introduced in South America, Australia, and Africa. It represents one of the largest and most variable species groups of the genus. SN, well known as "Black Nightshade" (the English name), is an herbal plant widely distributed throughout the world, extending from tropical regions to temperate regions [1].

In many developing countries, SN constitutes a minor food crop, with the shoots and berries not only used as vegetables and fruits but also for various medicinal and local uses [2]. SN serves mainly as vegetables for soup preparation in different parts of the world. Several studies have investigated the nutritive value of the 'vegetable black nightshade,' which put forward evidence that this species constitutes a nutritious vegetable [3]. This plant was chosen not only for being nutritive, but also for their folkloric reports of medicinal properties [4]. Studies document potential health benefits of different parts of this vegetable. SN leaves have been reportedly used in traditional medicine for the management of several diseases including seizure and epilepsy, pain, ulcer, inflammation, diarrhea, some eye infections, and jaundice [5,6].

In folklore medicine, the leaves are used to treat oral ulcers in India where an interesting pharmacological investigation has been performed by using an aqueous extract of SN leaves [7].

More recently, many research studies have reported that SN showed anti-cancer activity for hepatocellular carcinoma cells [8], human ovarian carcinoma cells [9], human colorectal carcinoma cells [10], and human endometrial carcinoma cells [11]. The leaves can provide appreciable amounts of protein and amino acids, minerals including calcium, iron, and phosphorus, vitamins A and C, fats and fibers, and appreciable amounts of methionine, which is an amino acid scarce in other vegetables. Other chemical constituents reported in leaves are steroidal glycosides [12]. Very recently, from the unripe berries, a previously undescribed steroidal alkaloids [13] and steroidal glycosides [14] were isolated. Those compounds showed a potent inhibitory activity against the lipopolysaccharide (LPS)-induced nitric oxide (NO) production.

Because medicinal plants are gaining popularity for the production of reliable and safe medicines suitable for human, many studies investigated the composition of extracts, their biological activities, and optimization of extraction procedures [15,16].

The extracts of the SN contain many polyphenolic compounds. The leaves are rich in polyphenols, including phenolic acids and flavones [17]. Zaidi et al. demonstrated that treatment of rats with SN leaves extract was able to reduce oxidative stress, and, in particular, they showed the potential of this extract in preventing/alleviating stress-induced diseases, involving oxidative damage to cellular constituents especially the brain [18]. Antioxidant activity might be due to the presence of the above-mentioned polyphenolic compounds on SN stems and leaves [19]. Sun et al. demonstrated that oxidative stress has been associated with pathological conditions, including Central Nervous System (CNS) diseases and physiological brain aging processes [20].

A very interesting study has shown that dietary inclusions of Solanum leaf could protect against cognitive and neurochemical impairments induced by scopolamine, and, hence, this vegetable could be used as a source of functional foods and nutraceuticals for the prevention and management of cognitive impairment-associated diseases such as Alzheimer's disease [21].

The formation and release of Radical Oxygen reactive Species (ROS) cause structural and functional alterations of cell membranes. Free radicals attack polyunsaturated fatty acids in bio-membranes and mitochondria begin the main source of ROS, when the mitochondrial respiratory chain is impaired. In these cases, a compound possessing antioxidant properties can be useful in stopping ROS production and limiting oxidative cell damages, which is particularly interesting if this activity is produced by a functional food [22]. Experimentally, ROS is well determined by using reduced glutathione (GSH), which is known as the most important scavengers of reactive species, and a reduced glutathione/oxidized glutathione ratio is used as a marker of oxidative stress [23].

At the central level, the oxidative stress may activate several calcium-dependent enzymes, causing mitochondria impairment, a decrease in adenosine triphosphate (ATP) levels, ROS production, and subsequent neuronal cell death [24]. A brief exposure to glutamate, which is a major excitatory neurotransmitter in the CNS, could determine several acute and chronic brain damages on differentiated astrocytes, which then causes cell swelling, whereas a prolonged incubation (excitotoxicity) induces cell damage [25,26]. This phenomenon causes alterations in glutamate transport, GSH depletion, and macromolecular synthesis [27].

Herein, we prepared two SN polar leaf extracts and assessed the total polyphenolic content and the concentration of phenolic acids and flavones compounds. Antioxidant activity in both an in vitro cellular free system and in vitro cellular system was evaluated. To verify whether SN1 and SN2 extracts were able to restore the oxidative status, which were modified by glutamate in primary cultures of astrocytes, GSH, ROS levels and the cytotoxicity of both extracts has been assessed.

2. Materials and Methods

2.1. Materials

Reference compounds (gallic acid, protocatechuic acid, chlorogenic acid, gentisic acid, caffeic acid, luteolin, apigenin) were purchased from Sigma (St. Louis, MO, USA). Acetonitrile, methanol, and water were chromatographic-grade and were bought from Carlo Erba (Milano, Italy). STable 2,2-diphenyl-1-picrylhydrazyl (DPPH) radical, 3(4,5-dimethyl-thiazol-2-yl)2,5-diphenyl-tetrazolium bromide (MTT), 2′,7′-dichlorofluorescein diacetate (DCFH-DA), 6-hydroxy-2,5,7,8-tetramethylchroman-2-carboxylic acid (Trolox), dimethylsulfoxide (DMSO), GYKI 52466, phosphate buffer solution (PBS), and other analytical chemicals were purchased from Sigma-Aldrich Chimica (Milan, Italy). Distilled and deionized water was used for the preparation of all samples and solutions and they were used after filtration through HA filters (0.45 µm Millipore, Bedford, MA, USA). All standards were diluted to the appropriate obtained concentration and were stored at +4 °C in amber vials in a dark place until analysis. Dulbecco's modified Eagle's medium (DMEM) and heat-inactivated Fetal Bovine Serum (FBS) were obtained from Invitrogen (Milan, Italy). The mouse monoclonal antibody against glial fibrillary acidic protein (GFAP) and anti-IgG polyclonal antibody were from Chemicon (Prodotti Gianni, Milan, Italy).

2.2. Methods

2.2.1. Plant Material and Extraction Procedures

SN leaves were collected around Catania (Sicily, Italy) in June 2018 and air-dried at room temperature (24 ± 2 °C) with no direct light. A voucher specimen of the plant has been deposited (N. 8234) in the herbarium of the Botany Department of the University of Catania (Italy). The dried and powdered leaves (5 g) were extracted by maceration for 24 h at room temperature with 150 mL each of methanol/water solution 80:20 v/v (SN1) and water (SN2), respectively. The extraction procedure was repeated three times, and the solvent extracts were combined and separated from the residue by filtration through Whatman N. 1 filter paper in a Buchner funnel under vacuum. The methanol was removed under a reduced pressure below 40 °C by using a rotary evaporator, and the aqueous phase remaining after evaporation of the organic phase was freeze-dried. The water extract was freeze-dried. The obtained dried extracts were 1.305 ± 0.24 g for SN1 and 0.975 ± 0.31 g for SN2, respectively. The obtained dried extracts were stored at −20 °C until undergoing assays.

2.2.2. Determination of Total Phenolic Content

The amount of total phenolic in the studied extracts was determined using a modified Folin-Ciocalteu method [28]. Extracts were prepared at a concentration of 0.5 mg/mL. An aliquot of a known dilution of the extract was mixed with 5 mL Folin–Ciocalteu reagent (previously diluted 10-fold with deionized water) and was allowed to react for 6 min. Then, 4 mL (70 g/L) of sodium carbonate solution was added to test tubes. The tubes were vortexed for 20 s and allowed to stand for 90 min for color development. Absorbance was measured at 760 nm by using the Perkin Elmer UV–Vis Lambda 25 spectrophotometer (Perkin Elmer Italia spa, Monza, Italy). Extract samples were evaluated at the final concentration of 1 mg/mL. The measurements were compared to a standard curve of prepared gallic acid solutions. The total phenolic content were 92.2 mg/g of extract for SN1 and 40.0 mg/g of extract for SN2, which was expressed as a gallic acid equivalent.

2.2.3. Chromatography

Chromatographic analysis was performed by using high-performance liquid chromatography (HPLC) (Perkin Elmer, Norwalk, CT, USA) Series 200 pump equipped with an LC-235C Diode Array Detector (DAD), auto-sampler, and column oven. Chromatographic data were processed by using a Turbochrom Workstation software, version 6.1.2 (Perkin Elmer, Norwalk, CT, USA). Separation and determinations were accomplished on a 5 µm Hypersil ODS RP-18 column (250 × 4.6 mm, Supelco, Bellefonte, PA, USA) fitted with a guard column (Hypersil ODS RP-18, 5 µm particles, 10 × 4.6 mm, Supelco, Bellefonte, PA, USA). All the samples were filtered, through 0.45 µm membrane filter, and degassed by an ultrasonic bath before the injection. The procedure was performed, as previously described [17].

2.2.4. Determination of Antioxidant Activity in an In Vitro Cellular Free System

The stable DPPH radical was used for the determination of free radical-scavenging activity of the extracts [29]. Because of its odd electron, the DPPH radical gives a strong absorption band at 517 nm in visible spectroscopy (deep violet color). As this electron becomes paired off in the presence of a free radical scavenger, the absorption vanishes, and the resulting decolorizing is stoichiometric with respect to the number of electrons taken up. The reaction mixture contained 86 µM DPPH radical and different concentrations of each extracts (0.025-0.5-0.1-0.2-0.4 mg/mL) in 1 mL of ethanol. After 10 min at room temperature, the absorbance at 517 nm was recorded [30]. Trolox (30 µM), water-soluble derivative of vitamin E, was used as a standard. The assay was performed in triplicates.

2.2.5. Determination of Antioxidant Activity within an In Vitro Cellular System

Primary cultures of astrocytes were prepared from new-born albino rat brains (from 1-day-old to 2-day-old Wistar strain rats) as described [31]. Cerebral tissues, after dissection and careful removal of the meninges, were mechanically dissociated through 82 µm pore sterile mesh (Nitex, Darmastadt, Germany). Isolated cells were suspended in DMEM, supplemented with 20% (v/v) FBS, 2 mM glutamine, streptomycin (50 mg/mL), and penicillin (50 U/mL), and plated at a density of 3×10^6 cells/100 mm dishes and of 0.5×10^5 cells/chamber of multi-chambered slides. Cells were maintained at 37 °C in a 5% CO_2 and 95% air humidified atmosphere for two weeks. The medium was exchanged every three days. The low initial plating density of dissociated cells was meant to favor the growth of astrocytes with only a very little oligodendroglial and microglial cells contamination. Astroglial cell cultures were characterized at 14 days in vitro (DIV), when it is confluent, by immunofluorescence staining with GFAP. All experiments conformed to the guidelines of the local Ethical Committee (University of Catania, Italy), and were carried out in accordance with the EC Directive 86/609/EEC for animal experiments.

Astrocytes at 14 DIV were treated with glutamate (500 µM) for 24 h, as previously described [32]. Those cultures were treated with different concentrations of SN1 and SN2 (0.5 and 1 mg/mL), which were added 20 min before glutamate exposure. Four replicates were carried out for each sample. In a subset of experiments, to assess the inhibition of glutamate effects, the astroglial cell cultures were incubated 20 min prior to glutamate exposure, with GYKI 52466 (100 µM), the specific AMPA/KA receptor antagonist.

Astroglial cell survival analysis was performed by the MTT reduction assay, which evaluated mitochondrial dehydrogenase activity, as previous reported [33,34]. Astrocytes were set up 0.5×10^5 cells per well of a 96-multiwell, flat-bottomed, 200-µL micro plate. They were maintained at 37 °C in a humidified 5% CO_2 and 95% air mixture [33]. At the end of treatment time, 20 µL of 0.5% MTT in (pH 7.4) PBS were added to each micro-well. After 1 h of incubation with the reagent, the supernatant was removed and replaced with of DMSO (200 µL). The optical density of each well was measured with a micro-plate spectrophotometer reader (Titertek Multiskan, Flow Laboratories, Helsinki, Finland) at $l = 570$ nm.

2.2.6. Glutathione Measurement

Astroglial cell cultures were scraped off and lysed in 50 µM sodium phosphate buffer (pH 7.4). The Bradford assay determined the protein concentration in cell extracts [32]. Then, a Hitachi U-2000 spectrophotometer (Hitachi, Tokyo, Japan) chemically determined the total glutathione intracellular content (GSH + GSSG), as described by Chen YH et al. [35].

2.2.7. ROS Levels Determination

Reactive species determination was performed by using DCFH-DA as a fluorescent probe. Furthermore, 100 μM of DCHF-DA was dissolved in 100% methanol, added to the cellular medium, and the cells were incubated at 37 °C for 30 min. Under these conditions, the acetate group was not hydrolyzed [12]. After incubation, astroglial cell cultures were lysed and centrifuged at 10,000× g for 10 min. The fluorescence corresponding to the radical oxidized species 2′,7′-dichlorofluorescein (DCF) was monitored by measuring the excitation (λ = 488 nm) and emission (λ = 525 nm), using an F-2000 spectrofluorometer (Hitachi).

Values are expressed as a percentage of fluorescence intensity per mg protein *versus* control (% I.F/mg prot vs. control). Protein concentration was measured, according to the Bradford assay applied by Li Volti et al [32].

2.3. Statistical Analysis

All values are presented as means ± S.D. of five separate measurements. Statistical analysis was performed using one-way ANOVA, which was followed by the Newman-Keuls post-hoc test. Differences were considered statistically significant at * $p < 0.05$ and ** $p < 0.001$.

3. Results

3.1. Analysis of SN Extracts

The total phenolic contents was assayed by a modified Folin-Ciocalteu method using gallic acid, and is reported in Table 1, with the physical characteristics and the dry weight yields of SN1 and SN2.

Table 1. Extraction yields and concentration of total phenolic content of SN1 and SN2 extracts (n = 3).

Extract	Solvent	Physical	Yield %	Total Phenolic Content (mg/g of Extract)
SN1	MeOH-H_2O 80–20 v/v	Brown solid	26.1 ± 3.9%	92.2 ± 4.8
SN2	H_2O	Brown solid	19.5 ± 5.0%	40.0 ± 6.9

The relative concentrations of phenolic acids and flavones in SN1 and SN2 were determined by HPLC analysis, according to Huang et al. [29]. The values were expressed in mg/g of dry extract. The results of the five phenolic acids determination (gallic, protocatechuic, chlorogenic, gentisic, and caffeic) and two flavones (luteolin and apigenin) in studied extracts are summarized in Table 2 while Figure 1 shows a representative chromatogram.

It was found that the major compound in both extracts was chlorogenic acid, whereas gentisic acid is the second. Luteolin is more abundant than apigenin, and less amounts of caffeic acid and protocatechuic acid were found. Gallic acid exists only in traces together with other unknown compounds.

Table 2. Contents of phenolic acid and flavones (mg/g of dry weight) in SN1 and SN2 leave extracts (n = 3).

Extract	Gallic Acid	Protocatechuic Acid	Chlorogenic Acid	Gentisic Acid	Caffeic Acid	Luteolin	Apigenin
SN1	0.09 ± 0.02	0.24 ± 0.91	2.77 ± 0.45	1.50 ± 0.66	0.64 ± 0.87	0.98 ± 0.33	0.16 ± 0.74
SN2	0.04 ± 0.05	0.19 ± 0.11	2.01 ± 0.98	1.81 ± 0.75	0.42 ± 0.54	0.8 ± 0.62	0.12 ± 0.02

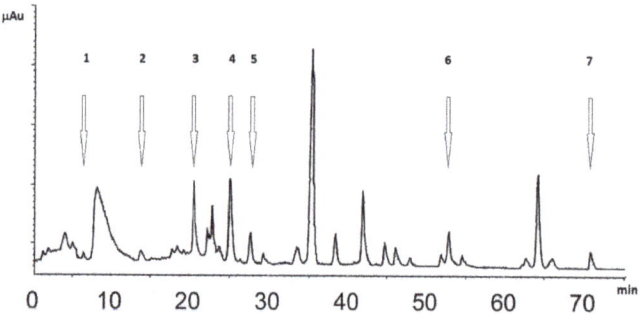

Figure 1. Representative chromatogram of the dry extract reporting the retention time (RT) of gallic acid (1, 0.65 min), protocatechuic acid (2, 13.85 min), chlorogenic acid (3, 20.5 min), gentisic acid (4, 25.1 min), caffeic acid (5, 27.5 min), luteolin (6, 52.9 min), and apigenin (7, 70.95 min). Axis: x label minutes, y label absorbance unit.

3.2. Antioxidant Activity in a Cellular Free System

The free radical-scavenging activity of SN1 and SN2 extracts was tested by their ability to bleach the stable DPPH radical. Both extracts were able to quench the DPPH-radical at all the concentrations used (0.025-0.5-0.1-0.2-0.4 mg/mL) in a dose-dependent manner. SN1 showed a more potent capacity than SN2. In addition, their effect appeared similar to Trolox (30 µM), which is a soluble analogous of vitamin E used as a standard. This experiment demonstrated that SN1 and SN2 possess antioxidant properties (Figure 2).

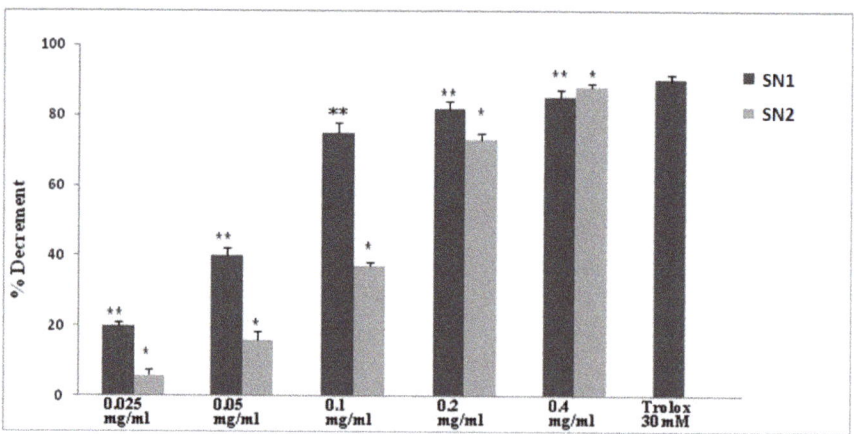

Figure 2. Quenching of DPPH of SN1 and SN2 extracts at different concentrations (0.025-0.5-0.1-0.2-0.4 mg/mL), compared to Trolox (30 mM). Axis: x label: concentration, y label: quenching of DPPH expressed as a percentage. (* $p < 0.05$ and ** $p < 0.001$).

3.3. Antioxidant Activity in the Cellular System

Primary rat astroglial cultures exposed to 500 µM glutamate for 24 h were used as an in vitro cellular model to assess the antioxidant effect of SN1 and SN2 extracts. Cells were characterized by immuno-fluorescent staining using GFAP as a marker [31].

The glutamate-evoked oxidative stress was evaluated by measuring the depletion of intracellular GSH levels (Figure 3) and ROS production (Figure 4). Glutamate (500 µM for 24 h) produced a significant decrease in the intracellular GSH levels and a significant increase of ROS levels, when compared to the untreated control ones. The pre-incubation of the cultures with SN1 and SN2 extracts

(0.5 and 1 mg/mL) was able to restore, in a dose-dependent manner, GSH and ROS levels. In particular, 1 mg/mL of the extracts showed values similar to untreated control values.

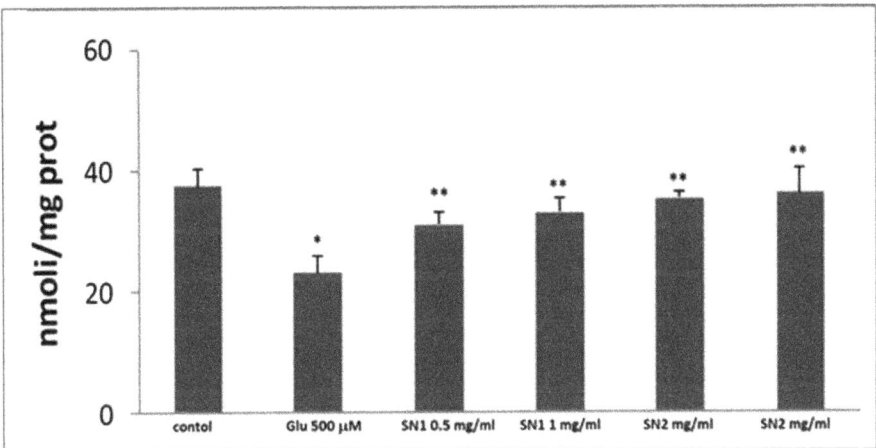

Figure 3. GSH levels in primary rat astroglial cell cultures at 14 DIV: exposed to glutamate 500 µM for 24 h. Bar 1: control. Bar 2: cell culture exposed 500 µM. Bar 3: cell culture exposed 500 µM plus SN1 0.5 mg/mL. Bar 4: cell culture exposed 500 µM plus SN1 1 mg/mL. Bar 5: cell culture exposed 500 µM plus SN2 0.5 mg/mL. Bar 6: cell culture exposed 500 µM plus SN2 1 mg/mL. Four replicates were carried out for each sample. (* $p < 0.05$ and ** $p < 0.001$). Axis: x label: concentration, y label: nmoli of GSH per mg of protein.

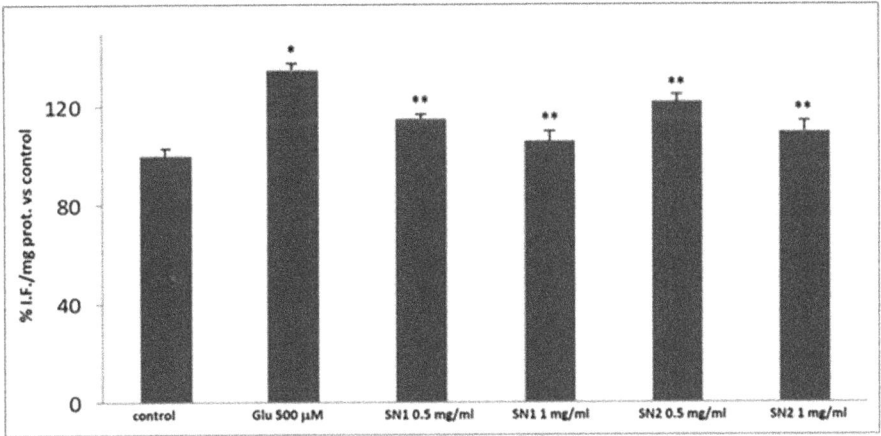

Figure 4. ROS levels in primary rat astroglial cell cultures at 14 DIV: exposed to glutamate 500 µM for 24 h. Bar 1: control. Bar 2: cell culture exposed 500 µM. Bar 3: cell culture exposed 500 µM plus SN1 0.5 mg/mL. Bar 4: cell culture exposed 500 µM plus SN1 1 mg/mL. Bar 5: cell culture exposed 500 µM plus SN2 0.5 mg/mL. Bar 6: cell culture exposed 500 µM plus SN2 1 mg/mL. Four replicates were carried out for each sample. (* $p < 0.05$ and ** $p < 0.001$). Axis: x label: concentrations, y label: percentage of fluorescence intensity per mg protein *versus* control.

4. Discussion

In this study, we assessed the antioxidant effect of SN1 and SN2 extracts of *Solanum nigrum* L. leaves, both in an in vitro cellular free system and in vitro cellular models.

Oxidative stress is the causative agent in a number of human diseases, such as atherosclerosis, ischemic reperfusion injury, inflammation, carcinogenesis, aging, and neurodegenerative diseases. Although there are many determinants in the development of these diseases, considerable experimental evidence links the production of ROS to biological damage that can potentially provide a mechanistic basis for their initiation and/or progression [36–39]. Moreover, because the ROS production is the fatal consequence of aerobic life, it is also an important component of the signaling network of plants [39], where polyphenols are the secondary metabolites produced as a defense mechanism against stress factors.

In this study, we exploited the sources, composition, and mechanisms of action of two polar *Solanum nigrum* L. leaf extracts natural products, and food, which represent a new frontier for therapy and a valuable tool to reduce the costs of health care systems.

In recent years, there has been great interest in the health effects of various natural products and in the in vivo protective function of natural antioxidants contained in dietary food against oxidative damage caused by ROS [40–45].

The free radical-scavenging activity is measured by the ability to bleach the stable DPPH radical. This assay provided information on the reactivity of test compounds with a stable free radical. SN1 and SN2 extracts were able to quench the DPPH-radical in a dose-dependent manner and showed comparable capacity. In fact, the two polar extracts content of different level phenolic components, but phenolic and flavones are not significantly different between SN1 and SN2. Only gentisic acid is more abundant in SN2. In addition, their effect appeared similar to Trolox. Then, this set of experiments demonstrate that SN1 and SN 2 possess comparable antioxidant properties.

Furthermore, we assessed the effect of the extracts in an in vitro cellular experimental model of excitotoxicity. In our previous research studies, using an experimental model of excitotoxicity, we demonstrated that glutamate exposure in primary cultures of astrocytes might be part of the biochemical response to oxidative stress induced by a prolonged exposure of astrocyte cultures to the neurotransmitter [46].

The antioxidant effect of the extracts SN1 and SN2 was also assessed in the cellular system using primary rat astroglial cell cultures exposed to the astroglial cell cultures in the presence of 500 μM glutamate for 24 hours. We used glutamate as a stressor because its high levels induce alterations in glutamate transport, mitochondria impairment, decrease ATP levels, GSH depletion, ROS production, macromolecular synthesis [35], and subsequent neuronal cell death [47,48].

Figure 2 shows the quenching of DPPH of SN1 and SN2 extracts at different concentrations, compared to Trolox, which shows a stronger activity at a lower concentration (0.025–0.5 and 0.1 mg/mL). In fact, we found that the extracts SN1 and SN2 were able to counteract the effect of glutamate, restoring, in a dose-dependent manner, the GSH and ROS levels similar to the control values.

The statistical analysis method in this study indicated high significance (* $p < 0.05$ and ** $p < 0.001$) when compared with the control group, as reported in Figures 3 and 4, where SN1 and SN2 extracts are compared with the cells exposed to glutamate 500 μM only.

The protective effect against glutamate toxicity of the extracts SN1 and SN2 appeared stronger than that of the synthetic antioxidant compounds used in our previous research studies [46].

Thus, these findings show that the extracts SN1 and SN2 possess antioxidant properties. Furthermore, it is possible to assume that the extracts of SN1 and SN2 of *Solanum nigrum* are able to counteract glutamate uptake-induced impairment of cystine/glutamate antiporter, which leads to depletion of the GSH content and biochemical alterations. This results in the delayed toxic effect for primary astroglial cell cultures [35]. Moreover, in a previous study, we reported that a pre-incubation with GYKI 52466, the selective inhibitor of AMPA/KA receptors, diminished glutamate effects, which indicates the involvement of receptor-linked events in GSH decrease and ROS increase levels [49].

Spectrophotometric and chromatographic analytical methods applied for estimation of total phenolic content and for determination of phenolic acids and flavones compounds in the examined extracts showed that these constituents are present in a valuable amount. Our results strongly suggest that phenolic compounds are important components of SN, and some of their pharmacological effects could be attributed to the presence of these compounds.

5. Conclusions

This study has provided some scientific rationale for these vegetables in the management of diseases, as obtained in folklore.

SN leave extracts were able to reduce oxidative stress, and, in particular, they showed the potential in quenching the radical in vitro free cellular system, and restoring the oxidative status among in vitro primary rat astroglial cell cultures exposed to glutamate, which possesses notable antioxidant properties and neuroprotective effects. Furthermore, preventing the increase in glutamate uptake and inhibiting glutamate excitotoxicity, SN1 and SN2 leave polar extracts may represent a new natural therapeutic strategy in the neuro-pathological conditions associated with excitotoxicity.

Therefore, these vegetables may serve as a potential source of natural phenolic antioxidants for functional foods and nutraceuticals for the prevention and management of neurodegenerative diseases.

Author Contributions: The CRediT taxonomy is listed below to define author contributions to primary research papers and the independent contributions of each author is provided, with the agreement of all co-authors. Each author on the paper had more CRediT contribution roles, as listed below: conceptualization, A.C., N.A.S.; methodology, A.C., R.A., M.R., N.A.S.; software, G.R.; validation, N.A.S., M.R.; formal analysis, M.R., G.R.; investigation, G.R., R.A.; resources, A.C., R.A., N.A.S., A.D.; data curation, R.A.; G.R., writing—original draft preparation, N.A.S., A.C., R.A.; writing—review and editing, M.R.; visualization, M.R.; supervision, A.C., N.A.S., R.A.; project administration, A.C., R.A., N.A.S.; funding acquisition, M.R.

Funding: This research received no external funding.

Conflicts of Interest: The authors declare no conflict of interest and the founder added the requested detail. The funder M.R. had roles in methodology, validation, formal analysis, visualization and the decision to publish the results.

Abbreviations

ATP	Adenosine Triphosphate
DCF	2′,7′-dichlorofluorescein
DCFH-DA	2′,7′-dichlorofluorescein diacetate
DIV	days in vitro
DMEM	Dulbecco's Modified Eagle's medium
DPPH	1,1-Diphenyl-2-Picrylhydrazyl radical
EDTA	ethylenediaminetetracetic acid
FBS	Heat Inactivated Fetal Bovine Serum
FITC	fluorescein isothiocyanate
GFAP	Glial Fibrillary Acidic Protein
GSH	reduced glutathione
HPLC	High performance liquid chromatography
MTT	3(4,5-dimethyl-thiazol-2-yl)2,5-diphenyl-tetrazolium bromide
NADH	β-Nicotinamide adenine dinucleotide
$O_2^{\bullet -}$	superoxide anion
$\bullet OH$	hydroxyl radical
$ROO\bullet$	peroxyl radical
ROS	reactive oxygen species
SOD	Superoxide dismutase
SN	Solanum Nigrum L
XO	Xanthine oxidase

References

1. Särkinen, T.; Poczai, P.; Barboza, G.E.; van der Weerden, G.M.; Baden, M.; Knapp, S. A revision of the Old World Black Nightshades (Morelloid clade of *Solanum* L., Solanaceae). *PhytoKeys* **2018**, *106*, 1–223.
2. Jagatheeswari, D.; Bharathi, T.; Sheik Jahabar Ali, H. Black Night Shade (*Solanum nigrum* L.)-An Updated Overview. *Int. J. Pharm. Biol. Arch.* **2013**, *4*, 288–295.
3. Akubugwo, I.E.; Obasi, A.N.; Ginika, S.C. Nutritional Potential of the Leaves and Seeds of Black Nightshade-Solanum nigrum L. Var virginicum from Afikpo-Nigeria. *Pak. J. Nutr.* **2007**, *6*, 323–326. [CrossRef]
4. Leporatti, M.L.; Ghedira, K. Comparative analysis of medicinal plants used in traditional medicine in Italy and Tunisia. *J. Ethnobiol. Ethnomed.* **2009**, *5*, 31. [CrossRef] [PubMed]
5. Jain, R.; Sharma, A.; Gupta, S.; Sarethy, I.P.; Gabrani, R. *Solanum nigrum*: Current perspectives on therapeutic properties. *Altern. Med. Rev.* **2011**, *16*, 78–85. [PubMed]
6. Wang, Z.; Li, J.; Ji, Y.; An, P.; Zhang, S.; Li, Z. Traditional herbal medicine: A review of potential of inhibitory hepatocellular carcinoma in basic research and clinical trial. *Evid. Based Complement. Altern. Med.* **2013**, *2013*, 268963. [CrossRef]
7. Patel, A.; Biswas, S.; Shoja, M.H.; Ramalingayya, G.V.; Nandakumar, K. Protective effects of aqueous extract of Solanum nigrum Linn. leaves in rat models of oral mucositis. *Sci. World J.* **2014**, *2014*, 345939.
8. Wang, C.K.; Lin, Y.F.; Tai, C.J. Integrated treatment of aqueous extract of *Solanum nigrum*-potentiated cisplatin- and doxorubicin-induced cytotoxicity in human hepatocellular carcinoma cells. *Evid. Based Complement. Alternat. Med.* **2015**, *2015*, 675270.
9. Wang, C.W.; Chen, C.L.; Wang, C.K. Cisplatin-, doxorubicin-, and docetaxel-induced cell death promoted by the aqueous extract of *Solanum nigrum* in human ovarian carcinoma cells. *Integr. Cancer Ther.* **2015**, *14*, 546–555. [CrossRef]
10. Tai, C.J.; Wang, C.K.; Chang, Y.J.; Lin, C.S.; Tai, C.J. Aqueous extract of *Solanum nigrum* leaf activates autophagic cell death and enhances docetaxel-induced cytotoxicity in human endometrial carcinoma cells. *Evid. Based Complement. Alternat. Med.* **2012**, *2012*, 859185.
11. Tai, C.J.; Wang, C.K.; Tai, C.J. Aqueous extract of *Solanum nigrum* leaves induces autophage and enhances cytotoxicity of cisplatin, doxoorubicin, docetaxel, and 5-flurouracil in human colorectal carcinoma cells. *Evid. Based Complement. Alternat. Med.* **2013**, *2013*, 514719.
12. Ikeda, T.; Tsumagari, H.; Nohara, T. Steroidal oligoglycosides from Solanum nigrum. *Chem. Pharm. Bull.* **2000**, *48*, 1062–1064. [CrossRef] [PubMed]
13. Gu, X.Y.; Shen, X.F.; Wang, L.; Wu, Z.W.; Li, F.; Chen, B.; Zhang, G.L.; Wang, M.K. Bioactive steroidal alkaloids from the fruits of Solanum nigrum. *Phytochemistry* **2018**, *147*, 125–131. [CrossRef] [PubMed]
14. Xiang, L.; Wang, Y.; Yi, X.; He, X. Anti-inflammatory steroidal glycosides from the berries of *Solanum nigrum* L. (European black nightshade). *Phytochemistry* **2018**, *148*, 87–96. [CrossRef] [PubMed]
15. Esmaeili, A.; Tahazadeh, A.R.; Ebrahimzadeh, M.A. Investigation of composition extracts, biological activities and optimization of *Solanum nigrum* L. extraction growing in Iran. *Pak. J. Pharm. Sci.* **2017**, *30*, 473–480. [PubMed]
16. Wang, H.C.; Chung, P.J.; Wu, C.H.; Lan, K.P.; Yang, M.Y.; Wang, C.J. *Solanum nigrum* L. polyphenolic extract inhibits hepatocarcinoma cell growth by inducing G2/M phase arrest and apoptosis. *J. Sci. Food Agric.* **2011**, *91*, 178–185. [CrossRef] [PubMed]
17. Huang, H.C.; Syu, K.Y.; Lin, J.K. Chemical Composition of *Solanum nigrum* Linn Extract and Induction of Autophagy by Leaf Water Extract and Its Major Flavonoids in AU565 Breast Cancer Cells. *J. Agric. Food Chem.* **2010**, *58*, 8699–8708. [CrossRef]
18. Zaidi, S.K.; Hoda, M.N.; Tabrez, S.; Ansari, S.A.; Jafri, M.A.; Khan, M.S.; Hasan, S.; Alqahtani, M.H.; Abuzenadah, A.M.; Banu, N. Protective Effect of *Solanum nigrum* Leaves Extract on Immobilization Stress Induced Changes in Rat's Brain. *Evid. Based Complement. Altern. Med.* **2014**, *2014*, 912450. [CrossRef]
19. Upadhyay, P.; Ara, S.; Prakash, P. Antibacterial and Antioxidant Activity of Solanum nigrum Stem and Leaves. *Chem. Sci.* **2015**, *4*, 1013–1017.
20. Sun, A.Y.; Wang, Q.; Simonyi, A.; Sun, G.Y. Botanical Phenolics and Neurodegeneration. In *Herbal Medicine Biomolecular and Clinical Aspects*, 2nd ed.; Benzie, I.F.F., Sissi Wachtel-Galor, S., Eds.; CRC Press Taylor & Francis Group: Boca Raton, FL, USA, 2011; pp. 315–325.

21. Ogunsuyi, O.B.; Ademiluyi, A.O.; Oboh, G.; Oyeleye, S.I.; Dada, A.F. Green leafy vegetables from two Solanum spp. (*Solanum nigrum* L and *Solanum macrocarpon* L.) ameliorate scopolamine-induced cognitive and neurochemical impairments in rats. *Food Sci. Nutr.* **2018**, *6*, 860–870. [CrossRef]
22. Lobo, V.; Patil, A.; Phatak, A.; Chndra, N. Free radicals, antioxidants and functional foods: Impact on human health. *Pharmacogn. Rev.* **2010**, *4*, 118–126. [CrossRef] [PubMed]
23. Zitka, O.; Skalickova, S.; Gumulec, J.; Masarik, M.; Adam, V.; Hubalek, J.; Trnkova, L.; Kruseova, J.; Eckschlager, T.; Kizek, R. Redox status expressed as GSH:GSSG ratio as a marker for oxidative stress in paediatric tumour patients. *Oncol Lett.* **2012**, *4*, 1247–1253. [CrossRef] [PubMed]
24. Liu, W.; Duan, X.; Fang, X.; Shang, W.; Tong, C. Mitochondrial protein import regulates cytosolic protein homeostasis and neuronal integrity. *Autophagy* **2018**, *14*, 1293–1309. [CrossRef] [PubMed]
25. Walls, A.B.; Waagepetersen, H.S.; Bak, L.K.; Schousboe, A.; Sonnewald, U. The glutamine glutamate/GABA cycle: Function, regional differences in glutamate and GABA production and effects of interference with GABA metabolism. *Neurochem. Res.* **2015**, *40*, 402–409. [CrossRef] [PubMed]
26. Devinsky, O.; Vezzani, A.; Najjar, S.; De Lanerolle, N.C.; Rogawski, M.A. Glia and epilepsy: Excitability and inflammation. *Trends Neurosci.* **2013**, *36*, 174–184. [CrossRef] [PubMed]
27. Schousboe, A.; Bak, L.K.; Waagepetersen, H.S. Astrocytic control of biosynthesis and turnover of the neurotransmitters glutamate and GABA. *Front. Endocrinol.* **2013**, *4*, 102. [CrossRef] [PubMed]
28. Agbor, G.A.; Vinson, J.A.; Donnelly, P.E. Folin-Ciocalteau Reagent for Polyphenolic Assay. *Int. J. Food Sci. Nutr. Diet.* **2014**, *3*, 147–156. [CrossRef]
29. Sak, K. Dependence of DPPH radical scavenging activity of dietary flavonoid quercetin on reaction environment. *Mini-Rev. Med. Chem.* **2014**, *14*, 494–504. [CrossRef] [PubMed]
30. Acquaviva, R.; Russo, A.; Galvano, F.; Galvano, G.; Barcellona, M.L.; Li Volti, G.; Vanella, A. Cyanidin and cyanidin 3-O-beta-D -glucoside as DNA cleavage protectors and antioxidants. *Cell Biol. Toxicol.* **2003**, *19*, 243–252. [CrossRef] [PubMed]
31. Campisi, A.; Caccamo, D.; Raciti, G.; Cannavò, G.; Macaione, V.; Currò, M.; Macaione, S.; Vanella, A.; Ientile, R. Glutamate-induced increases in transglutaminase activity in primary cultures of astroglial cells. *Brain Res.* **2003**, *978*, 24–30. [CrossRef]
32. Li Volti, G.; Ientile, R.; Abraham, N.G.; Vanella, A.; Cannavò, G.; Mazza, F.; Currò, M.; Raciti, G.; Avola, R.; Campisi, A. Immunocytochemical localization and expression of heme oxygenase-1 in primary astroglial cell cultures during differentiation: Effect of glutamate. *Biochem. Biophys. Res. Commun.* **2004**, *315*, 517–524. [CrossRef]
33. Acquaviva, R.; Campisi, A.; Murabito, P.; Raciti, G.; Avola, R.; Mangiamel, S.; Musumeci, I.; Barcellona, M.L.; Vanella, A.; Li Volti, G. Propofol attenuates peroxynitrite-mediated DNA damage and apoptosis in cultured astrocytes: An alternative protective mechanism. *Anesthesiology* **2004**, *101*, 1363–1371. [CrossRef] [PubMed]
34. Murphy, T.H.; Baraban, J.M. Glutamate toxicity in immature cortical neurons precedes development of glutamate receptor currents. *Brain Res. Dev. Brain Res.* **1990**, *57*, 146–150. [CrossRef]
35. Chen, Y.H.; Du, G.H.; Zhang, J.T. Salvianolic acid B protects brain against injuries caused by ischemia-reperfusion in rats. *Acta Pharmacol. Sin.* **2000**, *21*, 463–466. [PubMed]
36. Moloney, J.N.; Cotter, T.G. ROS signalling in the biology of cancer. *Semin. Cell Dev. Biol.* **2018**, *80*, 50–64. [CrossRef] [PubMed]
37. Schieber, M.; Chandel, N.S. ROS function in redox signaling and oxidative stress. *Curr. Biol.* **2014**, *24*, 453–462. [CrossRef] [PubMed]
38. Nistico, S.; Ventrice, D.; Dagostino, C.; Lauro, F.; Ilari, S.; Gliozzi, M.; Strongoli, M.C.; Vecchio, I.; Rizzo, M.; Mollace, V.; Muscoli, C.; et al. Effect of MN (III) tetrakis (4-benzoic acid) porphyrin by photodynamically generated free radicals on SODs keratinocytes. *J. Biol. Regul. Homeost. Agents* **2013**, *27*, 781–790. [PubMed]
39. Del Río, L.A. ROS and RNS in plant physiology: An overview. *J. Exp. Bot.* **2015**, *66*, 2827–2837. [CrossRef] [PubMed]
40. Carresi, C.; Musolino, V.; Gliozzi, M.; Maiuolo, J.; Mollace, R.; Nucera, S.; Maretta, A.; Sergi, D.; Muscoli, S.; Gratteri, S.; et al. Anti-oxidant effect of bergamot polyphenolic fraction counteracts doxorubicin-induced cardiomyopathy: Role of autophagy and c-kitposCD45negCD31neg cardiac stem cell activation. *J. Mol. Cell. Cardiol.* **2018**, *119*, 10–18. [CrossRef] [PubMed]
41. Sharma, A.; Kaur, M.; Katnoria, J.K.; Nagpal, A.K. Polyphenols in Food: Cancer Prevention and Apoptosis Induction. *Curr. Med. Chem.* **2018**, *25*, 4740–4757. [CrossRef]

42. Loffredo, L.; Perri, L.; Nocella, C.; Violi, F. Antioxidant and antiplatelet activity by polyphenol-rich nutrients: Focus on extra virgin olive oil and cocoa. *Br. J. Clin. Pharmacol.* **2017**, *83*, 96–102. [CrossRef] [PubMed]
43. Rizzo, M.; Ventrice, D.; Giannetto, F.; Cirinnà, S.; Santagati, N.A.; Procopio, A.; Mollace, V.; Muscoli, C. Antioxidant activity of oleuropein and semisynthetic acetyl-derivatives determined by measuring malondialdehyde in rat brain. *J. Pharm. Pharmacol.* **2017**, *69*, 1502–1512. [CrossRef] [PubMed]
44. Muscoli, C.; Lauro, F.; Dagostino, C.; Ilari, S.; Giancotti, L.A.; Gliozzi, M.; Costa, N.; Carresi, C.; Musolino, V.; Casale, F.; et al. Olea Europea-derived phenolic products attenuate antinociceptive morphine tolerance: An innovative strategic approach to treat cancer pain. *J. Biol. Regul. Homeost. Agents* **2014**, *28*, 105–116. [PubMed]
45. Zhang, Y.; Li, X.; Wang, Z. Antioxidant activities of leaf extract of Salvia miltiorrhiza Bunge and related phenolic constituents. *Food Chem. Toxicol.* **2010**, *48*, 2656–2662. [CrossRef] [PubMed]
46. Campisi, A.; Acquaviva, R.; Mastojeni, S.; Raciti, G.; Vanella, A.; De Pasquale, R.; Puglisi, S.; Iauk, L. Effect of berberine and Berberis aetnensis C. Presl. alkaloid extract on glutamate-evoked tissue transglutaminase up-regulation in astroglial cell cultures. *Phytother. Res.* **2001**, *25*, 816–820. [CrossRef]
47. Pierozan, P.; Biasibetti, H.; Schmitz, F.; Ávila, H.; Parisi, M.M.; Barbe-Tuana, F.; Wyse, A.T.; Pessoa-Pureur, R. Quinolinic acid neurotoxicity: Differential roles of astrocytes and microglia via FGF-2-mediated signaling in redox-linked cytoskeletal changes. *Biochim. Biophys. Acta* **2016**, *1863*, 3001–3014. [CrossRef] [PubMed]
48. Sandhu, J.K.; Pandey, S.; Ribecco-Lutkiewicz, M.; Monette, R.; Borowy-Borowski, H.; Walker, P.R.; Sikorska, M. Molecular mechanisms of glutamate neurotoxicity in mixed cultures of NT2-derived neurons and astrocytes: Protective effects of coenzyme Q10. *J. Neurosci. Res.* **2003**, *72*, 691–703. [CrossRef] [PubMed]
49. Campisi, A.; Caccamo, D.; Li Volti, G.; Currò, M.; Parisi, G.; Avola, R.; Vanella, A.; Ientile, R. Glutamate-evoked redox state alterations are involved in tissue transglutaminase upregulation in primary astrocyte cultures. *FEBS Lett.* **2004**, *578*, 80–84. [CrossRef] [PubMed]

© 2019 by the authors. Licensee MDPI, Basel, Switzerland. This article is an open access article distributed under the terms and conditions of the Creative Commons Attribution (CC BY) license (http://creativecommons.org/licenses/by/4.0/).

Article

Effect of a Milk-Based Fruit Beverage Enriched with Plant Sterols and/or Galactooligosaccharides in a Murine Chronic Colitis Model

Gabriel López-García [1], Antonio Cilla [1,*], Reyes Barberá [1], Amparo Alegría [1] and María C. Recio [2]

[1] Nutrition and Food Science Area, Faculty of Pharmacy, University of Valencia, Avda. Vicente Andrés Estellés/n, 46100 Burjassot (Valencia), Spain; gabriel.lopez@uv.es (G.L.-G.); reyes.barbera@uv.es (R.B.); amparo.alegria@uv.es (A.A.)
[2] Department of Pharmacology, Faculty of Pharmacy, University of Valencia, Avda. Vicente Andrés Estellés/n, 46100 Burjassot (Valencia), Spain; maria.c.recio@uv.es
* Correspondence: antonio.cilla@uv.es; Tel.: +34-963-544-972

Received: 7 March 2019; Accepted: 1 April 2019; Published: 4 April 2019

Abstract: The potential anti-inflammatory effect of plant sterols (PS) enriched milk-based fruit beverages (PS, 1 g/100 mL) (MfB) with/without galactooligosaccharides (GOS, 2 g/100 mL) (MfB-G) in an experimental mice model of chronic ulcerative colitis was evaluated. Beverages were orally administered to mice every day by gavage to achieve PS and GOS doses of 35 and 90 mg/kg, respectively, and experimental colitis was induced by giving mice drinking water *ad libitum* containing 2% (w/v) dextran sulphate sodium (DSS) for 7 days, alternating with periods without DSS up to the end of the study (56 days). MfB beverage showed significant reduction of symptoms associated to ulcerative colitis and improved the colon shortening and mucosal colonic damage, but it was not able to reduce the increase of myeloperoxidase levels produced by DSS. MfB-G showed higher incidence of bloody feces and loss of stool consistency than MfB, as well as high levels of immune cells infiltration in colon tissue and myeloperoxidase. Therefore, PS-enriched milk-based fruit beverage could be an interesting healthy food to extend the remission periods of the diseases and the need to evaluate, in a pre-clinical model, the anti-inflammatory effect of the combination of bioactive compounds in the context of a whole food matrix.

Keywords: milk-based fruit beverage; plant sterols; galactooligosaccharides; mice; chronic ulcerative colitis

1. Introduction

Ulcerative colitis (UC) is a chronic inflammatory disease, within intestinal bowel diseases (IBD), characterized by a tissue destruction of large intestine with relapsing phases followed by periods of remission [1]. The most common treatment strategies used include pharmacological therapy (steroidal and non-steroidal drugs) or, in the worst cases, removing portions of affected gastrointestinal tract through surgery. Although, pharmacological therapy has shown efficacies in ameliorating the severity and symptoms of IBD, typically does not lead to long periods of remission. Consequently, recent investigations have focused on the study of the impact of diet and healthy foods as potential alternative, and even the possible combination of nutraceuticals and healthy foods with pharmacological therapy in cases when the patient is not eligible for conventional therapy which could help extend remission periods and improve life quality of IBD patients [2–6].

A recent meta-analysis has associated a high intake of fruit and vegetables with a reduction of UC in European population, suggesting that the intake of their bioactive compounds (fiber, antioxidant vitamins and phytochemicals such as polyphenols, carotenoids, isoflavones) could explain this inverse association [7]. In murine colitis models, which mimic human pathology, it has been demonstrated the preventive effect on pro-inflammatory intestinal process associated to the intake of mango [8] or pineapple [9] juices, polyphenols-rich extracts from orange [10] or its sub-products [11], apple [12] and pomegranate [13]. Polyphenols seems to be the main bioactive compounds involved in the anti-colitic action of fruits, due to its ability to inhibit some pivotal pro-inflammatory mediators as nuclear transcriptional Factor-κB (NF-κB) and specific cytokines (Tumor Necrosis Factor-α (TNF-α), and interleukin 1-β (IL-1β)) and the induction of antioxidant defence systems. Furthermore, lipophilic compounds present in vegetables and fruits, such as fat-soluble vitamins (pro-vitamin A) [14], carotenoids (β-carotene) [15] and plant sterols (PS) [16–21], have displayed important anti-inflammatory and inmmunodulatory effects. In particular, PS could be a good strategy to prevent IBD, because they show a very little absorption in the intestine (0.5–2%) and reach the colon, where could exert anti-inflammatory effects. Moreover, the European Union authorized PS addition into milk-based fruit beverages [22], among other foods, to achieve the necessary amount of 1.5–3 g PS/day (not attainable with normal diet) for the well-known cholesterol-lowering effect [23]. PS have shown promising results in animal colitis models induced by trinitrobenzene sulfonic acid (TNBS) [16,17], dextran sodium sulphate (DSS) [18–24] or high-fat diet (40–60%) [19–21]. Doses between 20–150 mg/Kg/day help to mitigate some important parameters associated to UC such as symptoms, colon shortening and presence of edema, histopathology and myeloperoxidase (MPO) activity on colon tissue.

Other bioactive compounds that have gained attention are prebiotics, which stimulate selectively beneficial intestinal bacteria, helping to maintain intestinal mucosal barrier and enhance defense against pathogenic microorganism [25]. In this sense, galactooligosacharides (GOS) have been proposed as an active ingredient to prevent or alleviate symptoms associated with IBD [26]. Human clinical trials have demonstrated that consumption of GOS contained in chocolate or banana flavored chews (3.5–7 g/day) [27] and sachets (5.5 g/day) [28,29], reduce different pro-inflammatory markers such as calprotectin in feces, plasma C-reactive protein [28] and cytokines (IL-6, IL-1β and TNF-α) [29], with a concomitant increase in levels of fecal secretory immunoglobulin A (sIgA) [28] and reduction of common ulcerative colitis symptoms [27]. However, in animal studies the effect of GOS on colitis is controversial. GOS (4 g/Kg/day) administration for 10 days increase bifidobacterial number in colon in a TNBS colitis induced murine model, but it was not linked with a reduction of pro-inflammatory markers as MPO activity and colon damage [30]. On the other hand, in mice smad-3 feed with GOS (5 g/Kg/day) for 42 days an improvement of colitis severity and increase of natural killer activity were observed after colitis induction by *Helicobacter hepaticus* [31].

Enrichment of healthy foods with bioactive compounds could be an effective strategy to help mitigate the symptoms associated with IBD and extend remission periods [3]. In this sense, milk-based fruit beverages could be a suitable nutritional matrix for this purpose due to its low-fat content and good sources of bioactive compounds such as polyphenols, carotenoids and vitamins. Moreover, previous studies of our research group have demonstrated the beneficial effect of PS enriched milk-based fruit beverages on oxidative stress prevention and intestinal epithelia integrity maintenance in Caco-2 cells [32], as well as, their systemic anti-inflammatory properties, in hypercholesterolemic post-menopausal women, through the serum increase of anti-inflammatory cytokine IL-10 with concomitant reduction of IL-1β, cytokine which play an important role on the development of IBD [33].

The present study investigates for the first time, the effect of the daily intake of a healthy food, as milk-based fruit beverages, enriched with bioactive compounds (PS and/or GOS) on clinical symptoms and inflammatory process of UC, using a pre-clinical chronic murine model induced by DSS. The aim of this preliminary study is to evaluate if PS enriched milk-based fruit beverages with/without GOS are a suitable dietetic strategy to help to mitigate the health problems associated with UC.

2. Materials and Methods

2.1. PS Milk-Based Fruit Beverage Formulation

Two Ps-enriched (1 g/100 mL beverage) milk-based fruit beverages were used: MfB and MfB-G, without or with addition of GOS (1.8 g/100 mL beverage), respectively. Both beverages contained skimmed milk with the addition of milk fat and whey protein concentrate enriched with milk fat globule membrane (MFGM, Lacprodan® MFGM-10 from Arla Foods Ingredients, Sønderhøj, Denmark), mandarin juice from concentrate (45%) (Source of β-cryptoxanthin 205 µg/100 mL beverage) and banana puree (4%). PS were added during beverage formulation as microencapsulated free microcrystalline PS (β-sitosterol 80%; sitostanol 12%; campesterol; 7%; campestanol 1% and stigmasterol 0.7%) from tall oil in a powder form (Lipohytol® 146 ME Dispersible from Lipofoods, Barcelona, Spain). GOS syrup (Vivinal® GOS from Friesland Campina Ingredients, Amersfoort, Netherland) containing approximately 59% GOS, 21% lactose, 19% glucose and 1% galactose based on dry matter. In order to guarantee microbiological stability, beverages were subjected to pasteurization (90 °C for 30 s) and filled aseptically in 250-mL tetra bricks.

2.2. Animals and Treatment

Female C57BL/6 mice, aged 6–8 weeks with weights ranging between 18 to 20 g, provided by Harlan Interfauna Iberica (Barcelona, Spain) were used in this study. Animals were acclimatized for 7 days under a 12-h light/ dark cycle at 22 °C/60% humidity and fed with a standard laboratory rodent diet and water *ad libitum* before experiments. All experiments were approved by the Institutional Ethics Committee of the University of Valencia, Spain (2017/VSC/PEA/00143).

Chronic colitis induction was performed by repeated administration of a 2% (*w*/*v*) of DSS (in drinking water) for 7-day cycles according to Marín et al. [34] The DSS phases were interrupted by 7-days cycles of access to drinking water without DSS. MfB and MfB-G were orally administered everyday by gavage (0.15 mL) to achieve 35 and 95 mg/Kg body weight for PS and GOS, respectively (see Figure 1). Animals were randomly assigned into six intervention groups with 8 mice/group (*n* = 48): control (mice received water during all experiment), DSS (access to drinking water containing 2% DSS), DSS + MfB (drinking water containing 2% DSS + 0.15 mL of MfB), DSS + MfB-G (drinking water containing 2% DSS + 0.15 mL of MfB-G), MfB (received only 0.15 mL of MfB) and MfB-G (received only 0.15 mL of MfB-G).

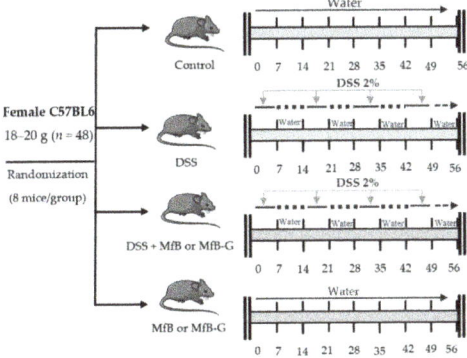

Figure 1. Experimental design of the chronic colitis induction by dextran sulphate sodium (DSS) (2%, *w*/*v*) and plant sterol enriched milk-based fruit beverages without (MfB) or with galactooligosacharides (MfB-G) administration in the procedures.

Water intake was monitored three times a week to guarantee equitative DSS intake among different colitis mice groups and the type of beverages did not affect drink behavior. No statistically significant differences ($p > 0.05$) in water intake was found among different mice group (ml/mouse/day); control (4.3 ± 0.5), DSS (4.2 ± 0.8), DSS + MfB (4.5 ± 0.7), DSS+MfB-G (4.8 ± 1.1), MfB (4.6 ± 0.4) and MfB-G (4.5 ± 0.5).

2.3. Disease Activity Index (DAI)

The disease activity index (DAI) was determined according to the parameters outlined in Table 1. Animal body weights, stool consistency and visible blood in feces were recorded three times a week during all experiments. Stool consistency and blood in feces parameters were evaluated checking fresh feces contained in each animal group cage and scored following the method described by Marín et al. [34]. Also, blood presence around mouse perianal area was taken as indicator of blood in feces. DAI was calculated by combining scores (WS: weight loss; SC: stool consistency; BF: bloody feces) following the next formula DAI: (WS + SC + BF)/3 according to the methodology described by Marín et al. [34].

Table 1. Scoring system to calculate the disease activity index (DAI *).

Score	Weight Loss (%)	Stool Consistency	Visible Blood in Feces
0	None	Normal	None
1	1–5		
2	6–10	Loose	Slight bleeding
3	11–20		
4	<20	Diarrhea	Gross bleeding

* DAI was calculated by combining scores (WS: weight loss; SC: stool consistency; BF: bloody feces).

2.4. Colon Shortening and Presence of Edema

Animals were sacrificed at day 56 by cervical dislocation and their colons (ileo-cecal junction to anal verge) were removed. Fecal residues were washed with cold phosphate buffered saline (PBS) and the length and weight were measured. Presence of edema in colonic tissue was evaluated working out the weight/length (mg/cm) ratio. Then, colons were cut in four portions of approximately 2 cm and frozen immediately at $-80\,^{\circ}\text{C}$ until use.

2.5. Histological Analysis

Portions of approximately 2 cm from the colon of every group of mice were fixed in freshly prepared 4% formaldehyde in PBS buffer (pH 7.2), embedded in paraffin, and sectioned at 4 µm. Sections were stained with hematoxylin and eosin for histopathological analysis and examined by light microscopy [35]. The slides were analysed and scored as previously described Cooper et al. [36] according to the criteria listed in Table 2. Scores were calculated by adding the score for the three parameters, giving a maximum score of 10.

Table 2. Scoring system used to calculate the histological score in dextran sulphate sodium (DSS)-induced colitis.

Histological Scoring System for DSS-Induced Colitis		
Feature	Score	Description
Severity of inflammation	0	None
	1	Mild
	2	Moderate
	3	Severe
Extent of inflammation	0	None
	1	Mucosa
	2	Mucosa and submucosa
	3	Transmural

Table 2. Cont.

Histological Scoring System for DSS-Induced Colitis		
Feature	Score	Description
Crypt damage	0	None
	1	1/3 damages
	2	2/3 damaged
	3	Crypts lost, surface and epithelium present
	4	Crypt and surface epithelium lost

Histological score is calculated by adding the score of three parameters up to a total score of 10 points. DSS: dextran sulphate sodium;

2.6. Myeloperoxidase (MPO) Assay

In the model of experimental colitis, neutrophil infiltration into the colon was assessed indirectly by measuring myeloperoxidase (MPO) following the method elaborated by Bradley et al. [37]. Colon samples were weighed and blended in homogenizer with phosphate buffer (0.22 M) containing 0.5% of hexadecyl-trimethyl-ammonium bromide. Samples were centrifuged at 12000 g/4 °C/20 min and supernatants were placed on 96-well plates. Estimation of MPO content in colonic tissues was evaluated spectrophotometrically (450 nm) by measuring the 3,3',5,5'-tetramethyl-benzidine (MPO substrate) oxidation in presence of H_2O_2 [34].

2.7. Statistical Analysis

The results are expressed as means ± standard deviation values ($n = 8$ for each group). One-way analysis of variance (ANOVA) followed by the Tukey post-hoc test was applied to determine differences among treatments. A significance level of $p < 0.05$ was adopted for all comparisons, and the Statgraphics Centurion XVI.I statistical package (Statpoint Technologies Inc., VA, USA) was used throughout.

3. Results

3.1. Evaluation of Clinical Symptoms of Induced Chronic Colitis in Mice

Mice exposed to DSS in drinking water (2% w/v) showed marked clinical symptoms (DAI values) during all experiment, being slightly lower in the first DSS cycle (7 days) (see Figure 2a). Administration of MfB significantly reduced DAI values between 24–67% compared to DSS group, with the exception of the first DSS cycle. MfB-G beverage attenuated DSS derived symptoms during the first cycles (7 and 21 days), reducing DAI values between 40–54% with respect to the DSS group. However, from the third DSS cycle (35 days) onwards, an increase of DAI (1.6 fold compared to the DSS group) was observed. With the aim to evaluate the effect of beverages on specific illness derived symptoms, Figure 2b,c show the evolution of mice weight, stool consistency and presence of blood in feces, respectively. In general, MfB and MfB-G administration did not have any differences in terms of weight loss compared to DSS group during the study, with the exception of MfB-G beverage, where a statistical ($p < 0.05$) higher weight loss at 10 day (20% vs. DSS group) was observed. Both beverages statistically ($p < 0.05$) reduced the presence of blood in feces (4 to 6-fold) during the acute phase of the illness (7 days) with respect to the DSS group (Figure 2c). However, in subsequent DSS cycles, this protective effect was not observed with both beverages and sometimes the incidence of blood in faces increased in the MfB-G mice group (third and fifth cycle). With respect to stool consistency, MfB prevented completely the appearance of diarrhea during the first two DSS cycles (7 and 21 days) and reduced up to 50% their incidence during the third DSS cycle, compared to DSS group, (see Figure 2c). However, after fourth (49 days) DSS cycle MfB did not show differences in the stool consistency in comparison to DSS group, although a slight beneficial effect was observed at the end of the study. On the other hand, MfB-G partially prevented the loss of stool consistency during the second DSS cycle, but this loss rapidly increased in the subsequent DSS cycles after which stool consistency remained constant for the rest of the study. The loss of stool consistency observed in the MfB-G group during the third cycle coincides with the increase of blood presence in feces. This suggests that the higher acute diarrhea observed is related to colonic epithelium erosion.

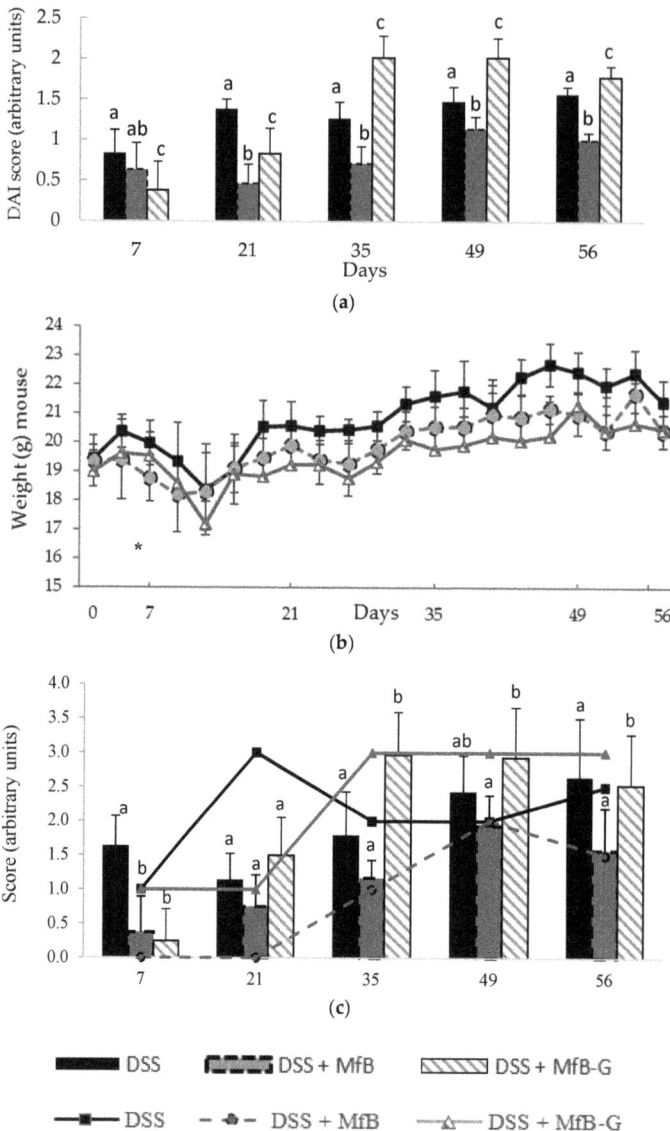

Figure 2. Effect of plant sterol enriched milk-based fruit beverages without (MfB) or with galactooligosacharides (MfB-G) on clinical symptoms in DSS-induced chronic colitis model. (**a**) Disease activity index (DAI) score after DSS (dextran sulphate sodium) administration in each cycle; (**b**) body weight per mouse during all the experiments; (**c**) stool consistency loss (lines) and presence of blood in feces (bars) at the end of each DSS administration cycle. Bars/markers show the mean ± standard deviation (n = 8). Different lowercase letters (a–c) indicate statistically significant differences ($p < 0.05$) among mice groups (DSS, DSS + MfB and DSS + MfB-G) within the same DSS cycle. (*) Denotes statistically significant differences ($p < 0.05$), compared to the DSS group, at the same day.

3.2. Colon Length Shortening and Presence of Edema

Figure 3a shows that mice exposed to DSS (6.10 ± 0.28 cm), DSS + MfB (7.20 ± 0.35 cm) and DSS + MfB-G (6.33 + 0.35 cm), suffered a significant shortening ($p < 0.05$) of colon length in comparison to the control (8.53 ± 0.29 cm). DSS + MfB mice showed longer colon (18%) than DSS mice, significantly preventing ($p < 0.05$) the colon shortening induced by DSS. Mice that received DSS showed higher presence of edema in colonic tissue than the control group (39–50 vs. 28 mg/cm, respectively) (see Figure 3b). Administration of beverages did not show any significant ($p > 0.05$) protective effect with respect to edema development. Mice that received MfB and MfB-G showed very similar colon length and tissue edema compared to control, which is according with the absence of clinical symptoms observed on these mice groups.

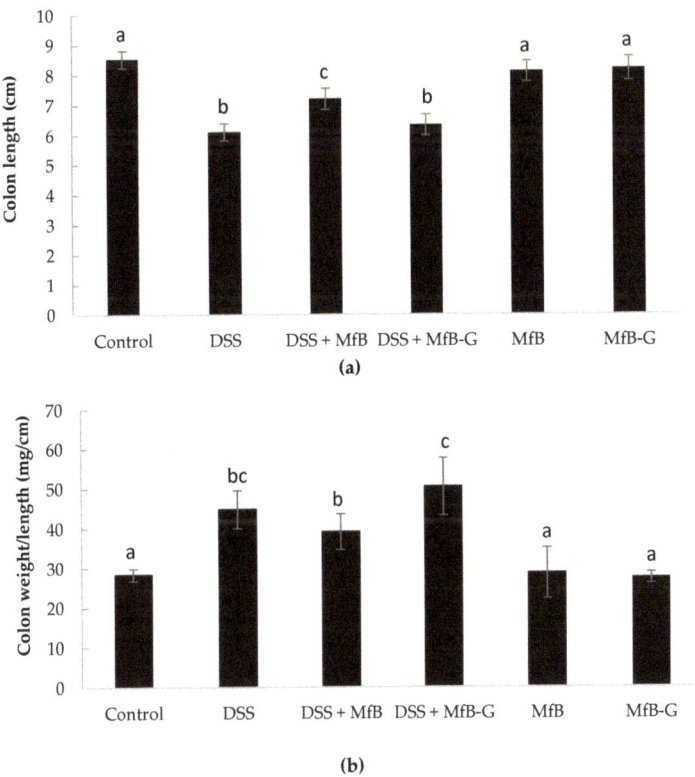

Figure 3. Evaluation of colon length (a) and presence of edema (b) after administration of plant sterol enriched milk-based fruit beverages without (MfB) or with galactooligosacharides (MfB-G) for 56 days on dextran sulphate sodium (DSS)- chronic colitis induction. Bars show the mean ± standard deviation ($n = 8$). Different lowercase letters (a–c) indicate statistically significant differences ($p < 0.05$) among samples.

3.3. Histopathological Analysis

As is shown in Figure 4A, control, MfB and MfB-G mice groups (Figure 4a,c,e, respectively) had a morphologically normal colon without signs, or very low levels of leucocyte infiltration. In contrast, DSS group (Figure 4b) suffered a marked inflammation characterized by a loss of epithelial architecture, reduction of number of globet cells and crypts, as well as a strong increase of immune cell infiltration compared to control (Figure 4a), what it was reflected in a high histological score (9/10 points). Treatment with DSS + MfB (Figure 4d) or DSS + MfB-G (Figure 4f) mice showed similar histopathology alteration compared to DSS group (b), with immune cell infiltration and distortion of crypts, although the epithelial architecture was more preserved resulting in a significant ($p < 0.05$) lower histological score (34 and 10%, respectively) values. However, DSS + MfB-G (Figure 4f) mice showed a higher histological score (25%) and damage in the epithelial architecture, with distortion of crypts and high presence of immune cell in comparison to the DSS + MfB group (Figure 4d).

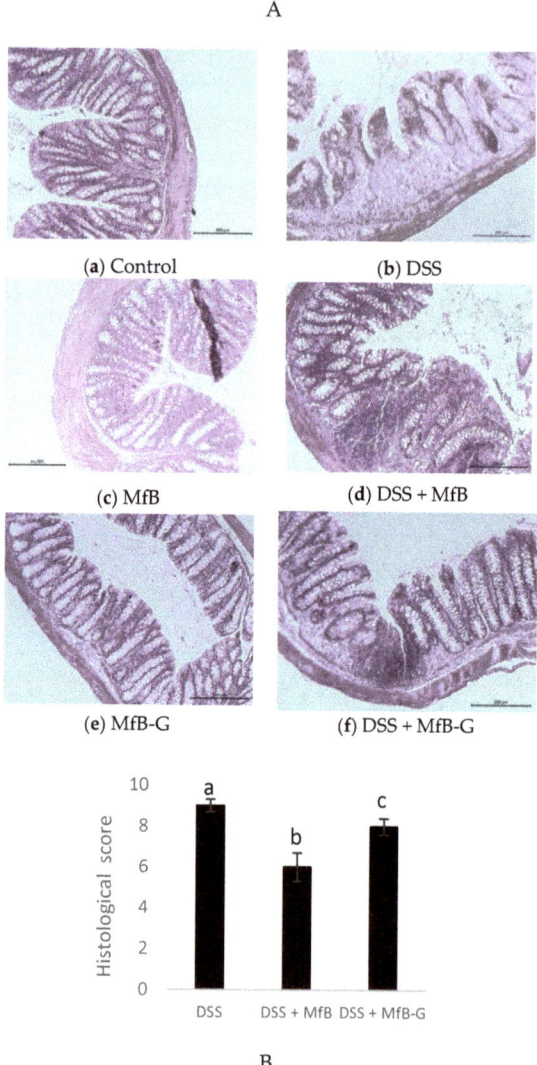

Figure 4. The representative photographs of hematoxylin and eosin staining (magnification ×200) (**A**) and the corresponding score (**B**) of colon tissues obtained after administration of plant sterol enriched milk-based fruit beverages without (MfB) or with galactooligosacharides (MfB-G) for 56 days on dextran sulphate sodium (DSS)- chronic colitis induction. Data are expressed as the means ± standard deviation of six mice in each group. Representative images of staining of colon tissues from each ($n = 3$) $p < 0.05$ vs. normal control. Different lowercase letters (a–c) indicate statistically significant differences ($p < 0.05$) among samples.

3.4. Presence of MPO in Colonic Tissue

Similar to that observed in the histopathological analysis, mice receiving DSS showed high levels of immune cells infiltration, raising from 2 to 3.5-fold MPO levels with respect to control (see Figure 5). Daily administration of beverages (DSS + MfB or DSS + MfB-G groups) failed to attenuate the increase of MPO levels induced by DSS. MfB and MfB-G mice groups had very similar MPO values with respect to the control, suggesting that both beverages apparently do not trigger an inflammatory response in the colon in the absence of colitis.

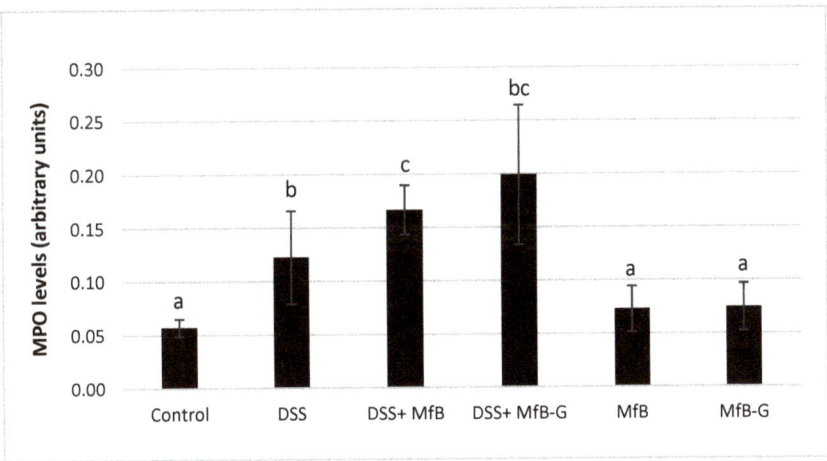

Figure 5. Presence of myeloperoxidase (MPO) levels in colonic tissue after administration of plant sterol enriched milk-based fruit beverages without (MfB) or with galactooligosacharides (MfB-G) for 56 days on dextran sulphate sodium (DSS)- chronic colitis induction. Bars show the mean ± standard deviation ($n = 8$). Different lowercase letters (a–c) indicate statistically significant differences ($p < 0.05$) among samples.

4. Discussion

To investigate the potential anti-inflammatory effects of PS enriched milk-based fruit beverages with or without GOS on UC, a model of chronic colitis induced by DSS in mice was selected. Although colitis animal models do not represent the complexity of human disease, they provide valuable information about factors involved in the inflammatory process and allow to evaluate different therapeutic strategies to improve life quality of patients with IBD [38]. In particular, the DSS-induced colitis model can easily reproduce the acute and chronic phases or the relapsing periods typical of UC, depending on the concentration and cycles of exposition to DSS in drinking water. Moreover, this model exhibits similar symptoms to those of human UC (diarrhea, bloody feces and body weight loss) and histological features such as mononuclear leucocyte infiltration, crypt architectural disruption and epithelial degeneration [39].

Our study showed that daily administration of MfB beverage resulted in a significant ($p < 0.05$) reduction of symptoms associated to UC, mainly preventing the presence of diarrhea or alleviating their increase during all experiment (vs. DSS group), as well as, protecting against presence of bloody feces in the acute phase of the disease (first DSS cycle). Reduction of clinical symptoms led to partially prevention of colon shortening induced by DSS cycles, showing longer colons (18%) compared to DSS group. Histopathological analysis of the colon revealed a slight improvement in the architecture of the colonic epithelium and a greater number of crypts compared to the DSS group, in agreement with the previous parameters mentioned. However, the distribution and morphology of the crypts were altered, as well as an increase of neutrophil infiltration and a high MPO level in the colonic tissue was observed. Stimulation of neutrophil activity or its migration into colon tissue, could be related with the prevention of colon shortening observed after MfB treatment. Neutrophils are considered the first line of defence against microorganisms and recently, it has been demonstrated that they can build a complex formed by chromatin and neutrophil proteins that act as immunomodulator and activate immune cells such as T cells, although their specific role on IBD has not been well described yet [40]. Beneficial effect of MfB upon DSS-induced colitis can also be related with its cytoprotective effect at intestinal level. In a previous in vitro study by our research group, it was demonstrated that milk-based fruit beverages had several beneficial effects against oxidative stress and prevention of cell dead, inhibiting some important pro-apoptotic events and preserving cell monolayer integrity in a differentiated colon cancer (Caco-2) cell model [32]. This fact could be important because DSS is a toxic compound to colonic epithelial cells that cause an increase of apoptotic cells and compromise the epithelial barrier integrity through the loss of some important proteins present in the tight junctions [41].

It is important to note our study design does not allow knowing which bioactive compound/s contained in the beverages has the anti-inflammatory effect on DSS-induced colitis. Nevertheless, it may be possible that the specific combination of all of them produce the effect observed in our study. The presence of mandarin juice (represents almost half of MfB composition), could contribute to the anti-inflammatory effect since contains important quantities of antioxidant phytochemicals such as flavonoids (mainly hesperidin, narirutin and vicenin-2) and β-cryptoxanthin (β-Cx) [42]. In this sense, flavonoid-rich extracts (containing mainly hesperidin, narirutin and vicenin-2) obtained from blood orange juice administrated (40 mg/kg/day) to CD1 mice with colitis-induced by dinitrobenzene sulfonic acid, have shown preventive effects against the colonic pathological tissue damage. Flavonoids acted mainly counteracting NF-κB signalling, decreasing expression of pro-apoptotic proteins (Bax) and restoring the redox balance in colonic tissue [10]. Similar effects were observed in a recent study with industrial orange by-products (citrus pectin and different sub-fractions obtained from orange after juice extraction) in DSS-treated mice [11]. Byproducts with high polyphenol total content and antioxidant capacity showed better anti-inflammatory effect in terms of clinical symptoms and reduction of pro-inflammatory mediators' expression (TNF-α, IL-1β, IL-6). On the other hand, the presence of β-Cx could contribute to the beneficial effects of the MfB, although its specific role on UC has not been studied yet. In steatohepatitis and insulin resistance murine models induced by high fat content diets, β-Cx administration (~2.5–7.5 mg/kg/day) lead to attenuation of lipotoxicity-induced inflammation, preventing hepatic tissue peroxidation (TBARS) and the macrophages activation [43], as well as the stimulation of antioxidant enzymes (catalase, superoxide dismutase and glutathione peroxidase) and inhibition of the expression of pro-inflammatory markers (NF-κB and TNFα) in liver [44]. Therefore, β-Cx is able to reduce the pro-inflammatory process through a direct and indirect anti-oxidant mechanism. Taking into account that oxidative stress has a pivotal role on UC, daily administration of MfB containing β-Cx (0.02 mg/kg/day) could help to mitigate the pro-inflammatory process.

Additionally, it is important to note that enrichment of MfB with PS could be a remarkable factor in its anti-colitic effect, since several studies using PS standard solutions (β-sitosterol, stigmasterol and γ-oryzanol) added to feed and administered at doses between 20–50 mg/kg/day (similar to our study 35 mg/kg/day) for 3-56 days have shown a marked anti-inflammatory effect independently of the colitis animal model used. In C57BL/6 mice, β-sitosterol (20 mg/kg/day) administration for 56 days prevented the colon shortening (~8%) and reduced MPO activity in colon tissue (~35%), what led to a lower level of pro-inflammatory cytokines (IL-1β, TNF-α and IL-6) after colitis induction by high fat diet (60 Kcal% from fat) [16]. Similarly, β-sitosterol administration (10 or 20 mg/kg) during 3 days to C57BL/6 mice with colon inflammation induced by TNBS prevented partially colon shortening (~3.4%) and improved the pro-inflammatory status, reducing pro-inflammatory cytokine levels (IL-1β, TNF-α and IL-6) and MPO activity in colon tissue (~42%), with a concomitant inhibition of NF-κB translocation into the nucleus [17]. In UC models induced by DSS, administration to C57BL/6 mice of γ-oryzanol (50 mg/Kg/day for 16 days), a mixture of phytosteryl ferulates derived from rice bran oil, mitigated clinical symptoms associated to UC and partially prevented colon shortening (~9%) and the pathophysiological activity during colonic inflammation through inhibition of NF-κB activation after 5 days of DSS at 3% (v/v) induction [24]. Differences observed in terms of MPO activity with respect to our study, could be attributed to the fact that PS are added into the beverages as a food ingredient composed of a complex mixture of PS (β-sitosterol, sitostanol, campesterol, campestanol and stigmasterol). As far as we know, there are no studies that evaluate the anti-inflammatory effect—in murine chronic colitis models—of some of the PS found in the food ingredient used in the PS enrichment of our beverages (sitostanol, campesterol and campestanol), which suppose 20% of the total PS. The small molecular structural differences among sitostanol, campesterol or campestanol, in comparison to β-sitosterol, could explain the lack of beneficial effect observed on the pro-inflammatory process. Feng et al. [21] observed that β-sitosterol or stigmasterol administration (0.4% w/w) in C57BL/6J mice similarly mitigate inflammation severity and macroscopic damage (colon shortening and histopathology score) induced by DSS (1.5% w/v, for 5 days), but only stigmasterol was able to reduce the expression of cyclooxygenase-2. Authors indicate that the presence of a double-bond in the side chain in stigmasterol could be responsible for their additional anti-inflammatory effect compared to β-sitosterol. In this sense, different behaviour of β-sitosterol, campesterol and stigmasterol at the same concentration (24 µM for 48 h) and cell system (macrophages) have been observed. Stigmasterol suppressed cytokine secretion into the supernatant, while β-sitosterol promoted it and campesterol did not have any effect [45]. Moreover, it is also possible that some specific compounds present in our beverages hidden the beneficial effects of PS on colitis. Llewellyn et al. [46] observed that administration of a high casein diet (41% w/w) for 14 days promoted intestinal barrier damage and increased colonic cytokines levels (IL-6 and TNF-α) in DSS induced model in mice (3% w/v for 7 days). Although, the casein intake is around 850-fold higher than our study, the longer administration time of beverage (14 vs. 56 days) and DSS (7 vs. 56 days) exposition could explain our results. Besides, the potential impact of other compounds present in the beverage on the colitis process cannot be ruled out.

Regarding the potential beneficial effect of GOS on DSS-induced colitis, results show that MfB-G administration during all the experiment does not confer additional beneficial effects with respect to MfB. DSS + MfB-G mice suffered a dramatic increase of clinical symptoms from the third DSS cycle, remaining constant up to the end of the study. This fact could explain the absence of a beneficial effect of MfB-G on colon shortening (12% shorter than MfB), the higher colonic mucosa alteration (distortion of crypts and high immune cells infiltration) and MPO level compared to MfB. The effect of GOS and mechanisms which could improve UC in murine colitis models have been poorly studied and currently are controversial. Holma et al. [30] reported that administration to rats of two kinds of GOS (whey and lactose derived from cow milk) at 4 g/kg/day for 10 days increased notably bifidobacterial number in feces, which are high producers of anti-inflammatory compounds, such as short-chain fatty acids. However, the increase of bifidobacterial number was not correlated with an improvement of colonic damage and they did not prevent immune cell infiltration (MPO activity) and edema induced by TNBS during 72 h. In contrast, in mice deficient in *smad-3* (phenotype characterized by colon moderate inflammatory response) infected by *Helicobater hepaticus*, GOS supplementation (5 g/kg/day) for 42 days reduced colitis severity preserving colon architecture by modulating the function and trafficking of natural killer cells [31]. The use of different animal species, agents of induction of colitis, treatment times and type of GOS could explain the differences found between the studies. However, it has been reported previously the harmful effect of fructooligosacharides (FOS) administration (at 6 g/day), a kind of soluble fibre similar to GOS, in rats with colitis induced by DSS (at 3% for seven days). Similar to our results, FOS slightly prevented symptoms associated to UC (13–45%, DAI value) at the beginning of the study but quickly worsened without showing differences with respect to DSS group up to the end of the study. FOS treatment did not prevent the colon shortening and exacerbated colon histological damage severity and MPO activity, as well as reduced crypt cell proliferation in the distal colon, which is an integral part of the colon repair process. Authors indicated that FOS delayed the onset of repair, promoting the harmful effect of DSS on colon epithelia [47]. Other possible hypothesis could be related with the bifidogenic effect of FOS, what lead to the increase of bifidobacterial genus in the colon. It has been indicated that a quick rate of FOS fermentation in the cecum produce an overproduction of organic acids as lactic and acetic acids [48], which can damage the intestinal epithelium, leading to an increase of colonic permeability [49]. Similar to FOS, it has been indicated that GOS fermentation (72 h) lead to a quick increase and accumulation of lactic and acetic acids in the colon, compared to propionic and butyric acids, in a dynamic in vitro colon model (TIM-2). These increases were correlated with an increase of bifidobacteria and lactobacilli genus [50]. Then, we hypothesized that MfB-G administration could produce a change in intestinal microbiota by increasing the organic acids that produce bacteria, which could exacerbate the loss of the epithelial barrier integrity and the colonic mucosal permeability induced by DSS. Moreover, the potential inhibitory effect of GOS on the colonic repair mechanism cannot be ruled out.

In summary, PS-enriched milk-based fruit beverage (MfB) shows a moderate anti-inflammatory effect, helping to alleviate the clinical symptoms associated to colitis, the colon shortening and colonic damage in a DSS-induced mice model of chronic colitis. Neutrophil infiltration in colonic tissue and MPO level remained higher after MfB treatment, suggesting that they could be involved in its anti-inflammatory action as a compensatory mechanism trying to overcome the damage induced by DSS. GOS addition to the MfB did not show any additional beneficial effects in comparison with MfB and even exacerbated the pro-inflammatory action induced by DSS. The higher DAI value, colonic mucosa damage, immune cell infiltration and MPO level, suggest that presence of GOS in colon compromise colonic epithelial permeability or delay the reparation colonic mechanism, promoting the cytotoxic effect of DSS on colon cells.

Our results demonstrate the importance to evaluate the biological effects of bioactive compounds in the context of a complex food matrix. PS-enriched foods could be a suitable strategy to extend remission periods and improve the quality of life of patients with UC, but further investigation is needed to confirm the beneficial role of PS in a food matrix on the UC.

Author Contributions: Conceptualization and funding acquisition: A.C., R.B., and A.A.; methodology: M.C.R.; analysis and experiments: G.L.-G., and M.C.R.; writing, review and editing: G. L-G., A.C., R.B., A.A., and M.C.R.

Funding: This research was funded by Spanish Ministry of Economy and Competitiveness, through National Project AGL2015-68006-C2-1-R (MINECO-FEDER). G.L.-G. holds a grant (ACIF/2016/449) from the Generalitat Valenciana (Spain).

Acknowledgments: The authors want to express their acknowledgement to the Central Service for Experimental Research (SCSIE) of the University of Valencia, in particular to Animal Production and Microscopy sections, for their support.

Conflicts of Interest: The authors declare no conflict of interest.

References

1. Vanga, R.; Long, M.D. Contemporary management of ulcerative colitis. *Gastroenterol. Rep.* **2018**, *20*, 12. [CrossRef]
2. Charlebois, A.; Rosenfeld, G.; Bressler, B. The impact of dietary interventions on the symptoms of inflammatory bowel disease: A systematic review. *Crit. Rev. Food Sci. Nutr.* **2016**, *56*, 1370–1378. [CrossRef] [PubMed]
3. Al Mijan, M.; Lim, B.O. Diets, functional foods, and nutraceuticals as alternative therapies for inflammatory bowel disease: Present status and future trends. *World J. Gastroenterol.* **2018**, *24*, 2673–2685. [CrossRef] [PubMed]
4. Santini, A.; Novellino, E.; Armini, V.; Ritieni, A. State of the art of Ready-to Use Therapeutic Food: A tool for nutraceuticals addition to foodstuff. *Food Chem.* **2013**, *140*, 843–849. [CrossRef] [PubMed]
5. Daliu, P.; Santini, A.; Novellino, E. From pharmaceuticals to nutraceuticals: Bridging disease prevention and management. *Expert Rev. Clin. Pharmacol.* **2019**, *12*, 1–7. [CrossRef] [PubMed]
6. Santini, A.; Cammarata, S.M.; Capone, G.; Ianaro, A.; Tenore, G.C.; Pani, L.; Novellino, E. Nutraceuticals: Opening the debate for a regulatory framework. *Br. J. Clin. Pharmacol.* **2018**, *84*, 659–672. [CrossRef]
7. Li, F.; Liu, X.; Wang, W.; Zhang, D. Consumption of vegetables and fruit and the risk of inflammatory bowel disease: A meta-analysis. *Eur. J. Gastroenterol. Hepatol.* **2015**, *27*, 623–630. [CrossRef] [PubMed]

8. Kim, H.; Banerjee, N.; Barnes, R.C.; Pfent, C.M.; Talcott, S.T.; Dashwood, R.H.; Mertens-Talcott, S.U. Mango polyphenolics reduce inflammation in intestinal colitis—Involvement of the miR-126/PI3K/AKT/mTOR axis in vitro and in vivo. *Mol. Carcinog.* **2017**, *56*, 197–207. [CrossRef] [PubMed]
9. Hale, L.P.; Chichlowski, M.; Trinh, C.T.; Greer, P.K. Dietary supplementation with fresh pineapple juice decreases inflammation and colonic neoplasia in IL-10-deficient mice with colitis. *Inflamm. Bowel Dis.* **2010**, *16*, 2012–2021. [CrossRef] [PubMed]
10. Fusco, R.; Cirmi, S.; Gugliandolo, E.; Di Paola, R.; Cuzzocrea, S.; Navarra, M. A flavonoid-rich extract of orange juice reduced oxidative stress in an experimental model of inflammatory bowel disease. *J. Funct. Foods* **2017**, *30*, 168–178. [CrossRef]
11. Pacheco, M.T.; Vezza, T.; Diez-Echave, P.; Utrilla, P.; Villamiel, M.; Moreno, F.J. Anti-inflammatory bowel effect of industrial orange by-products in DSS-treated mice. *Food Funct.* **2018**, *9*, 4888–4896. [CrossRef] [PubMed]
12. D'Argenio, G.; Mazzone, G.; Tuccillo, C.; Ribecco, M.T.; Graziani, G.; Gravina, A.G.; Caserta, S.; Guido, S.; Fogliano, V.; Caporaso, N.; et al. Apple polyphenols extract (APE) improves colon damage in a rat model of colitis. *Dig. Liver Dis.* **2012**, *44*, 555–562. [CrossRef] [PubMed]
13. Rosillo, M.A.; Sánchez-Hidalgo, M.; Cárdeno, A.; Aparicio-Soto, M.; Sánchez-Fidalgo, S.; Villegas, I.; de la Lastra, C.A. Dietary supplementation of an ellagic acid-enriched pomegranate extract attenuates chronic colonic inflammation in rats. *Pharmacol. Res.* **2012**, *66*, 235–242. [CrossRef] [PubMed]
14. Okayasu, I.; Hana, K.; Nemoto, N.; Yoshida, T.; Saegusa, M.; Yokota-Nakatsuma, A.; Song, S.Y.; Iwata, M. Vitamin A inhibits development of dextran sulfate sodium-induced colitis and colon cancer in a mouse model. *Biomed. Res. Int.* **2016**. [CrossRef]
15. Trivedi, P.P.; Jena, G.B. Mechanistic insight into beta-carotene-mediated protection against ulcerative colitis-associated local and systemic damage in mice. *Eur. J. Nutr.* **2015**, *54*, 639–652. [CrossRef]
16. Mencarelli, A.; Renga, B.; Palladino, G.; Distrutti, E.; Fiorucci, S. The plant sterol guggulsterone attenuates inflammation and immune dysfunction in murine models of inflammatory bowel disease. *Biochem. Pharmacol.* **2009**, *78*, 1214–1223. [CrossRef] [PubMed]
17. Lee, I.A.; Kim, E.J.; Kim, D.H. Inhibitory effect of β-sitosterol on TNBS-induced colitis in mice. *Planta Med.* **2012**, *78*, 896–898. [CrossRef]
18. Aldini, R.; Micucci, M.; Cevenini, M.; Fato, R.; Bergamini, C.; Nanni, C.; Cont, M.; Camborata, C.; Spinozzi, S.; Montagnani, M.; et al. Antiinflammatory effect of phytosterols in experimental murine colitis model: Prevention, induction, remission study. *PLoS ONE* **2014**, *9*, e108112. [CrossRef]
19. Kim, K.A.; Lee, I.A.; Gu, W.; Hyam, S.R.; Kim, D.H. β-Sitosterol attenuates high-fat diet-induced intestinal inflammation in mice by inhibiting the binding of lipopolysaccharide to toll-like receptor 4 in the NF-κB pathway. *Mol. Nutr. Food Res.* **2014**, *58*, 963–972. [CrossRef]
20. Te Velde, A.A.; Brüll, F.; Heinsbroek, S.E.; Meijer, S.L.; Lütjohann, D.; Vreugdenhil, A.; Plat, J. Effects of dietary plant sterols and stanol esters with low-and high-fat diets in chronic and acute models for experimental colitis. *Nutrients* **2015**, *7*, 8518–8531. [CrossRef]
21. Feng, S.; Dai, Z.; Liu, A.; Wang, H.; Chen, J.; Luo, Z.; Yang, C.S. β-Sitosterol and stigmasterol ameliorate dextran sulphate sodium-induced colitis in mice fed a high fat Western-style diet. *Food Funct.* **2017**, *8*, 4179–4186. [CrossRef]
22. European Commission. Decision 2004/336/EC of 31 March 2004 authorizing the placing on the market of yellow fat spreads, milk based fruit drinks, yoghurt type products and cheese type products with added phytosterols/phytostanols as novel foods or novel food ingredients under Regulation (EC) No 258/97 of the European Parliament and of the Council. *Off. J. Eur. Union* **2004**, *L105*, 49–51.
23. European Commission. Regulation (EU) No 686/2014 of 20 June 2014 amending Regulations (EC) No 983/2009 and (EU) No 384/2010 as regards the conditions of use of certain health claims related to the lowering effect of plant sterols and plant stanols on blood LDL-cholesterol. *Off. J. Eur. Union* **2014**, *L182*, 27–30.
24. Islam, M.S.; Murata, T.; Fujisawa, M.; Nagasaka, R.; Ushio, H.; Bari, A.M.; Hori, M.; Ozaki, H. Anti-inflammatory effects of phytosteryl ferulates in colitis induced by dextran sulphate sodium in mice. *Br. J. Pharmacol.* **2008**, *154*, 812–824. [CrossRef] [PubMed]
25. Ioannidis, O.; Varnalidis, I.; Paraskevas, G.; Botsios, D. Nutritional modulation of the inflammatory bowel response. *Digestion* **2011**, *84*, 89–101. [CrossRef] [PubMed]

26. Laurell, A.; Sjöberg, K. Prebiotics and symbiotic in ulcerative colitis. *Scand. J. Gastroenterol.* **2017**, *52*, 477–485. [CrossRef] [PubMed]
27. Silk, D.B.A.; Davis, A.; Vulevic, J.; Tzortzis, G.; Gibson, G.R. Clinical trial: The effects of a trans-galactooligosaccharide prebiotic on faecal microbiota and symptoms in irritable bowel syndrome. *Aliment. Pharmacol. Ther.* **2009**, *29*, 508–518. [CrossRef] [PubMed]
28. Vulevic, J.; Juric, A.; Tzortzis, G.; Gibson, G.R. A Mixture of trans-galactooligosaccharides reduces markers of metabolic syndrome and modulates the fecal microbiota and immune function of overweight adults. *J. Nutr.* **2013**, *143*, 324–331. [CrossRef] [PubMed]
29. Vulevic, J.; Drakoularakou, A.; Yaqoob, P.; Tzortzis, G.; Gibson, G.R. Modulation of the fecal microflora profile and immune function by a novel trans-galactooligosaccharide mixture (B-GOS) in healthy elderly volunteers. *Am. J. Clin. Nutr.* **2008**, *88*, 1438–1446.
30. Holma, R.; Juvonen, P.; Asmawi, M.Z.; Vapaatalo, H.; Korpela, R. Galacto-oligosaccharides stimulate the growth of bifidobacteria but fail to attenuate inflammation in experimental colitis in rats. *Scand. J. Gastroenterol.* **2002**, *37*, 1042–1047. [CrossRef]
31. Gopalakrishnan, A.; Clinthorne, J.F.; Rondini, E.A.; McCaskey, S.J.; Gurzell, E.A.; Langohr, I.M.; Gardner, E.M.; Fenton, J.I. Supplementation with galacto-oligosaccharides increases the percentage of NK cells and reduces colitis severity in smad3-deficient mice. *J. Nutr.* **2012**, *142*, 1336–1342. [CrossRef]
32. López-García, G.; Cilla, A.; Barberá, R.; Alegría, A. Protective effect of antioxidants contained in milk-based fruit beverages against sterol oxidation products. *J. Funct. Foods* **2017**, *30*, 81–89. [CrossRef]
33. Alvarez-Sala, A.; Blanco-Morales, V.; Cilla, A.; Silvestre, R.A.; Hernández-Álvarez, E.; Granado-Lorencio, F.; Barberá, R.; Garcia-Llatas, G. A positive impact on the serum lipid profile and cytokines after the consumption of a plant sterol-enriched beverage with a milk fat globule membrane: A clinical study. *Food Funct.* **2018**, *9*, 5209–5219. [CrossRef] [PubMed]
34. Marín, M.; Giner, R.M.; Ríos, J.L.; Recio, M.C. Intestinal anti-inflammatory activity of ellagic acid in the acute and chronic dextrane sulfate sodium models of mice colitis. *J. Ethnopharmacol.* **2013**, *150*, 925–934. [CrossRef] [PubMed]
35. Giner, E.; Recio, M.C.; Ríos, J.L.; Cerdá-Nicolás, J.M.; Giner, R.M. Chemopreventive effect of oleuropein in colitis-associated colorectal cancer in C57BL/6 mice. *Mol. Nutr. Food Res.* **2016**, *60*, 242–255. [CrossRef] [PubMed]
36. Cooper, H.S.; Murthy, S.N.; Shah, R.S.; Sedergran, D.J. Clinicopathologic study of dextran sulfate sodium experimental murine colitis. *Lab. Investig.* **1993**, *69*, 238–249. [PubMed]
37. Bradley, P.P.; Priebat, D.A.; Christensen, R.D.; Rothstein, G. Measurement of cutaneous inflammation: Estimation of neutrophil content with an enzyme marker. *J. Investig. Dermatol.* **1982**, *78*, 206–209. [CrossRef] [PubMed]
38. Almero, J. Modelos experimentales in vivo de enfermedad inflamatoria intestinal y cáncer colorrectal: Conceptos, modelos actuales y aplicabilidad. *Nutr. Hosp.* **2007**, *22*, 178–189.
39. Perŝe, M.; Cerar, A. Dextran sodium sulphate colitis mouse model: Traps and tricks. *J. Biomed. Biotechnol.* **2012**. [CrossRef]
40. Gottlieb, Y.; Elhasid, R.; Berger-Achituv, S.; Brazowski, E.; Yerushalmy-Feler, A.; Cohen, S. Neutrophil extracellular traps in paediatric inflammatory bowel disease. *Pathol. Int.* **2018**, *68*, 517–523. [CrossRef]
41. Eichele, D.D.; Kharbanda, K.K. Dextran sodium sulphate colitis murine model: An indispensable tool for advancing our understanding of inflammatory bowel diseases pathogenesis. *World J. Gastroenterol.* **2017**, *23*, 6016–6029. [CrossRef]
42. Granado-Lorencio, F.; Largarda, M.J.; García-López, F.J.; Sánchez-Siles, L.M.; Blanco-Navarro, I.; Alegría, A.; Pérez-Sacristán, B.; Garcia-Llatas, G.; Donoso-Navarro, E.; Silvestre-Mardomingo, R.A.; et al. Effect of β-cryptoxanthin plus phytosterols on cardiovascular risk and bone turnover markers in postmenopausal women: A randomized crossover trial. *Nutr. Metab. Cardiovasc. Dis.* **2014**, *24*, 1090–1096. [CrossRef] [PubMed]
43. Kobori, M.; Ni, Y.; Takahashi, Y.; Watanabe, N.; Sugiura, M.; Ogawa, K.; Nagashimada, M.; Kaneko, S.; Naito, S.; Ota, T. β-Cryptoxanthin alleviates diet-induced nonalcoholic steatohepatitis by suppressing inflammatory gene expression in mice. *PLoS ONE* **2014**, *9*, e98294. [CrossRef] [PubMed]

44. Sahin, K.; Orhan, C.; Akdemir, F.; Tuzcu, M.; Sahin, N.; Yılmaz, I.; Juturu, V. β-Cryptoxanthin ameliorates metabolic risk factors by regulating NF-κB and Nrf2 pathways in insulin resistance induced by high-fat diet in rodents. *Food Chem. Toxicol.* **2017**, *107*, 270–279. [CrossRef] [PubMed]
45. Sabeva, N.S.; McPhaul, C.M.; Li, X.; Cory, T.J.; Feola, D.J.; Graf, G.A. Phytosterols differentially influence ABC transporter expression, cholesterol efflux and inflammatory cytokine secretion in macrophage foam cells. *J. Nutr. Biochem.* **2011**, *22*, 777–783. [CrossRef] [PubMed]
46. Llewellyn, S.R.; Britton, G.J.; Contijoch, E.J.; Vennaro, O.H.; Mortha, A.; Colombel, J.F.; Grispan, A.; Clemente, J.C.; Merad, M.; Faith, J.J. Interactions between diet and the intestinal microbiota alter intestinal permeability and colitis severity in mice. *Gastroenterology* **2018**, *154*, 1037–1046. [CrossRef] [PubMed]
47. Geier, M.S.; Butler, R.N.; Giffard, P.M.; Howarth, G.S. Prebiotic and symbiotic fructooligosaccharide administration fail to reduce the severity of experimental colitis in rats. *Dis. Colon Rectum* **2007**, *50*, 1061–1069. [CrossRef]
48. Campbell, J.M.; Fahey, G.C., Jr.; Wolf, B.W. Selected indigestible oligosaccharides affect large bowel mass, cecal and fecal short-chain fatty acids, pH and microflora in rats. *J. Nutr.* **1997**, *127*, 130–136. [CrossRef] [PubMed]
49. Bruggencate, S.J.T.; Bovee-Oudenhoven, I.M.; Lettink-Wissink, M.L.; Van der Meer, R. Dietary fructooligosaccharides increase intestinal permeability in rats. *J. Nutr.* **2005**, *135*, 837–842. [CrossRef]
50. Maathuis, A.J.; van den Heuvel, E.G.; Schoterman, M.H.; Venema, K. Galacto-oligosaccharides have prebiotic activity in a dynamic in vitro colon model using a 13C-labeling technique. *J. Nutr.* **2012**, *142*, 1205–1212. [CrossRef]

© 2019 by the authors. Licensee MDPI, Basel, Switzerland. This article is an open access article distributed under the terms and conditions of the Creative Commons Attribution (CC BY) license (http://creativecommons.org/licenses/by/4.0/).

Article

Anti-Obesity Effect of Extract from *Nelumbo Nucifera* L., *Morus Alba* L., and *Raphanus Sativus* Mixture in 3T3-L1 Adipocytes and C57BL/6J Obese Mice

Wan-Sup Sim [1], Sun-Il Choi [1], Bong-Yeon Cho [1], Seung-Hyun Choi [1], Xionggao Han [1], Hyun-Duk Cho [2], Seung-Hyung Kim [3], Boo-Yong Lee [4], Il-Jun Kang [5], Ju-Hyun Cho [2,*] and Ok-Hwan Lee [1,*]

[1] Department of food science and biotechnology, Kangwon National University, Chuncheon 24341, Korea; simws9197@naver.com (W.-S.S.); docgotack89@hanmail.net (S.-I.C.); bongyeon.cho92@gmail.com (B.-Y.C.); zzaoszz@naver.com (S.-H.C.); xionggao414@hotmail.com (X.H.)
[2] Haram Co. Ltd. Jeungpyeong 27914, Korea; hansol305@naver.com
[3] Institute of Traditional Medicine and Bioscience, Daejeon University, Daejeon 34520, Korea; sksh518@dju.kr
[4] Department of Food Science and Biotechnology, CHA University, Seongam, Gyeonggi 13488, Korea; bylee@cha.ac.kr
[5] Department of Food Science and Nutrition, Hallym University, Chuncheon 24252, Korea; ijkang@hallym.ac.kr
* Correspondences: dusvnd608@hanmail.net (J.-H.C.); loh99@kangwon.ac.kr (O.-H.L.); Tel: +82-43-217-1077 (J.-H.C.); +82-33-250-6454 (O.-H.L.); Fax: +82-43-217-1088 (J.-H.C.); +82-33-259-5561 (O.-H.L.)

Received: 3 May 2019; Accepted: 17 May 2019; Published: 19 May 2019

Abstract: The antioxidant and anti-adipogenic activities of a mixture of *Nelumbo nucifera* L., *Morus alba* L., and *Raphanus sativus* were investigated and their anti-obesity activities were established in vitro and *in vivo*. Among the 26 different mixtures of extraction solvent and mixture ratios, ethanol extract mixture no. 1 (EM01) showed the highest antioxidant (α,α-Diphenyl-β-picrylhydrazyl, total phenolic contents) and anti-adipogenic (Oil-Red O staining) activities. EM01 inhibited lipid accumulation in 3T3-L1 adipocytes compared to quercetin-3-O-glucuronide. Furthermore, body, liver, and adipose tissue weights decreased in the high-fat diet (HFD)-EM01 group compared to in the high-fat diet control group (HFD-CTL). EM01 lowered blood glucose levels elevated by the HFD. Lipid profiles were improved following EM01 treatment. Serum adiponectin significantly increased, while leptin, insulin growth factor-1, non-esterified fatty acid, and glucose significantly decreased in the HFD-EM01 group. Adipogenesis and lipogenesis-related genes were suppressed, while fat oxidation-related genes increased following EM01 administration. Thus, EM01 may be a natural anti-obesity agent.

Keywords: anti-obesity; *Nelumbo nucifera* L., *Morus alba* L., *Raphanus sativus*; 3T3-L1 adipocyte; C57BL/6J mice

---

1. Introduction

Obesity is defined as excessive weight gain, particularly inordinate fat accumulation in the body [1]. Due to rapid economic growth and westernized diets, the number of patients with obesity has increased [2], resulting in increased rates of diseases such as type 2 diabetes [3], hypertension [4], hypercholesterolemia [5], and hyperlipidemia [6].

Obesity occurs because of a combination of overeating, lack of exercise, and neurotransmitters, drugs, and genetic factors [7]. To improve obesity, pharmacological treatment has been applied.

However, because these treatments involve diverse side effects such as diarrhea and vomiting, the development of anti-obesity materials from natural products with few side effects is required [8,9].

Nelumbo nucifera L contains several flavonoids and alkaloids and has recently been used as a plain or blended tea to treat obesity in China [10]. *Morus alba* L is abundant in polyphenols such as rutin and quercetin and has been used to treat dyslipidemia, diabetes, fatty liver disease, and hypertension [11]. *Raphanus sativus* is an annual herb belonging to the cruciferous family and has been reported to be effective for treating hyperlipidemia by decreasing blood sugar level [12]. Some researchers evaluated the anti-obesity effects of *Eriobotrya japonica* and *N. nucifera* in adipocytes and obese mice using a single material or mixture, but few studies have examined mixtures of three natural materials [13]. Because each material has a different physiological activity, it is necessary to confirm the physiological activities of each according to the type and mixing ratio. These materials may exert synergistic effects, but there might be also negative effects [14,15].

In this study, we blended *N. nucifera* L., *M. alba* L., and *R. sativus* in different mixing ratios and selected the optimal mixed material based on antioxidant and anti-adipogenic experiments. The anti-obesity effects of optimal mixed material were evaluated in 3T3-L1 adipocytes and C57BL/6J obese mice.

2. Materials and Methods

2.1. Sample Preparation

Nelumbo nucifera leaves, *M. alba* leaves, and Dried *R. sativus* root were supplied by Haram Co. Ltd. (Jeungpyeong, Korea) and freeze-dried and mixed in 13 ratios. The mixing ratio of *N. nucifera* L, *M. alba* L, *R. sativus* is 80:20:0 (M1), 70:30:0 (M2), 60:40:0 (M3), 50:50:0 (M4), 80:0:20 (M5), 70:0:30 (M6), 70:20:10 (M7), 60:30:10 (M8), 60:20:20 (M9), 50:30:20 (M10), 100:0:0 (M11), 0:100:0 (M12), 0:0:100 (M13). These materials were extracted by boiling and reflux with hot water (WM) and 70% ethanol (EM) at 60 °C for 12 h. Then they were filtered, concentrated by a vacuum evaporator (Rotavapor R-200; Buchi, Flawil, Switzerland), and freeze-dried (Bondiro; Il Shin Lab Co. Ltd., Seoul, Korea) to obtain a powder [16].

2.2. Antioxidant Activity Analysis

We performed a DPPH (α,α-Diphenyl-β-picrylhydrazyl) assay as described previously with some modifications [17]. First, 0.8 mL of 0.4 mM DPPH solution was added to 0.2 mL of the sample, and the mixture was incubated in the dark for 10 min. Next, it was measured at the 517 nm absorbance using a microplate reader (Molecular Devices, Sunnyvale, CA, USA).

$$\text{DPPH radical scavenging activity (\%)} = (1 - (A_{experiment}/A_{control})) \times 100 \tag{1}$$

$A_{experiment}$: Absorbance of experimental group, $A_{control}$: Absorbance of control group.

The total phenolic contents were determined using Folin-Ciocalteu's colorimetric method [18]. First, 1 mL of sample was added to 10% Folin-Ciocalteu reagent and 2% Na_2CO_3 reagent in order and incubated for 1 h in the dark. It was measured at the 750 nm absorbance using a microplate reader. Gallic acid was used as a standard and the total phenolic contents were calculated from the standard calibration curve ($y = 19.12x - 0.0261$, $R^2 = 0.9992$).

2.3. Cell Culture and Differentiation

3T3-L1 preadipocytes were obtained from American Type Culture Collection (CL-173, ATCC, Manassas, VA, USA) and grown in culture plates containing Dulbecco's modified Eagle's medium (Lonza, Basel, Switzerland) supplemented with 10% bovine serum (Gibco, Grand Island, NY, USA) and 1% penicillin/streptomycin (P/S; Gibco) kept at 37 °C and 5% CO_2 incubator condition. 3T3-L1 preadipocytes were differentiated to adipocytes at the 2 day after confluence by exchanging with MDI medium (DMEM containing 10% fetal bovine serum (Gibco), 1% P/S,

0.5 mM 3-isobutyl-1-methylxanthine (Sigma-Aldrich, Saint Louis, MO, USA), 1 µM dexamethasone (Sigma-Aldrich), and 1 µg/mL insulin (Gibco)). Every 2 days during incubation, the culture medium was changed to DMEM containing 10% fetal bovine serum and 1 µg/mL insulin with extracts until day 6 [19].

2.4. Cell Viability Assay

3T3-L1 preadipocytes were plated into 96-well plates (1×10^6 cells/well) and differentiated with MDI medium along with 100 µg/mL extracts since the 2 day after confluence for 6 days. At the day 6, Mixed XTT (2,3-bis(2-methoxy-4-nitro-5-sulfophenyl)-2H-tetrazolium-5-carboxanilide) and PMS (N-methyl dibenzopyrazine methyl sulfate reagents) (WelGene, Seoul, Korea) were included into the medium and incubated for 4 h at 37 °C. Then the soluble formazan salt generated in the medium was measured at 450 nm against 690 nm using a microplate reader [20].

2.5. Oil-Red O Staining Assay

The amount of lipid accumulation of 3T3-L1 cells differentiated on a 24-well plate for 6 days was determined using Oil-Red O staining. In brief, the cells were washed with phosphate-buffered saline (PBS; Gibco) and fixed in 10% formaldehyde in distilled water for 1 h. Next, the cells were dried with 60% isopropanol and stained with Oil-Red O (Sigma-Aldrich) solution for 1 h, and then washed with distilled water. The stained lipids were eluted with 100% isopropanol and measured using a microplate reader at 490 nm.

2.6. Animal Experiment Design

Eight-week-old male C57BL/6J mice were obtained from Korea BioLink (Eumseong, Korea) and maintained under standard light (12:12-h light/dark cycle), temperature (22 ± 2 °C) and humidity (55 ± 15%) conditions. The diets included a normal diet (AIN-76A; Research Diets, Inc., New Brunswick, NJ, USA) and high-fat diet (HFD; D12492; Research Diets, Inc.), and obesity was induced in the mice fed an HFD for 2 weeks. After 2 weeks, the mice were randomly divided into eight groups; C57BL/6J normal group, 60% kcal HFD control group, positive control group (*Garcinia cambogia*, GC_245 mg/kg), EM11_100 mg/kg group, EM12_100 mg/kg group, EM01_50 mg/kg group, EM01_100 mg/kg group, quercetin-3-O-glucuronide (Q3OG) 10 mg/kg group; $n = 10$). Meanwhile, *Garcinia cambogia* is an ingredient in dietary supplements for weight loss [19]. These test substances were taken orally with saline solution once a day for 8 weeks. The mice had free access to food and water, and their body weight and calorie intake were measured weekly. Biochemical measurements were analyzed after treatment of the candidate materials for 8 weeks and glucose tolerance test were performed 3 days later. Experiments were carried out with the approval of the Institutional Animal Care and Use Committee of Kangwon University for the ethical and scientific feasibility study and effective management of animal experiments (Permit No. KIACUC-12-0140).

2.7. Glucose Tolerance Test

After administering the anti-obesity candidate materials for 8 weeks, mice were fasted for 12 h and the blood glucose level was measured at 0, 15, 30, 45, 60, and 75 min after glucose (1 g/kg) peritoneal administration. Blood for glucose measurement was obtained from the tail vein of mice and measured using a serum analyzer (Accutrend plus GCTL Cobas Roche, Basel, Switzerland).

2.8. Serum Biochemical Parameter Analysis

After administering the anti-obesity candidate materials for 8 weeks, blood samples were collected by cardiac puncture using centrifugation (2000× *g*, 4 °C, 15 min) and stored at −74 °C. Alanine aminotransferase (ALT), aspartate aminotransferase (AST), and creatinine, which are indices of liver and kidney function, and total cholesterol (TC), high-density lipoprotein cholesterol (HDL-cholesterol), low-density lipoprotein cholesterol (LDL-cholesterol), triglyceride (TG), non-esterified fatty acid (NEFA), and glucose, which are indices of lipid content, were measured using a biochemical automatic analyzer (Hitachi-720, Hitachi Medical, Tokyo, Japan).

2.9. Serum Adipokine Analysis

Also, adiponectin, leptin, and insulin-like growth factor-1 (IGF-1) were isolated in blood samples collected by cardiac puncture after administration of anti-obesity candidate materials for 8 weeks. Each antibody was diluted in coating buffer and coated on a microwell for overnight incubation at 4 °C. Each well was washed three times with washing buffer, and 100 μL of serum was dispensed, incubated at room temperature for 1 h, and washed twice with washing buffer. Thereafter, 100 μL of the antibody avidin-horseradish peroxidase conjugate was added and incubated at room temperature for 1 h, and then washed again. The TMB substrate was dispensed in 100 μL, incubated in the dark for 30 min, and treated with 50 μL of stop solution, after which absorbance was measured at 450 nm.

2.10. Real-Time Polymerase Chain Reaction (RT-PCR)

Total cellular RNA was extracted from the liver, epididymal, and abdominal subcutaneous adipose tissue using a homogenizer and Trizol reagent (Sigma-Aldrich). Total RNA was used for cDNA synthesis with the One-step SYBR Green PCR kit (AB Science, Avenue George V, France). The Applied Biosystems 7500 Real-Time PCR system (Applied Biosystems, Foster City, CA, USA) used for real-time quantitative PCR. The probes containing the fluorescence reporter dye 6-carboxy-fluorescein (Applied Biosystems) was used to indicate mRNA gene expression. The mouse glyceraldehyde-3-phosphate dehydrogenase probe (VIC/MGB probe, primer limited, 4352339E, Applied Biosystems) was used as an internal standard. The sequence of the used primer was as follows; Leptin sense (5′-AACCCTTACTGAACTCAGATTGTTAG-3′) and antisense (5′-TAAGTCAGTTTAAATGCTTAGGG-3′); PPARγ sense (5′-ATGCCATTCTGGCCCACCAACTT-3′) and antisense (5′-CCCTTGCATCCTTCACAAGCATG-3′); PPARα sense (5′-GCCTGTCTGTCGGATGT-3′) and antisense (5′-GGCTTCGTGGATTCTCTTG-3′); Adiponectin sense (5′-TTCAAATGAGATTGTGGGAAAAT-3′) and antisense (5′-ACCGATACAGTACAGTACAGTA-3′); UCP1 sense (5′-CGACTCAGTCCAAGAGTACTTCTCTTC-3′) and antisense (5′-GCCGGCTGAGATCTTGTTTC-3′); UCP2 sense (5′-TTCAAATGAGATTGTGGGAAAAT-3′) and antisense (5′-ACCGATACAGTACAGTACAGTA-3′); ACX1 sense (5′-CAGGAAGAGCAAGGAAGTGG-3′) and antisense (5′-CCTTTCTGGCTGATCCCATA-3′); DGAT1 sense (5′-TGCTACGACGAGTTCTTGAG-3′) and antisense (5′-CTCTGCCACAGCATTGAGAC-3′); SCD1 sense (5′-CATCGCCTGCTCTACCCTTT-3′) and antisense (5′-GAACTGCGCTTGGAAACCTG-3′); SREBP1c/ADD1 sense (5′-AGCCTGGCCATCTGTGAGAA-3′) and antisense (5′-CAGACTGGTACGGGCCACAA-3′); ACS1 sense (5′-TCCTACAAAGAGGTGGCAGAACT-3′) and antisense (5′-GGCTTGAACCCCTTCTGGAT-3′); CPT1b sense (5′-GTCGCTTCTTCAAGGTCTGG-3′) and antisense (5′-AAGAAAGCAGCACGTTCGAT-3′); FAS sense (5′-CTGAGATCCCAGCACTTCTTGA-3′) and antisense (5′-GCCTCCGAAGCCAAATGAG-3′); GAPDH VIC (5′-TGCATCCTGCACCACCAACTGCTTAG-3′). The PCR conditions were 50 °C for 2 min, 94 °C for 10 min, 95 °C for 15 s, and 45 °C for 1 min of 40 cycles.

2.11. Statistical Analysis

The results are expressed as the mean ± standard deviation of triplicate experiments. An analysis of variance and Duncan's multiple range tests were used to determine the significance at the $p < 0.05$ level.

3. Results

3.1. Effects of 26 Extracts by Mixture Ratio of N. Nucifera L., M. alba L., R. Sativus on the Antioxidant and Anti-Adipogenic Activities

α,α-Diphenyl-β-picrylhydrazyl (DPPH) is a very stable free radical and representative reaction substance used to measure antioxidant capacity. DPPH radical scavenging activity assay is a method that utilizes the principle that a purple compound is discolored as yellow when radicals are eliminated through hydrogen donation in a phenolic compound or flavonoid with a hydroxyl radical (-OH) [21]. For the extraction method, the radical scavenging activities of ethanol extracts were superior to those of hot water extracts. For the mixing ratio, EM05 (84.30%), EM06 (83.00%), and EM01 (81.65%) were superior in order, but there was no significant difference between these samples (Figure 1A).

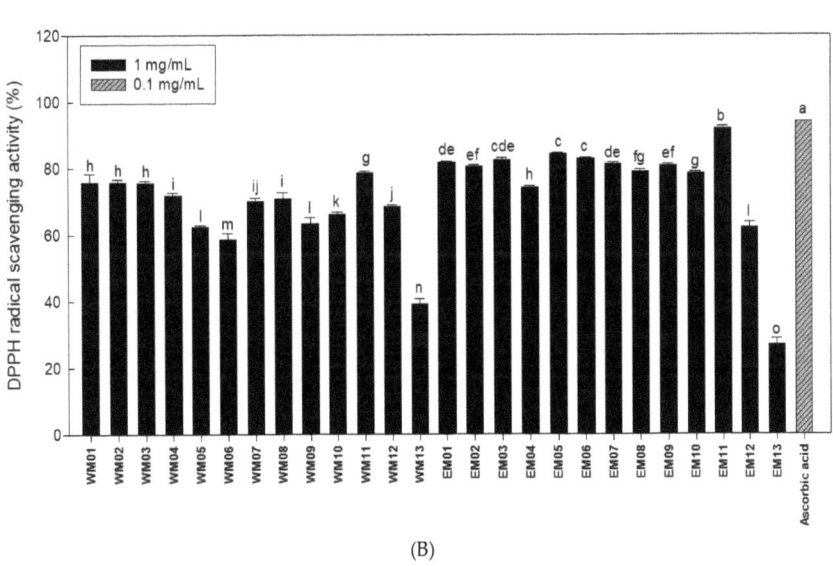

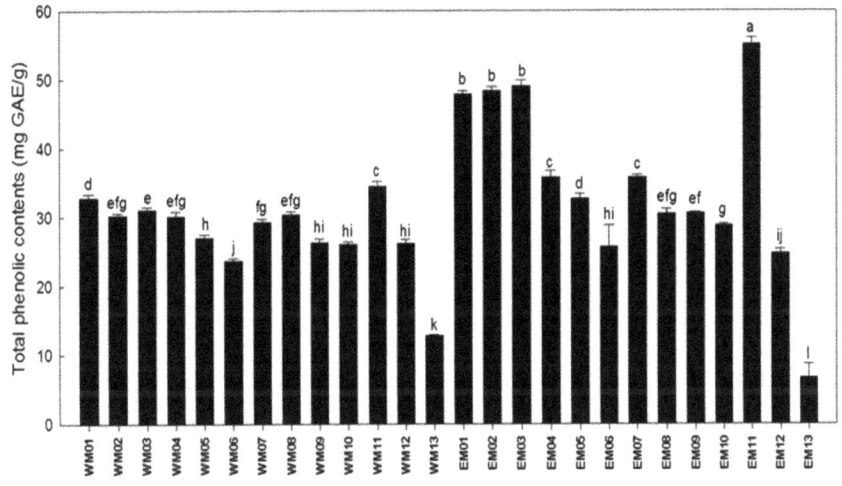

Figure 1. Cont.

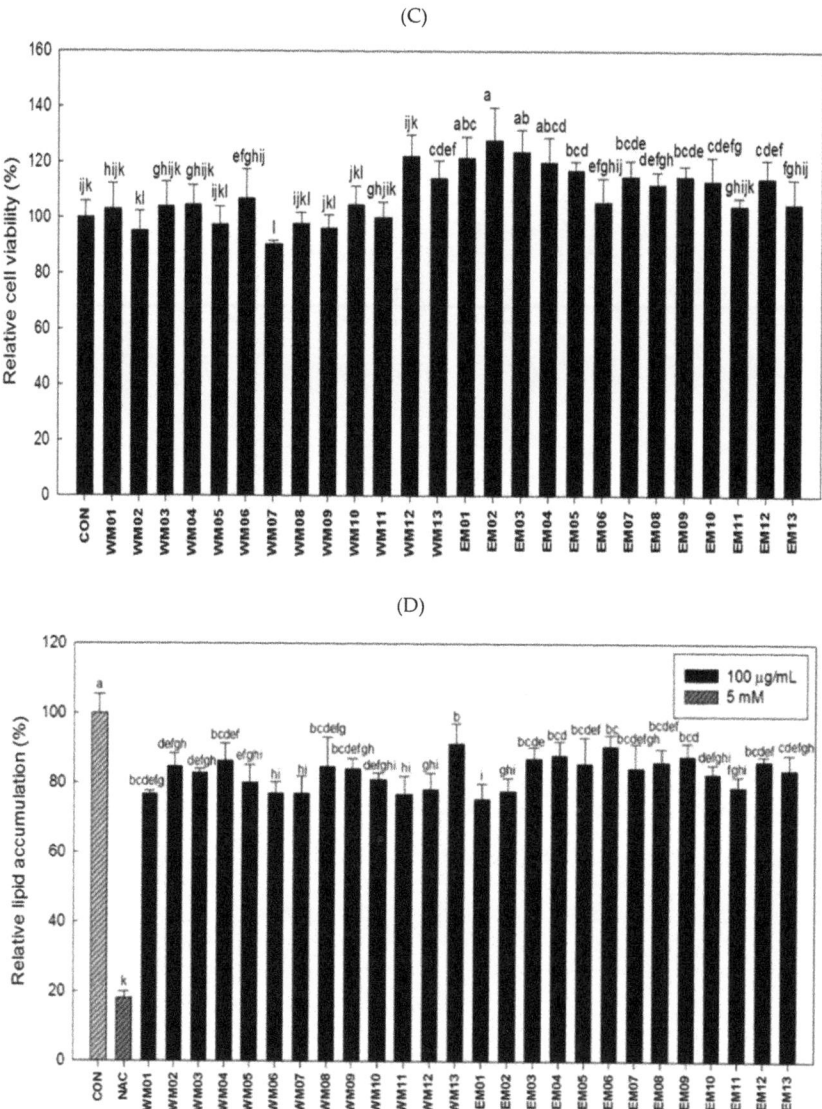

Figure 1. Effects of 26 extracts by mixture ratio of *Nelumbo nucifera* L., *Morus alba* L., and *Raphanus sativus* on antioxidant activity, cell viability, lipid accumulation. (**A**) DPPH radical scavenging activity. (**B**) Total phenolic contents. (**C**) Post-confluent 3T3-L1 preadipocytes were differentiated along with the treatment of each extracts for 6 days. XTT (2,3-bis(2-methoxy-4-nitro-5- sulfophenyl)-2H-tetrazolium-5-carboxanilide) and PMS (*N*-methyl dibenzopyrazine methyl sulfate reagents) mixture was added to the medium. After 4 h of incubation, cell viability was found out by calculating the absorbance at 450 nm against 690 nm. (**D**) Stained lipids were extracted and quantified by calculating the absorbance at 490 nm. Data are presented as the mean ± SEM (n = 3). Means with different letters on bars indicate that there is a significant difference at $p < 0.05$ by Duncan's multiple range test.

The content of phenol, a representative antioxidant, was highest in EM03 (49.00 mg GAE/g), EM02 (48.40 mg GAE/g), and EM01 (47.90 mg GAE/g), with no significant difference between these samples (Figure 1B).

No cytotoxicity or changes in morphology were observed at the concentration of 100 μg/mL mixtures by XTT assay (Figure 1C). Next, the anti-adipogenic activity was measured by Oil-Red O staining, and EM01 (75.30%) was found to be the most effective mixture for inhibiting lipid accumulation among the 26 mixtures (Figure 1D). Therefore, we selected EM01 as an optimal mixture and carried out anti-obesity experiments using in vitro and in vivo models.

3.2. Effect of EM01 on Lipid Accumulation

We previously found that quercetin-3-O-glucuronide is a bioactive compound in mixed materials [16]. In antioxidant and anti-obesity experiments, EM01 (100 μg/mL) was measured to determine its anti-obesity activity with quercetin-3-O-glucuronide (7.8 μM). As shown in Figure 2, EM01 (80.73%) showed better lipid accumulation inhibitory activity than the single materials, EM11 (84.92%), EM12 (86.39%), and single bioactive compound treatment (84.65%). There may have been a synergistic interaction in EM01 between quercetin-3-O-glucuronide and other compounds.

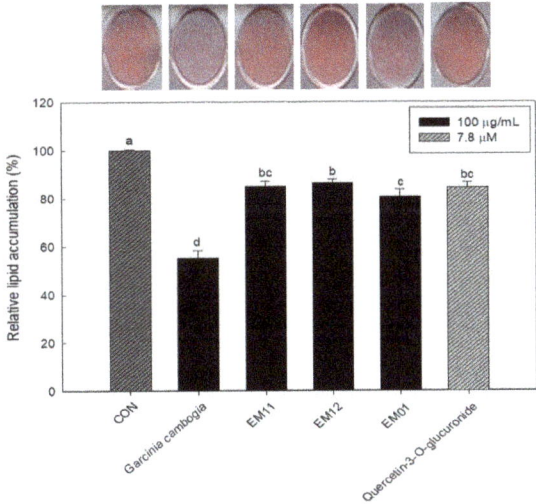

Figure 2. Effect of EM01 and its bioactive compound on lipid accumulation. Post-confluent 3T3-L1 preadipocytes were differentiated along with the treatment of each extracts and its bioactive compound for 6 days. Stained lipids were eluted and quantified by calculating the absorbance at 490 nm. Data are presented as the mean ± SEM ($n = 3$). Means with different letters on bars indicate that there is a significant difference at $p < 0.05$ by Duncan's multiple range test.

3.3. Effects of EM01 on Body Weight, Food Intake, FER, Organ Weight, and Adipose Tissue Weight in HFD-Induced Obese Mice

There was no significant difference in the initial body weight between experimental groups, but the final body weights in different groups were significantly different. The body weight increased significantly in the HFD-CTL group compared to in the ND group, suggesting that obesity was induced by the HFD. The HFD-EM01 group (100 μg/mL) showed a higher weight loss rate than the HFD-CTL group (Figure 3A). Food intake was not significantly different between groups except for the ND group, and the food efficiency ratio (FER) was significantly decreased in the HFD-EM01 group (100 μg/mL) (Figure 3B,C).

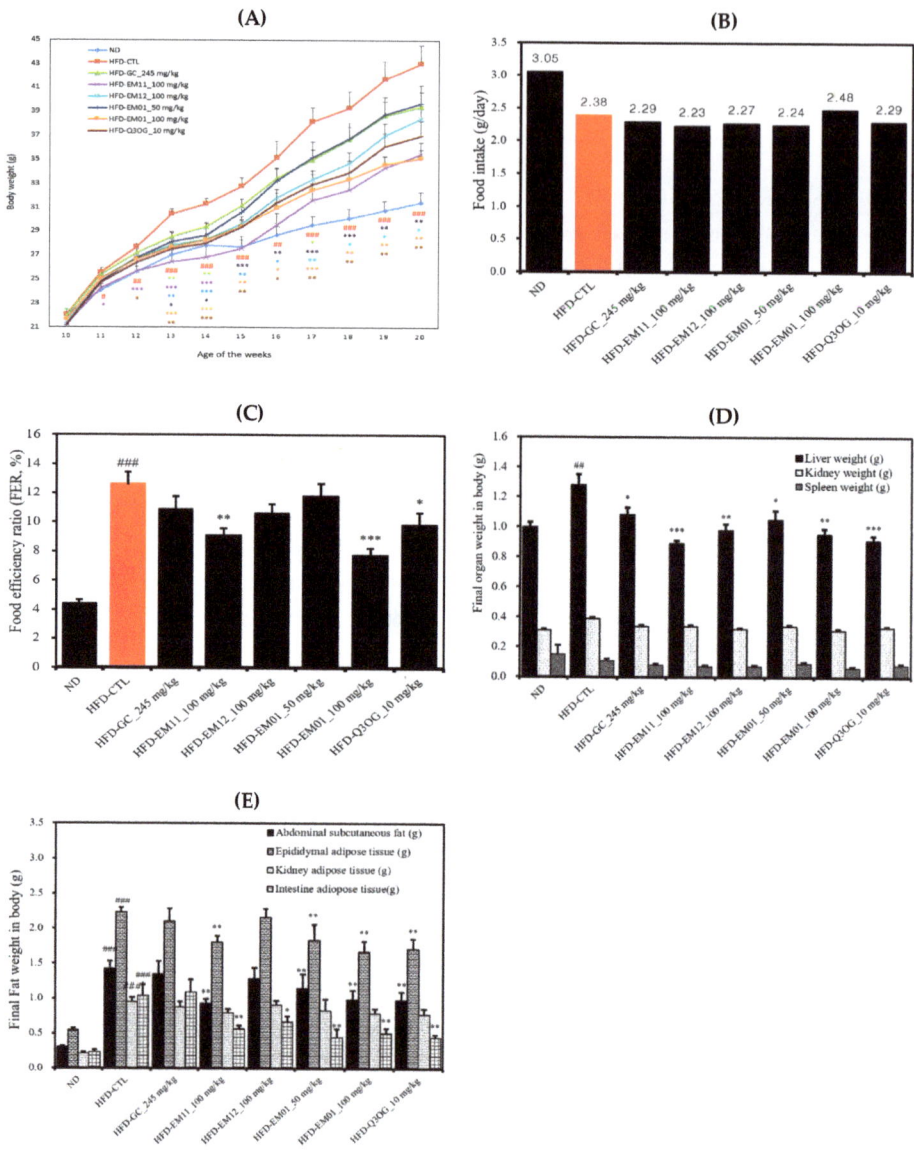

Figure 3. Effects of EM01 on (**A**) body weight, (**B**) food intake, (**C**) FER, (**D**) organ weight, and (**E**) adipose tissue fat weight in high-fat diet (HFD)-induced obese mice. GC, *Garcinia cambogia* extract; HFD, mice were fed a high-fat diet (60% kcal fat); ND, mice were fed a normal diet (10% kcal fat); FER, food efficiency ratio (total weight gain/total food intake × 100). Data are presented as the mean ± SEM ($n = 12$); # $p < 0.05$, ## $p < 0.01$, ### $p < 0.001$ vs. ND; * $p < 0.05$, ** $p < 0.01$ and *** $p < 0.001$ vs. HFD-CTL.

The kidney and spleen weights were not significantly different among groups (Figure 3D). Liver weight increased significantly in the HFD-CTL group compared to in the ND group, and weight decreased significantly in the HFD-EM01 and HFD-Q3OG groups compared to in the HFD-CTL group. According to Figure 3E, the weight of the kidney adipose tissue significantly increased in the HFD-CTL group compared to in the ND group, but no significant inhibitory effect was observed in any group. The weight of abdominal subcutaneous fat, epididymal, and intestine adipose tissue increased significantly in the HFD-CTL group compared to in the ND group, but there was an inhibitory effect on fat accumulation in the HFD-EM01 and HFD-Q3OG group compared to the positive control (HFD-GC group).

3.4. Effects of EM01 on glucose tolerance in HFD-induced obese mice

When glucose tolerance occurs, blood glucose levels do not rise despite glucose administration, which is common in obese patients [22]. To investigate the effect of EM01 administration on glucose-induced hyperglycemia, glucose was orally administered, and the glucose tolerance test was performed over time. Blood glucose levels in all groups increased after 30 min of glucose injection. After 60 min of glucose injection, blood glucose levels in EM01 and Q3OG administrated groups dropped markedly even close to ND-group. On the other hand, it was confirmed that blood glucose levels did not decrease rapidly in EM11 and EM12 administrated groups (single material) (Figure 4).

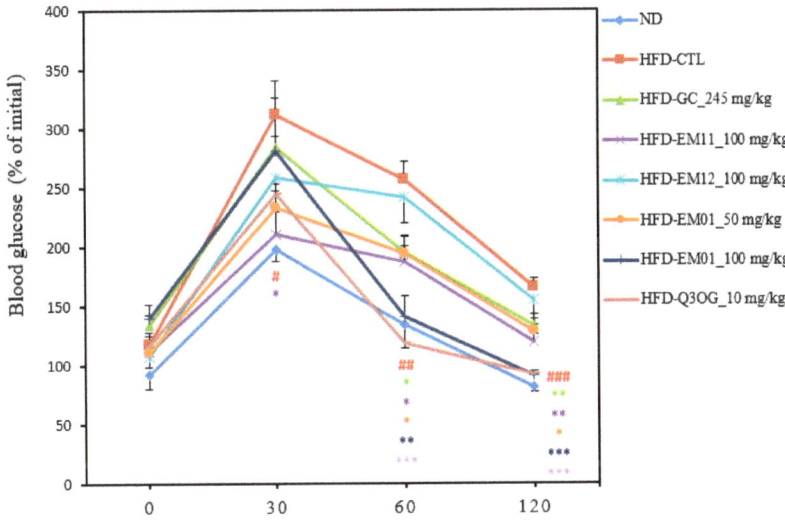

Figure 4. Effects of EM01 on glucose tolerance in HFD-induced obese mice. Data are presented as the mean ± SEM ($n = 12$); # $p < 0.05$, ## $p < 0.01$, ### $p < 0.001$ vs. ND; * $p < 0.05$, ** $p < 0.01$ and *** $p < 0.001$ vs. HFD-CTL.

3.5. Effects of EM01 on the Serum Lipid Profile in HFD-Induced Obese Mice

The HFD-CTL group showed significant increases in all parameters of the serum lipid profile compared to the ND group. The TC level was significantly decreased in the HFD-EM01 and HFD-Q3OG groups compared to in the HFD-CTL group (Figure 5A). HDL-cholesterol was lowered by more than LDL-cholesterol in these groups (Figure 5B,C). TC may decrease when LDL-cholesterol levels, often referred to as bad hormones, are suppressed [23]. The TG level was higher in the HFD-CTL group than in the ND group, but there was no significant difference between all groups compared to HFD-CTL (Figure 5D). AST, ALT, and creatinine are used as liver toxicity marker [24] and their levels were significantly lowered by EM01 administration (Figure 5E,F). This suggests that the 8-week oral administration did not affect liver and kidney toxicity in obese mice.

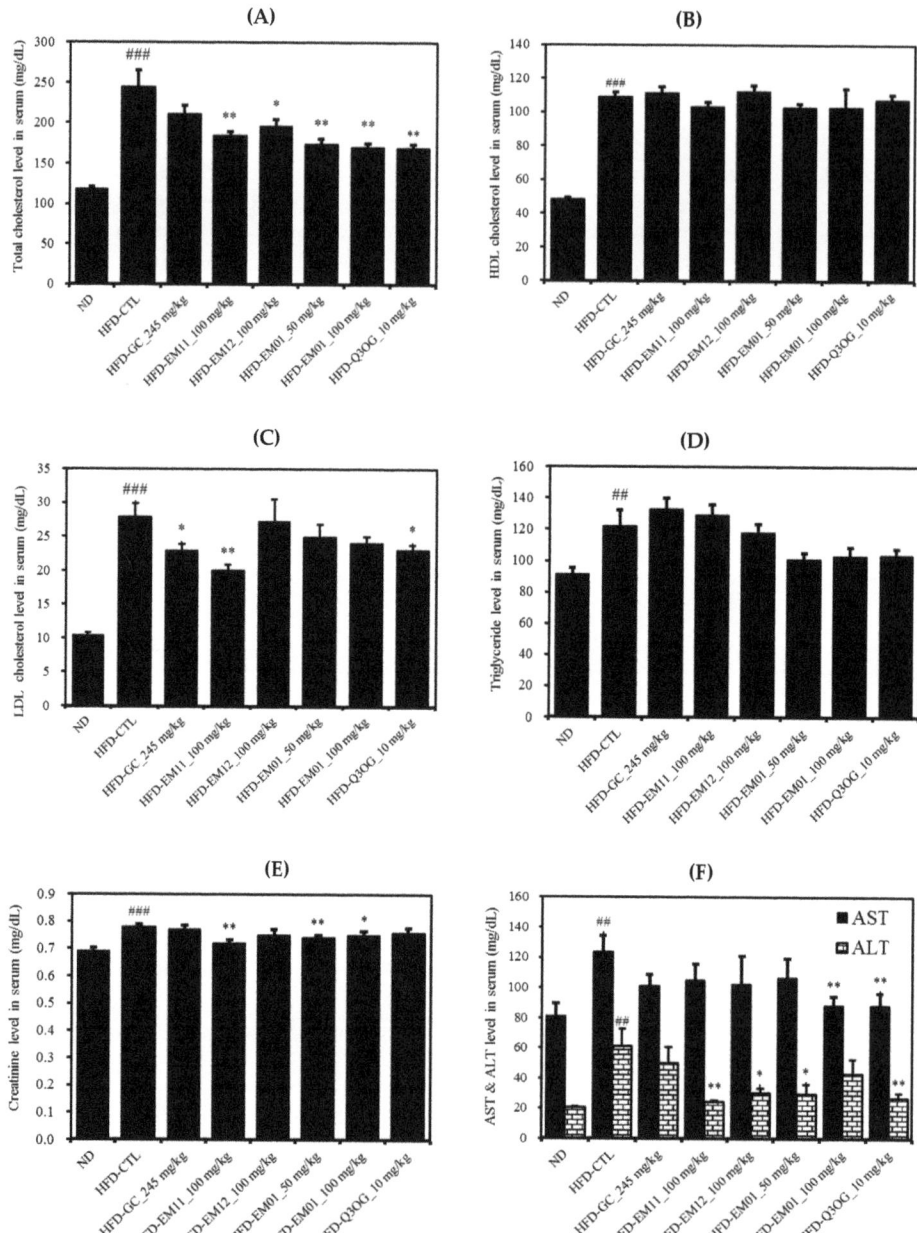

Figure 5. Effects of EM01 on serum lipid profile in HFD-induced obese mice. (**A**) Total cholesterol. (**B**) HDL cholesterol. (**C**) LDL cholesterol. (**D**) Triglyceride. (**E**) Creatinine. (**F**) AST & ALT. Data are presented as the mean ± SEM ($n = 12$); # $p < 0.05$, ## $p < 0.01$, ### $p < 0.001$ vs. ND; * $p < 0.05$, ** $p < 0.01$ and *** $p < 0.001$ vs. HFD-CTL. AST, aspartate aminotransferase; ALT, alanine aminotransferase.

3.6. Effects of EM01 on the Energy Balancing Metabolism in HFD-Induced Obese Mice

Adipokine is a hormone specifically secreted from adipose tissue that affects endocrine system function. We analyzed adiponectin, leptin, IGF-1, which plays an important role in normal growth and health maintenance, NEFA, and glucose, which is associated with energy homeostasis [25]. Adiponectin levels in the serum were significantly decreased in the HFD-CTL group compared to in the ND group, but there was a significant increase in the HFD-EM01 and HFD-Q3OG groups compared to in the HFD-CTL group (Figure 6A). Leptin, IGF-1, NEFA, and glucose levels in the serum were significantly increased in the HFD-CTL group compared to in the ND group, while HFD-EM01 and HFD-Q3OG group were significantly decreased compared to the HFD-CTL group (Figure 6B–E).

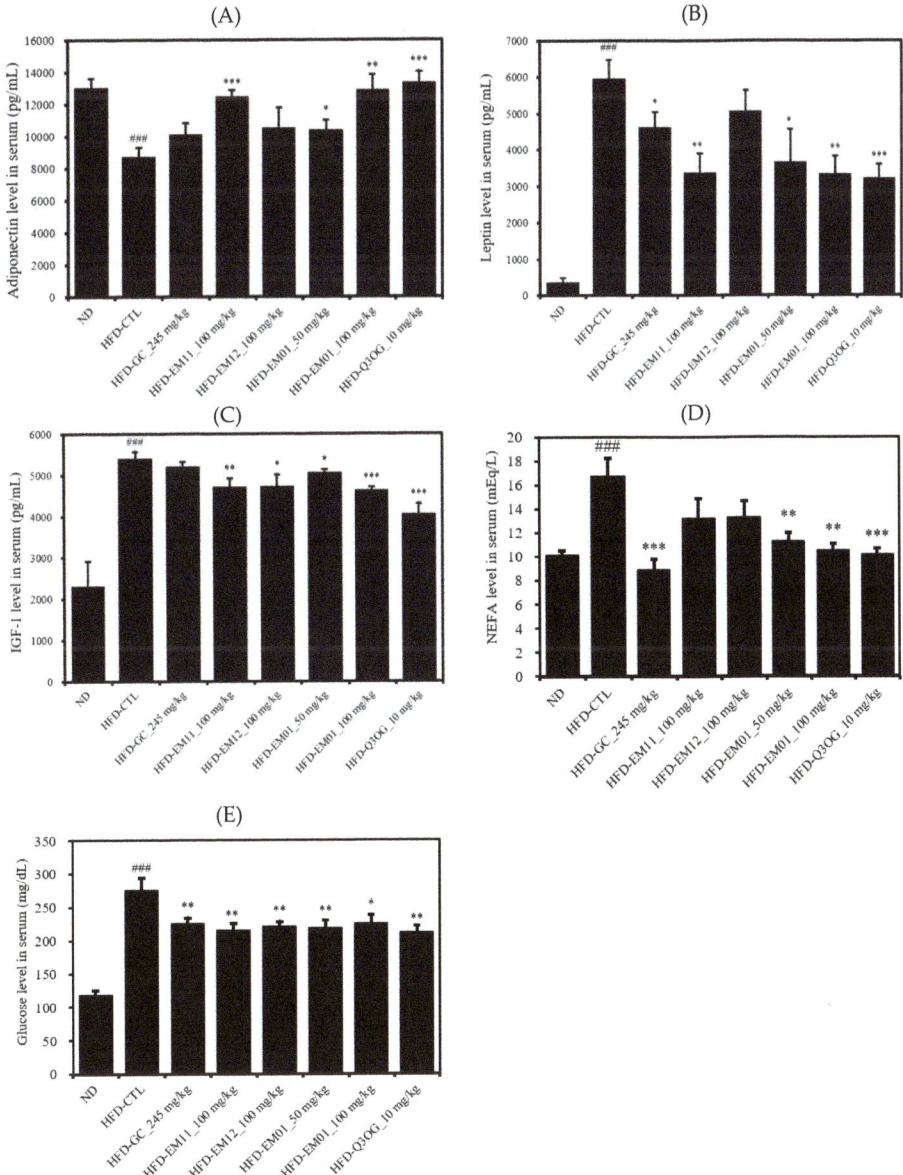

Figure 6. Effects of EM01 on the energy balancing metabolism in HFD-induced obese mice. (**A**) Adiponectin. (**B**) Leptin. (**C**) IGF-1. (**D**) NEFA. (**E**) Glucose. Data are presented as the mean ± SEM ($n = 12$); # $p < 0.05$, ## $p < 0.01$, ### $p < 0.001$ vs. ND; * $p < 0.05$, ** $p < 0.01$ and *** $p < 0.001$ vs. HFD-CTL. IGF-1, insulin-like growth factor-1; NEFA, non-esterified fatty acid.

3.7. Effects of EM01 mRNA Expression Level of Lipid Metabolism-Related Genes in HFD-Induced Obese Mice

We analyzed the mRNA levels of adipogenesis, lipogenesis, and fatty acid oxidation-related genes in the liver, epididymal adipose tissue, and abdominal subcutaneous fat after 8 weeks of EM01 administration (Figure 7).

As shown in Figure 7A, the mRNA levels of FAS, DGAT1, SCD-1, leptin, SREBP1c, PPARγ in the liver were significantly increased in the HFD-CTL group but decreased in the HFD-EM01 and HFD-Q3OG groups. The mRNA levels of COX1, adiponectin, UCP2, and PPARα increased in the HFD-EM01 and HFD-Q3OG groups compared to in the HFD-CTL group.

As shown in Figure 7B, the mRNA levels of the FAS, leptin in the epididymal adipose tissue increased in the HFD-CTL group, but markedly decreased in HFD-EM01 and HFD-Q3OG groups. The mRNA levels of the ACS1, ACOX1, CPT1b, UCP2, adiponectin, PPARα increased in HFD-EM01 and HFD-Q3OG groups compared to those in the HFD-CTL group.

As shown in Figure 7C, the mRNA level of UCP1 in abdominal subcutaneous fat significantly increased in the HFD-EM01 and HFD-Q3OG groups compared to in the HFD-CTL group.

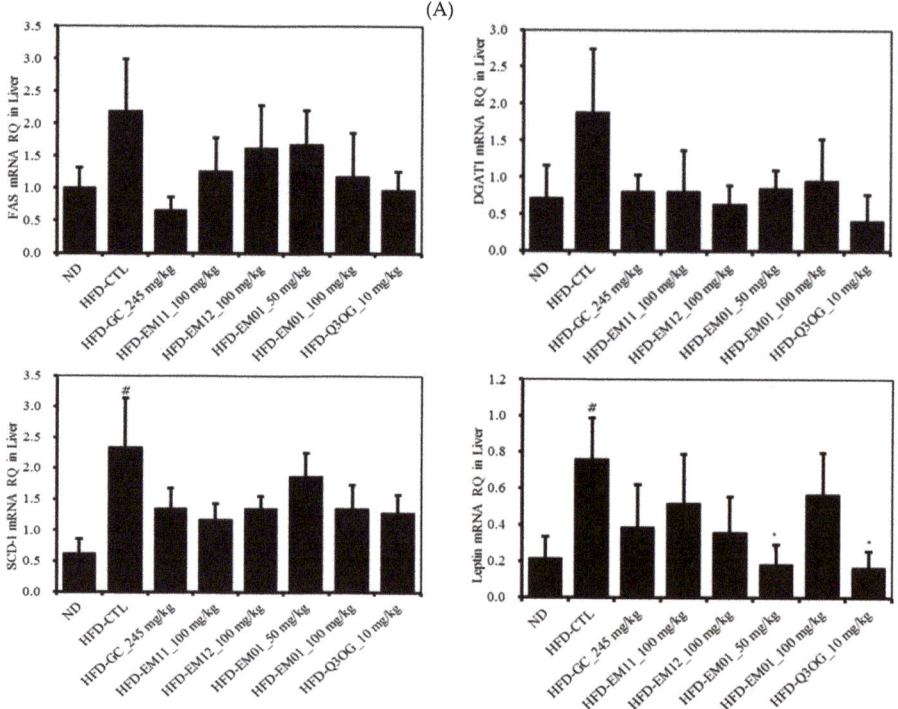

Figure 7. *Cont.*

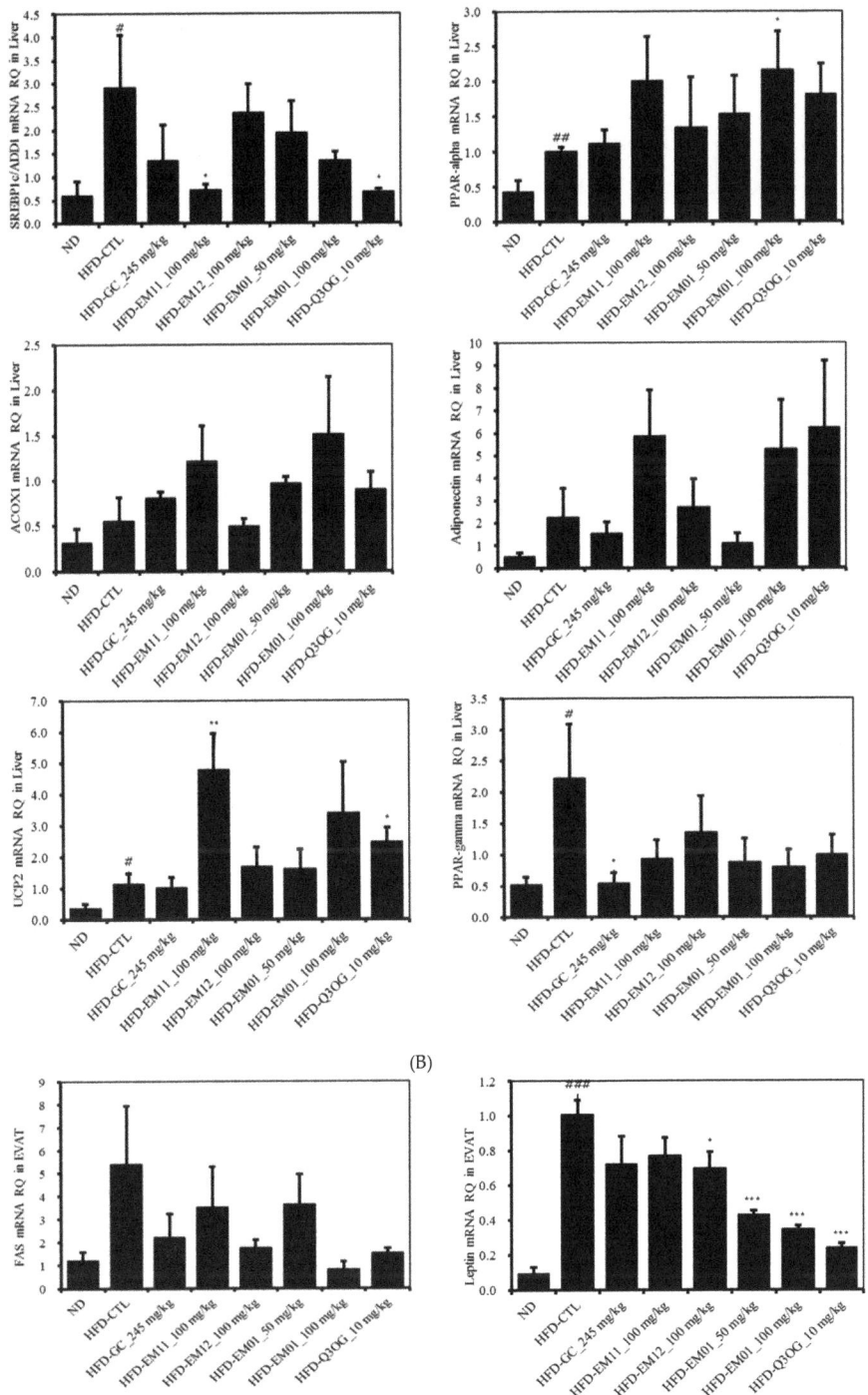

Figure 7. Cont.

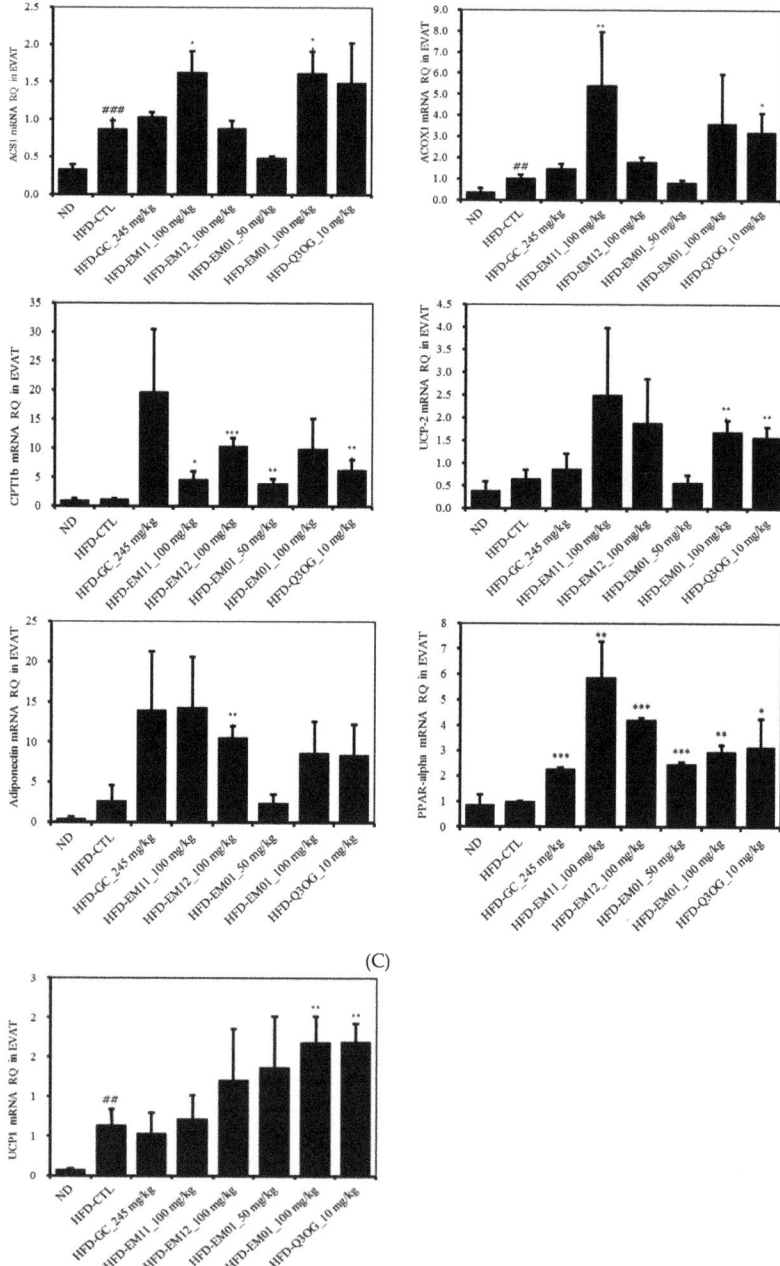

Figure 7. Effects of EM01 on mRNA expression level of lipid metabolism-related genes in HFD-induced obese mice. (**A**) Liver. (**B**) Epididymal adipose tissue. (**C**) Abdominal subcutaneous fat. Data are presented as the mean ± SEM ($n = 12$); # $p < 0.05$, ## $p < 0.01$, ### $p < 0.001$ vs. ND; * $p < 0.05$, ** $p < 0.01$ and *** $p < 0.001$ vs. HFD-CTL. FAS, fatty acid synthase; SCD-1, stearoyl-CoA desaturase-1; SREBP-1c, sterol regulatory element binding protein-1c; PPARγ, peroxisome proliferator-activated receptor γ; DGAT1, diglyceride acyltransferase; UCP, mitochondrial uncoupling proteins; ACOX1, peroxisomal acyl-coenzyme A oxidase 1; PPARα, peroxisome proliferator-activated receptor α; ACS1, acetyl CoA synthetase 1; CPT1b, carnitine palmitoyltransferase 1b.

4. Discussion

In recent years, many side effects have been reported for drugs used for therapeutic purposes, and numerous clinical studies have examined natural products and natural product-derived compounds [26,27]. In this study, we confirmed the anti-obesity effect of three natural materials. In previous studies based on these materials, *N. nucifera* L was found to be a medicinal plant that is not only useful for treating gastritis, bleeding, diarrhea, hemorrhoids, and enuresis, but also contains various biologically active components such as polyphenolics, flavonoids, and tannic acid [28]. SOD, CAT, GST played as defence means against the reactive oxygen species in biological systems. TBARS are formed as a byproduct of lipid peroxidation. This material was reported to have anti-oxidative activity by increasing the levels of SOD, CAT, GST and decreasing TBARS level in liver [29] and anti-obesity activity in HFD-induced mice or anti-obesity activity for inducing apoptosis in 3T3-L1 adipocytes, but few studies have examined these effects [30]. *Morus alba* L, which is prepared for medicinal use such as treating headache, fever, and cough, shows pharmacological activities, particularly antidiabetic effects on lowering blood sugar [31]. *Raphanus sativus* is an herbaceous plant belonging to the cruciferous family and has been reported to have excellent anti-obesity efficacy [32]. Recent studies have demonstrated the anti-obesity activity of a single material, but the mechanism action of this material in a mixture has not been evaluated.

We selected EM01, which contains a high content of *Nelumbo nucifera* [16], among the 26 mixtures (13 kinds of hot water extracts and 13 ethanol extracts) through the antioxidant and anti-obesity experiments at in vitro model. EM01 treatment has an anti-obesity effect at in vivo model using C57BL/6J obese mice. The weight of HFD-induced mice was effectively lower and the weight of the adipose tissue was significantly decreased compared with the control.

Type 2 diabetes is one of the most common disabilities caused by obesity. Obesity directly affects insulin function, which causes glucose to enter the cell membrane for energy metabolism, resulting in insulin resistance and type 2 diabetes [33]. A glucose tolerance test was conducted to investigate the glucose processing ability. EM01 effectively lowered the blood glucose level. This result was similar to those of the glucose tolerance test using neferine an alkaloid compound extracted from *N. nucifera* seed [34].

EM01 reduced the serum lipid profile (TC, HDL-cholesterol, LDL-cholesterol, TG) of HFD-induced obese mice. It may be useful for preventing dyslipidemia induced by obesity [35]. AST, ALT, and creatinine are toxicity indicators of the liver and kidney [24]. EM01 reduced the levels of these markers in the serum and improved the function of each organ.

Adipocytes not only play a role as energy reservoirs but also regulate endocrine function. Adipokine, a hormone specifically expressed and secreted from adipocytes such as adiponectin and leptin, is involved in fatty acid and glucose metabolism. IGF-1 plays an important role in regulating adipose tissue growth. These adipokines interact closely with each other and regulate the hormonal system in the body [36]. EM01 significantly increased the adiponectin level and reduced leptin and IGF-1 level in serum. According to a previous study [37], glucose and lipid metabolism are regulated by insulin signaling. Glucose and fatty acid enter cells and are metabolized for glycogenesis and β-oxidation, respectively. Based on these results, our study confirms that EM01 improves the process of bringing glucose and fatty acid into the cells, resulting in anti-obesity and anti-diabetic effects.

Adipogenesis is the process of cell differentiation by which pre-adipocytes become adipocytes. PPARγ is a key factor in adipogenic transcription [1]. In this study, it was found that mRNA expression of PPARγ in the liver tissue was reduced by administration of EM01. Also, the expression levels of FAS, DGAT1, SCD-1, leptin, SREBP1c were decreased, while the expression levels of UCP2, PPARα, ACOX1, and adiponectin increased in the EM01, Q3OG treatment group in the liver. These results suggest that EM01 treatment inhibits adipogenesis of 3T3-L1 adipocytes and improves fatty acid oxidation. EM01 regulated lipid metabolism in epididymal adipose tissue and abdominal subcutaneous fat. Our results demonstrate that EM01 has an anti-obesity effect in 3T3-L1 adipocytes and C57BL/6J obese mice.

5. Conclusions

In conclusion, we observed the anti-obesity activity of EM01, which is an optimal mixed material containing *N. nucifera* L., *M. alba* L., and *R. sativus* in vitro and in vivo models. EM01 reduced lipid accumulation and weight gain, fat mass, serum lipid concentration, and mRNA expression levels of lipid-metabolism-related genes in HFD-induced obese mice. Therefore, our findings show that EM01 has potential as an effective material for anti-obesity treatment.

Author Contributions: W.-S.S., S.-I.C., B.-Y.C., S.-H.C., X.H., H.-D.C., J.-H.C., S.-H.K., B.-Y.L., I.-J.K. and O.-H.L. designed research; W.-S.S., S.-H.K. performed research and analyzed the data; W.-S.S., O.-H.L. wrote the paper. All authors read and approved the final manuscript.

Funding: This research was funded Technology Development Program (C1013823-01-01) funded by the Ministry of SMEs and Startups (MSS, Korea).

Acknowledgments: This work was supported by the Technology Development Program (C1013823-01-01) funded by the Ministry of SMEs and Startups (MSS, Korea); NRF (National Research Foundation of Korea) Grant funded by the Korean Government (NRF-2018-Fostering Core Leaders of the Future Basic Science Program/Global Ph.D. Fellowship Program); The basic Science Research Program through the National Research Foundation of Korea (NRF) funded by the Ministry of Education (NRF-2017R1D1A3B06028469) and has been worked with the support of a research grant from Kangwon National University in 2017.

Conflicts of Interest: The authors declare no conflict of interest.

References

1. Moon, J.; Do, H.J.; Kim, O.Y.; Shin, M.J. Antiobesity effects of quercetin-rich onion peel extract on the differentiation of 3T3-L1 preadipocytes and the adipogenesis in high fat-fed rats. *Food Chem. Toxicol.* **2013**, *58*, 347–354. [CrossRef]
2. Mu, M.; Xu, L.-F.; Hu, D.; Wu, J.; Bai, M.-J. Dietary patterns and overweight/obesity: A review article. *Iran J. Public Health.* **2017**, *46*, 869–876. [PubMed]
3. Kahn, S.E.; Hull, R.L.; Utzschneider, K.M. Mechanisms linking obesity to insulin resistance and type 2 diabetes. *Nature* **2006**, *444*, 840–847. [CrossRef] [PubMed]
4. Rahmouni, K.; Correia, M.L.; Haynes, W.G.; Mark, A.L. Obesity-associated hypertension: New insights into mechanisms. *Hypertension.* **2005**, *45*, 9–14. [CrossRef] [PubMed]
5. Zahid, N.; Claussen, B.; Hussain, A. High prevalence of obesity, dyslipidemia and metabolic syndrome in a rural area in Pakistan. *Diabetes Metab. Syndr: Clinical Res. Rev.* **2008**, *2*, 13–19. [CrossRef]
6. Lee, H.S.; Nam, Y.; Chung, Y.H.; Kim, H.R.; Park, E.S.; Chung, S.J.; Kim, J.H.; Sohn, U.D.; Kim, H.C.; Oh, K.W.; et al. Beneficial effects of phosphatidylcholine on high-fat diet-induced obesity, hyperlipidemia and fatty liver in mice. *Life Sci.* **2014**, *118*, 7–14. [CrossRef]
7. Kopelman, P.G. Obesity as a medical problem. *Nature* **2000**, *404*, 635–643. [CrossRef] [PubMed]
8. Padwal, R.S.; Majumdar, S.R. Drug treatments for obesity: Orlistat, sibutramine, and rimonabant. *The Lancet* **2007**, *369*, 71–77. [CrossRef]
9. Weigle, D.S. Pharmacological therapy of obesity: Past, present, and future. *J. Clin. Endocrinol. Metab.* **2003**, *88*, 2462–2469. [CrossRef] [PubMed]
10. Ono, Y.; Hattori, E.; Fukaya, Y.; Imai, S.; Ohizumi, Y. Anti-obesity effect of *Nelumbo nucifera* leaves extract in mice and rats. *J. Ethnopharmacol.* **2006**, *106*, 238–244. [CrossRef]
11. Ann, J.Y.; Eo, H.; Lim, Y. Mulberry leaves (*Morus alba* L.) ameliorate obesity-induced hepatic lipogenesis, fibrosis, and oxidative stress in high-fat diet-fed mice. *Genes Nutr.* **2015**, *10*, 46–59. [CrossRef]
12. Chaturvedi, P. Inhibitory response of *Raphanus sativus* on lipid peroxidation in albino rats. *J. Evid. Based Complementary Altern Med.* **2008**, *5*, 55–59. [CrossRef] [PubMed]
13. Sharma, B.R.; Oh, J.; Kim, H.A.; Kim, Y.J.; Jeong, K.S.; Rhyu, D.Y. Anti-obesity effects of the mixture of *Eriobotrya japonica* and *Nelumbo nucifera* in adipocytes and high-fat diet-induced obese mice. *Am. J. Chin. Med.* **2015**, *43*, 681–694. [CrossRef]
14. Zheng, G.; Sayama, K.; Okubo, T.; Juneja, L.R.; Oguni, I. Anti-obesity effects of three major components of green tea, catechins, caffeine and theanine, in mice. *In Vivo* **2004**, *18*, 55–62.

15. Nadeem, S. Synergistic effect of *Commiphora mukul* (gum resin) and *Lagenaria siceraria* (fruit) extracts in high fat diet induced obese rats. *Asian Pac. J. Trop. Dis.* **2012**, S883–S886. [CrossRef]
16. Jang, G.W.; Park, E.Y.; Choi, S.H.; Choi, S.I.; Cho, B.Y.; Sim, W.S.; Han, X.; Cho, H.D.; Lee, O.H. Development and validation of analytical method for wogonin, quercetin, and quercetin-3-O-glucuronide in extracts of *Nelumbo nucifera*, *Morus alba* L., and *Raphanus sativus* mixture. *J. Food Hyg. Saf.* **2018**, *33*, 289–295. [CrossRef]
17. Cho, M.; Ko, S.B.; Kim, J.M.; Lee, O.H.; Lee, D.W.; Kim, J.Y. Influence of extraction conditions on antioxidant activities and catechin content from bark of *Ulmus pumila* L. *Appl. Biol. Chem.* **2016**, *59*, 329–336. [CrossRef]
18. Tawaha, K.; Alali, F.Q.; Gharaibeh, M.; Mohammad, M.; El-Elimat, T. Antioxidant activity and total phenolic content of selected Jordanian plant species. *Food Chem.* **2007**, *104*, 1372–1378. [CrossRef]
19. Cho, B.Y.; Park, M.R.; Lee, J.H.; Ra, M.J.; Han, K.C.; Kang, I.J.; Lee, O.H. Standardized *Cirsium setidens* Nakai ethanolic extract suppresses adipogenesis and regulates lipid metabolisms in 3T3-L1 adipocytes and C57BL/6J mice fed high-fat diets. *J. Med. Food.* **2017**, *20*, 763–776. [CrossRef] [PubMed]
20. Lee, J.H.; Cho, B.Y.; Choi, S.H.; Jung, T.D.; Choi, S.I.; Lim, J.H.; Lee, O.H. Sulforaphane attenuates bisphenol A-induced 3T3-L1 adipocyte differentiation through cell cycle arrest. *J. Funct. Foods.* **2018**, *44*, 17–23. [CrossRef]
21. Bondet, V.; Brand-Williams, W.; Berset, C. Kinetics and mechanisms of antioxidant activity using the DPPH. free radical method. *LWT-Food Sci. Technol.* **1997**, *30*, 609–615. [CrossRef]
22. Sinha, R.; Fisch, G.; Teague, B.; Tamborlane, W.V.; Banyas, B.; Allen, K.; Savoye, M.; Rieger, V.; Taksali, S.; Barbetta, G.; et al. Prevalence of impaired glucose tolerance among children and adolescents with marked obesity. *N. Engl. J. Med.* **2002**, *346*, 802–810. [CrossRef] [PubMed]
23. Barter, P.; Gotto, A.M.; LaRosa, J.C.; Maroni, J.; Szarek, M.; Grundy, S.M.; Kastelein, J.P.; Bittner, V.; Fruchart, J.C. HDL cholesterol, very low levels of LDL cholesterol, and cardiovascular events. *N. Engl. J. Med.* **2007**, *357*, 1301–1310. [CrossRef] [PubMed]
24. Xiong, Y.; Shen, L.; Liu, K.J.; Tso, P.; Xiong, Y.; Wang, G.; Woods, S.C.; Liu, M. Anti-obesity and anti-hyperglycemic effects of ginsenoside Rb1 in rats. *Diabetes* **2010**, *59*, 2505–2512. [CrossRef] [PubMed]
25. Pardina, E.; Ferrer, R.; Baena-Fustegueras, J.A.; Lecube, A.; Fort, J.M.; Vargas, V.; Catalan, R.; Peinado-Onsurbe, J. The relationships between IGF-1 and CRP, NO, leptin, and adiponectin during weight loss in the morbidly obese. *Obes. Surg.* **2010**, *20*, 623–632. [CrossRef] [PubMed]
26. Butler, M.S. Natural products to drugs: Natural product-derived compounds in clinical trials. *Nat. Prod. Rep.* **2008**, *25*, 475–516. [CrossRef]
27. Pauwels, E.; Stoven, V.; Yamanishi, Y. Predicting drug side-effect profiles: A chemical fragment-based approach. *BMC bioinformatics.* **2011**, *12*, 169–182. [CrossRef]
28. Choe, J.H.; Jang, A.; Choi, J.H.; Choi, Y.S.; Han, D.J.; Kim, H.Y.; Lee, M.A.; Kim, H.W.; Kim, C.J. Antioxidant activities of lotus leaves (*Nelumbo nucifera*) and barley leaves (*Hordeum vulgare*) extracts. *Food Sci. Biotechnol.* **2010**, *19*, 831–836. [CrossRef]
29. Huang, B.; Ban, X.; He, J.; Tong, J.; Tian, J.; Wang, Y. Hepatoprotective and antioxidant activity of ethanolic extracts of edible lotus (*Nelumbo nucifera* Gaertn.) leaves. *Food Chem.* **2010**, *120*, 873–878. [CrossRef]
30. Bin, X.; Jin, W.; Wenqing, W.; Chunyang, S.; Xiaolong, H.; Jianguo, F. *Nelumbo nucifera* alkaloid inhibits 3T3-L1 preadipocyte differentiation and improves high-fat diet-induced obesity and body fat accumulation in rats. *J. Med. Plant Res.* **2011**, *5*, 2021–2028.
31. Hunyadi, A.; Martins, A.; Hsieh, T.J.; Seres, A.; Zupkó, I. Chlorogenic acid and rutin play a major role in the in vivo anti-diabetic activity of *Morus alba* leaf extract on type II diabetic rats. *PLOS One.* **2012**, *7*, e50619. [CrossRef] [PubMed]
32. Vivarelli, F.; Canistro, D.; Sapone, A.; De Nicola, G.R.; Marquillas, C.B.; Iori, R.; Antonazzo, I.C.; Gentilini, F.; Paolini, M. *Raphanus sativus* cv. sango sprout juice decreases diet-induced obesity in Sprague Dawley rats and ameliorates related disorders. *PLoS ONE* **2016**, *11*, e0150913. [CrossRef] [PubMed]
33. Kahn, B.B.; Flier, J.S. Obesity and insulin resistance. *Clin. Investig.* **2000**, *106*, 473–481. [CrossRef]
34. Pan, Y.; Cai, B.; Wang, K.; Wang, S.; Zhou, S.; Yu, X.; Xu, B.; Chen, L. Neferine enhances insulin sensitivity in insulin resistant rats. *J. Ethnopharmacol.* **2009**, *124*, 98–102. [CrossRef]
35. Zhou, T.; Luo, D.; Li, X.; Luo, Y. Hypoglycemic and hypolipidemic effects of flavonoids from lotus (*Nelumbo nuficera* Gaertn) leaf in diabetic mice. *J. Med. Plant. Res.* **2009**, *3*, 290–293.

36. Winsz-Szczotka, K.; Kuźnik-Trocha, K.; Komosińska-Vassev, K.; Kucharz, E.; Kotulska, A.; Olczyk, K. Relationship between adiponectin, leptin, IGF-1 and total lipid peroxides plasma concentrations in patients with systemic sclerosis: Possible role in disease development. *Int. J. Rheum Dis.* **2016**, *19*, 706–714. [CrossRef]
37. Saltiel, A.R.; Kahn, C.R. Insulin signalling and the regulation of glucose and lipid metabolism. *Nature* **2001**, *414*, 799–806. [CrossRef] [PubMed]

© 2019 by the authors. Licensee MDPI, Basel, Switzerland. This article is an open access article distributed under the terms and conditions of the Creative Commons Attribution (CC BY) license (http://creativecommons.org/licenses/by/4.0/).

Article

Use of Phycobiliproteins from Atacama Cyanobacteria as Food Colorants in a Dairy Beverage Prototype

Alexandra Galetović [1,*], Francisca Seura [2], Valeska Gallardo [2], Rocío Graves [2], Juan Cortés [1], Carolina Valdivia [3], Javier Núñez [3], Claudia Tapia [3], Iván Neira [3], Sigrid Sanzana [3] and Benito Gómez-Silva [1]

1. Departamento Biomédico, Laboratorio de Bioquímica, Facultad de Ciencias de la Salud and Centre for Biotechnology and Bioengineering (CeBiB), Universidad de Antofagasta, Avenida Angamos N° 601, Antofagasta 1270300, Chile; juancortesgonzalez.96@gmail.com (J.C.); benito.gomez@uantof.cl (B.G.-S.)
2. Departamento de Ciencias de los Alimentos y Nutrición, Universidad de Antofagasta, Avenida Angamos N° 601, Antofagasta 1270300, Chile; fran.ncisca@hotmai.com (F.S.); nutricionistavale.gd@gmail.com (V.G.); Rociograves@hotmail.com (R.G.)
3. Departamento de Tecnología Médica, Universidad de Antofagasta/Avenida Angamos N° 601, Antofagasta 1270300, Chile; carovaldiviab@gmail.com (C.V.); javier.ignacio.ng@gmail.com (J.N.); claudiafran.t@gmail.com (C.T.); ivan.neira@uantof.cl (I.N.); sigrid.sanzana@uantof.cl (S.S.)
* Correspondence: alexandra.galetovic@uantof.cl; Tel.: +56-55-2637054

Received: 25 January 2020; Accepted: 3 February 2020; Published: 24 February 2020

Abstract: The interest of the food industry in replacing artificial dyes with natural pigments has grown recently. Cyanobacterial phycobiliproteins (PBPs), phycoerythrin (PE) and phycocyanin (PC), are colored water-soluble proteins that are used as natural pigments. Additionally, red PE and blue PC have antioxidant capabilities. We have formulated a new food prototype based on PBP-fortified skim milk. PBPs from Andean cyanobacteria were purified by ammonium sulfate precipitation, ion-exchange chromatography, and freeze-drying. The stability of PE and PC was evaluated by changes in their absorption spectra at various pH (1–14) and temperature (0–80 °C) values. Purified PBPs showed chemical stability under pH values of 5 to 8 and at temperatures between 0 and 50 °C. The antioxidant property of PBP was confirmed by ABTS (2,2'-Azino-bis (3-ethylbenzothiazoline-6-sulfonic acid) diammonium salt radical ion scavenging, and FRAP (Ferric Antioxidant Power) assays. The absence of PBP toxicity against *Caenorhabditis elegans* was confirmed up to 1 mg PBP/mL. Skim milk fortified with PE obtained a higher score after sensory tests. Thus, a functional food based on skim milk-containing cyanobacterial PBPs can be considered an innovative beverage for the food industry. PBPs were stable at an ultra-high temperature (138 °C and 4 s). PBP stability improvements by changes at its primary structure and the incorporation of freeze-dried PBPs into sachets should be considered as alternatives for their future commercialization.

Keywords: cyanobacteria; phycobiliproteins; natural pigment; phycoerythrin; phycocyanin; food colorant

1. Introduction

Cyanobacteria are a large, diverse and ancient group of ubiquitous Gram-negative prokaryotes that are found in terrestrial or aquatic habitats. They perform oxygenic photosynthesis and colonize freshwater, marine and brackish waters and soils and rocks from drylands. They also are found at extreme environments that are subjected to high ultraviolet radiation, high or low temperatures, desiccation, and nutrient deficiencies [1,2]. Cyanobacteria can be found as free-living unicellular or filamentous microorganisms and as photobionts in symbiotic association with fungi. Members of some filamentous genera (e.g., *Anabaena* and *Nostoc*) have heterocysts—cells specialized in atmospheric nitrogen fixation. Cyanobacteria are an important source of natural products with interest from the pharmaceutical and biotechnological industries [3,4]. Particularly, cyanobacterial pigments have attracted attention for their use in the food, textile and cosmetic industries [5,6].

Three types of cyanobacterial water-soluble phycobiliproteins (PBPs), C-allophycocyanin (APC), phycocyanin (PC), and C-phycoerythrin (PE) are organized in phycobilisomes in the photosynthetic apparatus. PE, PC and APC act as antennae pigments with absorption maxima at 562, 615, and 652 nm, respectively [7]. PBPs contain linear tetrapyrrolic chromophores (bilins) that are covalently bound to apoproteins via cysteine residues [7,8]. The ability of PBPs to act as free radical scavengers has been demonstrated to be centered on their tetrapyrrolic systems, supporting their use in the food, cosmetics and pharmaceutical industries, as well as their use as fluorochromes in biomedical research. In addition, PBPs isolated from various cyanobacterial species also have beneficial effects as anticancer, neuroprotective, anti-inflammatory, anti-allergic and hepatoprotective biomolecules [9–15].

Color in food has an important impact on consumers since it is one of the first characteristics we perceive from a product. Additionally, colored food is attractive, and color allows for better identification and selection among similar products. Coloration helps to relate water with food; for example, yogurt, animal or vegetal milk (e.g., soybean, coconut, or almonds) with fruits such as strawberries (reddish), blue berries (purple), and melon (green). Then, consumers would consider ingesting these types of food as beneficial to their well-being even though they may not have fruits [16].

Artificial dyes are stable at different temperatures, pH and light regimes, maintaining their coloration for long periods. Some synthetic dyes have been approved for their use in the food industry; however, some of them have been reported to be neurotoxic, mutagenic, and genotoxic (lemon-yellow tartrazine), damaging in the liver and the kidney (brilliant blue), and triggering biochemical changes and cancer in the thyroid gland (cherry-red erythrosine) [17–19]. As such, the demand for natural dyes in the food industry has grown in recent years due to the toxicity of artificial colorants [20]. There is great interest in finding non-harmful alternative pigments, especially blue pigments such as PC, which has the capability to scavenge hydroxyl ions in order to avoid lipoperoxidation [21]. Natural dyes can be considered renewable and sustainable bioresources with minimal environmental impact [22]. These eco-friendly, presumably mostly non-toxic, natural colorants could have applications in other industrial sectors like the cosmetics or pharmacological industries [23–26]. Other studies have shown that some natural colorants from plants have health-promoting properties in the human diet, like natural carotenoids that provide beneficial biological effects such as antioxidant and anticancer properties [27,28].

Cyanobacterial PBPs are natural pigments used as colorants in some food products; for example, aqueous extracts of non-purified blue PC from *Spirulina* have been added in ice creams, yogurts, isotonic beverages, confectionery, and jellies [29]. Particularly, bright blue PC has been selected over others less-bright natural colorants such as gardenia blue and indigo in confectionery production [30–32]. Red PE has been mostly used as a fluorescent probe in biomedical studies, rather than in the food industry.

Spirulina (*Arthrospira platensis*) has been used as a source of PBPs for additions to food products; however, this microorganism mostly produces PC. The Andes wetlands in northern Chile are a source of microbial communities that include PC and PE producing cyanobacteria. Isolated strains *Nostoc* sp. Caquena (CAQ-15), LLA-10 and *Nostoc* sp. Llayta (LLC-10) from Andes wetlands, above 3000 m of altitude, predominantly accumulate red PE, blue PC, and a purple fluorescent mixture of both, respectively. In addition, the CAQ-15 strain modulates its PBP content by complementary chromatic adaptation [33]. Here, we report the use of these isolated *Nostocaceae* strains for the purification of PBPs and their application as colorants and antioxidant molecules in the formulation of dairy functional prototypes. Additionally, we report results on the stability, antioxidant capabilities and toxicity of the purified PBPs, as well as sensory tests of the final prototypes. This work represents the first step in the use of PBPs from Andean cyanobacteria as ingredients for the food industry.

2. Results and Discussion

2.1. PBPs Stability under Different Temperature Regimes

The stability of the purified PBPs from the cyanobacterial strains LLC-10 and CAQ-15 were measured as the concentration of the remaining non-denatured PBPs after incubation at various temperature and pH regimes. Nearly 80% of the PC and PE proteins from both strains were stable after 72 h of incubation at temperatures from 10 to 21 °C. The denaturation of the PC and PE proteins increased to nearly 50% after 24 h of incubation at 26 to 53 °C (Table 1 and Figure 1). These proteins showed total denaturation after 48 h of incubation at temperatures over 55 °C (Table 1 and Figure 1). PC from both strains had a similar response to temperature; however, PE from the LLC-10 strain reached 50% denaturation at 35 °C, while PC required a lower temperature (26 °C) to reach the same level of denaturation. Additionally, PE was a thermally more stable protein (Table 1). Purified PE from the LLC-10 strain and PC from the LLA-10 strain were subjected to incubations at 138 °C for 4 s in order to evaluate changes in stability after this heat treatment. Based on changes in their visible absorption spectra, only 10% and 15% denaturation were observed at the PE and PC solutions, respectively.

The thermal stability shown by PBPs from the Atacama cyanobacterial strains was consistent with the denaturation profiles that are expected for mesophilic proteins and also for more stable proteins from thermophilic cyanobacteria [34]. PC denaturation from *Spirulina platensis* occurred after incubations over 45 °C at pH 7 [35–37]; the bleaching of PBPs from mesophilic cyanobacteria and algae was observed at temperatures between 60 and 65 °C [38]. Additionally, PC from *Anabaena fertilissima* PUPPCC 410.5 was unstable at 42 °C with a 50% loss after 4 days; this protein was very stable and maintained its antioxidant properties at 4 °C over 6–9 days [39,40]. Moreover, PC from *Spirulina fusiformis* was denatured at temperatures above 70 °C [41,42]. Though the PBPs from *Phormidium rubidium* A09DM were stable at temperatures from 4 to 40 °C, their corresponding absorptions at their maximal wavelengths decreased 2–4 fold at 60 to 80 °C [41]. Comparatively, PBPs from thermophilic cyanobacteria had better temperature stability than PBPs from mesophilic cyanobacteria; for example, PE from *Leptolyngbya* sp. KC 45 maintained 80% of its antioxidant activity after exposure to 60 °C for 30 min [43]. Additionally, PC from *Thermosynechococcus elongatus* TA-1 was stable between 4 and 60 °C at a pH range of 4 to 9, but PC denaturation occurred at temperatures over 75 °C [44]. Likewise, the thermostable PC from the *Synechococcus lividus* PCC 6715 strain that was isolated from a hot spring maintained 90% and 70% stability at 50 °C for 5 h and two weeks, respectively [45]. Additionally, PC from the thermophilic red algae *Cyanidioschyzon merolae* showed a midpoint denaturation at 83 °C and pH 5, with a half-life of 40 min [46].

Future work will consider the isolation of thermophilic cyanobacteria from a thermal spring in the Atacama region and the purification of their phycobiliproteins. The biochemical properties of these PBPs and relevant genetic studies would provide new unidentified protein resources for biotechnological applications.

Table 1. Interpolation of temperatures to evaluate protein stability. The remaining non-denatured phycobiliproteins content was expressed as percentage (0%, 50% or 80%) of the control condition at 0 °C. Each value shown represents the interpolated temperature at 24, 48 and 72 h of incubation. The standard deviation for LLC-10 (PC), LLC-10 (PE), CAQ-15 (PC) and CAQ-15 (PE) were 9.3 to 13.4, 13.6 to 14.2, 15.1 to 16.4, and 15.72 to 18.6, respectively.

Strain	PBP	Percentage of Remaining Phycobiliproteins at Different Interpolated Temperatures (°C)								
		0%			50%			80%		
		24 h	48 h	72 h	24 h	48 h	72 h	24 h	48 h	72 h
LLC-10	PC	84.7	54.3	52.8	42.3	27.1	26.4	16.9	10.9	10.6
	PE	106.9	76.9	70.4	53.4	38.5	35.2	21.4	15.4	14.1
CAQ-15	PC	68.1	57.8	49.5	34.0	28.9	24.7	13.6	11.6	9.9
	PE	82.7	62.5	52.5	41.4	31.3	26.2	16.5	12.5	10.5

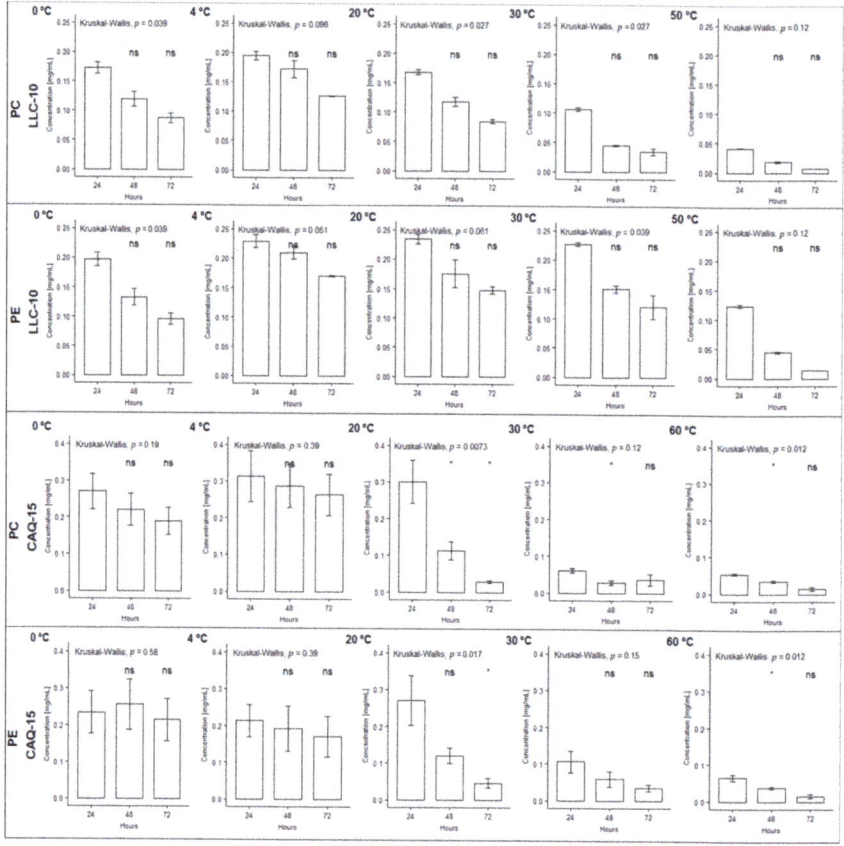

Figure 1. Effect of the temperature on the stability of phycocyanin (PC) and C-phycoerythrin (PE) phycobiliproteins from the cyanobacterial *Nostoc* sp. Llayta (LLC-10) and *Nostoc* sp. Caquena (CAQ-15) strains. The stability of the proteins was expressed as mg/mL of the remaining native phycobiliproteins after 24 to 72 h of incubation. (*) significant, ns: no significant with respect to the reference group at 24 h.

2.2. *PBPs Stability under Different pH Regimes*

Based on their absorbance spectra, PE and PC from the LLC-10 strain were stable at pH 5 to 8. At pH 5, PE and PC showed absorbance maximum at 565 and 620 nm, respectively (Figure 2). At an acidic pH (1 to 3), a PBP precipitation was observed (Figure 2a, lower left), showing a wide non-characteristic profile of their absorption spectrum (Figure 2a, upper left). Additionally, the incubation of PBPs at an alkaline pH (9 to 14) rendered uncolored solutions with a change in their absorption spectra due to protein denaturation (Figure 2b, lower right).

Both, acidic or alkaline pH values alter the electrostatic and hydrogen bond interactions among amino acid residues in proteins; in PBP, this effect translates into structural changes in chromophores and the apoproteins [41]. The stability of Atacama PBPs from the LLC-10 strain at pH 5–8 was similar to PC from *Spirulina platensis* at pH 4–6 [35–37]. In addition, the addition of preservative molecules such as citric acid, sugars and calcium chloride improve PBP stability [47–49]. Blue PC from *Spirulina platensis* increased stability in the presence of citric acid at 35 °C over 15 days [50]. Then, non-toxic PBP stabilizers should be explored in depth to expand the use of these proteins in the food industry.

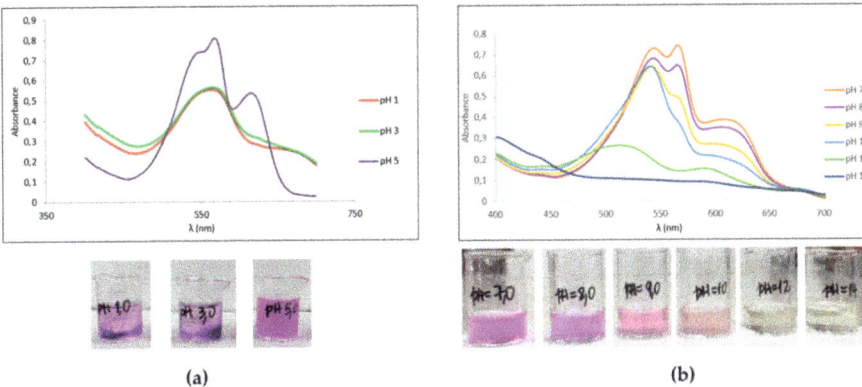

Figure 2. Effect of pH on phycobiliprotein stability. The stability of mixed solutions of PE plus PC from the LLC-10 strain was evaluated at an acidic (**a**), neutral and basic pH (**b**) range. Changes in absorption spectra, coloration and solubility are presented. These experiments were run in triplicate. The information provided in this figure corresponds to one complete experiment. The application of the Kolmogorov–Smirnov goodness-of-fit test showed that the inequality hypothesis was significant ($p < 0.05$) among the distribution functions for the absorption of each pH condition.

2.3. Antioxidants Activity

The antioxidant capabilities of purified PE and PC have been previously demonstrated. Additionally, the antioxidant power of PC has been related to its ability to sequester hydroxyl ions avoiding lipo-peroxidation [9,21,51–54].

The antioxidant activity of the purified Atacama PBPs was evaluated by two assays, ABTS and FRAP, and the results are shown in Table 2. The methanol extracts from the cyanobacterial LLA-10, CAQ-15 and LLC-10 strains showed antioxidant activity values of 195 ± 38 to 717 ± 60 µmoles Trolox equivalent (TE)/100 g fresh mass by the ABTS assay (Table 2). Purified phycocyanin (PC-LLA-10), phycocyanin (PC-CAQ-15), and phycoerythrin (PE-CAQ-15) showed antioxidant activities between 2 and 3 µmoles TE/100 mg pigment (Table 2). These results indicate that PBPs and the methanol extracts from Atacama cyanobacteria have an antioxidant power comparable to fruits such as mulberry, pineapple and passion-fruit [55]. Further support was obtained from the FRAP assay (Table 2). Consequently, cyanobacteria from the Atacama Desert are an innovative source of functional natural antioxidants that have a potential protective role against oxidative stress and biotechnological applications in the food, pharmaceutical and cosmetic industries.

Table 2. Antioxidant activity of purified phycobiliproteins and methanol extracts from the Atacama native cyanobacterial strains CAQ-15, LLC-10 and LLA-10. The antioxidant capabilities were evaluated by the ABTS and FRAP assays and expressed as TEAC (Trolox equivalent antioxidant capacity). TE: Trolox equivalents; PE: Phycoerythrin; and PC: Phycocyanin. The assays were conducted in triplicate, and the results are shown as the mean values with the corresponding standard deviation.

Sample	ABTS	FRAP
Methanol extract	µmoles TE/100 g fresh biomass	
CAQ-15	717 ± 61	50.23 ± 1.64
LLC-10	641 ± 95	12.63 ± 0.73
LLA-10	195 ± 38	13.14 ± 1.52
Phycobiliprotein	µmoles TE/100 mg phycobiliprotein	
PE-CAQ-15	198 ± 45	0.92 ± 0.15
PC-CAQ-15	312 ± 56	1.55 ± 0.10
PC-LLA-10	205 ± 41	2.50 ± 0.15

2.4. Toxicity of Phycobiliproteins against C. elegans

The nematode *Caenorhabditis elegans* offers several advantages as an emerging model in environmental toxicology. It is easy and inexpensive to culture in the laboratory, it has a short life cycle that allows for short-time span experiments, and there is increasing evidence on its genetic and physiological similarity with mammals, so results related to its use have the potential to predict possible effects in higher animals [56,57]. Our work showed that PBPs that were purified from the LLC-10 strain (genus *Nostoc*) were not toxic to *C. elegans*; the nematode survival was 100% at all concentrations used; comparatively, ivermectin, a nematicidal drug, showed a 100% mortality (Figure 3), which is in agreement with the information provided by Ju et al. (2014) on other cyanobacterial pigments [58].

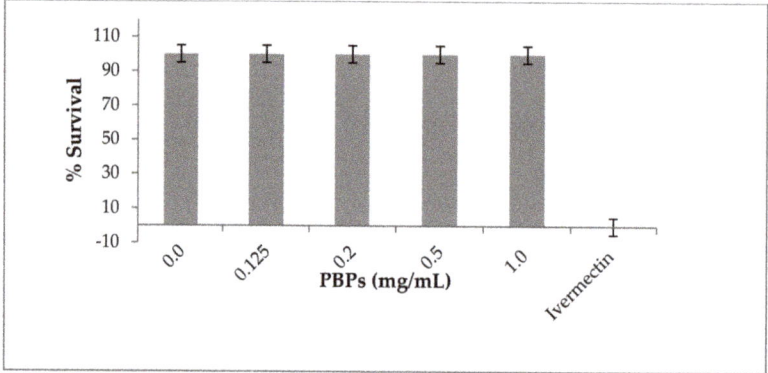

Figure 3. Toxicity of phycobiliproteins against the nematode *Caenorhabditis elegans*. Toxicity test of a mixed solution of PE plus PC from the LLC-10 strain was evaluated at a phycobiliprotein (PBP) concentration from 0.125 to 1.0 mg/mL. Ivermectin (0.3 mg/mL) was used as a nematicidal control drug. The nematode M-9 buffer was used as a control without phycobiliproteins. Quadruplicate tests were carried out for 24 h at 18 °C.

2.5. Sensory Test

Prototypes of skim milk that were fortified with the phycobiliproteins PC or PE purified from two Atacama cyanobacterial strains were the functional foods that were tested by a volunteer team by using an acceptability hedonic scale (Figure 4). The results of the sensory evaluation showed that there were no statistically significant differences between the prototypes. However, the parameters' appearance (related to the color reached by the prototype) and texture were the best valued by judges. The appearance of the prototypes had a good acceptance (mean score 3.7) that was only surpassed by texture. The highest score at the sensory test was obtained by the skim milk that was fortified with PE (prototype N°2).

Several reports have shown a wide acceptance for food products that are supplemented with microalgal natural pigments, given the improvements in color and antioxidant properties, e.g., chlorophyll and carotenoids from *Chlorella vulgaris* and *Haematococcus pluvialis* [59,60]. In addition, microalgae have been incorporated into dairy products as a source of bioactive and coloring compounds, with good acceptability, particularly in texture and appearance [61]. In this study, texture and appearance stood out among all parameters tested. Therefore, a PE-fortified food would provide health benefits to consumers.

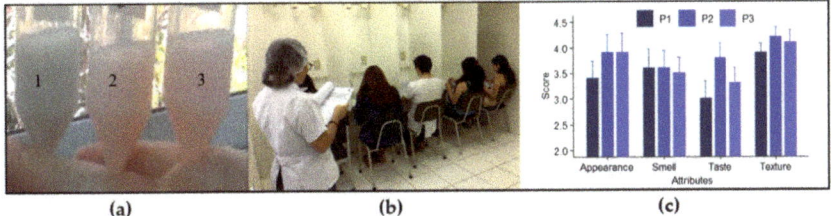

Figure 4. Sensory test for a PBP-containing dairy product. (**a**) Three skim milk prototypes were fortified with PE and PC: Prototype N° 1 (PC from the LLC-10 strain at 120 mg%); Prototype N° 2 (PE from the CAQ-15 strain at 13 mg%; Prototype N° 3 (PE from the LLC-10 strain at 140 mg%). (**b**) Sensory test by a volunteer team evaluating four sensory factors (appearance, smell, taste and texture) that used a consumer acceptability 5-point hedonic scale: 1—dislike extremely; 2—dislike slightly; 3—neither like nor dislike; 4—like slightly; and 5—like extremely) (**c**) Final sensory evaluation scores for four attributes of prototypes P1, P2 and P3.

2.6. Microbiological Assays

All prototypes were subjected to a microbiological control to evaluate the potential presence of *Enterobacteriaceae*. All prototypes were free of microbial contaminant such as coliforms, *Salmonella* and *Shigella*.

3. Materials and Methods

3.1. Strains and Culture Conditions of Cyanobacteria

Diazotrophic cyanobacteria *Nostoc* sp. Llayta (LLC-10) and *Nostoc* sp. Caquena (CAQ-15), were isolated from wetlands at the XV Region in northern Chile. The strains were cultured in an Arnon liquid medium [62] without additions of combined nitrogen at 30 °C under continuous white fluorescent light irradiation (180 µE m^{-2} s^{-1}) and continuous aeration with air enriched with 1% (v/v) CO_2. Cultures were harvested by centrifugation (10 min; 6000 rpm, rotor Sorvall SS-34) at the exponential phase of growth (15 days). The liquid cultures rendered approximately 1.0 g of wet weight per 160 mL of culture.

3.2. Phycobiliprotein Purification

PBP purification was conducted according to the method of Ranjithak and Kaushik [63] with modifications. Briefly, cyanobacterial cell suspensions were digested overnight with lysozymes at a final concentration of 1.0 mg/mL in a 50 mM phosphate buffer of pH 7.2 at 37 °C. Next, the cell suspensions were sonicated in a water–ice bath for 2 min at 15-s intervals (Microson Ultrasonic Cell Disruptor XL, Farmingdale, NY, USA). One volume of the sonicated extract was diluted with one volume of phosphate buffer and centrifuged at 10,000 rpm for 15 min (RC 5B Plus Sorvall SS-34, Newtown, USA). The PBP-rich supernatants were recovered, and 500 µL of aliquots were used to obtain the corresponding 280–700 nm UV/VIS absorption spectrum. The selective precipitation of proteins from the remaining supernatants was conducted by adding ammonium sulfate, with continuous stirring at 4 °C to reach a 0%–35% saturation and a 35%–60% saturation to enrich precipitates with PE and PC, respectively. After 24 h, each protein precipitate was resuspended in 1.0 mL of phosphate buffer and dialyzed in 12,000 Da cut-off cellulose dialysis tubing (Sigma, Steimheim, Germany) against distilled water at 4 °C. Aliquots of the dialyzed PBP solutions were used to obtain the corresponding 280–700 nm absorption spectra. The absorption maxima 620 nm (PE), 565 nm (PC) and 650 nm (APC) were used to compute their concentration according to the following equations:

$$[PC] \text{ (mg/mL)} = ([(A615 - A730) - 0.476 (A652 - A730)] \times 1/5.34)$$

$$[APC] \text{ (mg/mL)} = ([(A652 - A730) - 0.208 (A615 - A730)] \times 1/5.09)$$

$$[PE]] \text{ (mg/mL)} = ([((A562 - A730) - 0.241 [PC] - 0.849 [APC]] \times 1/9.62)$$

PBPs were further purified by ion exchange chromatography by using a DEAE (Diethylaminoethyl-cellulose) column (7.0 by 1.5 cm). Two or three milliliters of a PBP-dialyzed solution was loaded in the column and eluted with a 0.03–1.0 M NaCl gradient at room temperature. One-milliliter fractions were collected, and the colored fractions were pooled and scanned to obtain the UV (Ultraviolet)–Vis (Visible) absorption spectra. Concentration and purity were inferred by using the corresponding equations. The collected purified PBPs were lyophilized and stored at −20 °C until used.

3.3. PBPs Stability

Triplicates of each purified PBP solution were incubated in a 50 mM phosphate buffer of pH 7.2 at temperatures from 0 to 80 °C over 72 h. Protein stability was evaluated in triplicate assays at room temperature in a 50 mM sodium phosphate buffer adjusted to pH 1 to 14. Changes in PBP concentration were obtained from the corresponding UV–Vis absorption spectra every 24 h. In addition, purified PE from the LLC-10 strain and PC from the LLA-10 strain were subjected to incubations at 138 °C for 4 s; visible absorption spectra were obtained before and after the heat treatment to evaluate changes in stability.

3.4. PBPs Antioxidant Capabilities

The biomass from cyanobacterial strains *Nostoc* sp. Llayta (LLC-10) and *Nostoc* sp. Caquena (CAQ-15) at the exponential growth phase was recovered by centrifugation (4000 rpm, 10 min; rotor Sorvall SS-34) and washed twice with a 0.9% NaCl solution. The washed cell pellets were extracted with 4.0 mL of 70% methanol and vortexed at maximal speed for 2 min in a Genic2 multitube holder (Daigger Sci. Ind., model G560E, Bohemia, NY, USA). The methanol extracts were sonicated on a water–ice bath (Microson Ultrasonic Cell Disruptor XL, Farmingdale, NY, USA) for 3 min at 10-s intervals and clarified by centrifugation (4000 rpm, 10 min), and then the supernatants were saved. The methanol extracts were filtered through 0.2 μm SFCA (Surfactant-Free Cellulose Acetate) syringe filters (Ultra Cruz) and stored at −20 °C. The antioxidant capacity of the methanol extracts and the purified PBP was performed by ABTS and FRAP assays according to Re et al. (1999) and Benzie and Strain (1996), respectively [64,65]. Results were expressed as μmoles of Trolox equivalents per 100 g of fresh mass or milligram of pigment.

3.5. Toxicity Assays

Wild-type *C. elegans* var. Bristol-N2 and the *Escherichia coli* strain OP50 were obtained from the *Caenorhabditis* Genetics Center (St. Paul, MN, USA). The nematode was plate-propagated as previously described [66]. *C. elegans* var. Bristol-N2 was cultured on nematode growth agar that was seeded with either *E. coli* OP50 or *E. coli* strain B, following the procedure reported by Brenner [67]. Gravid worms were gently shaken at room temperature in 10 volumes of a fresh 1% NaClO/0.5 M NaOH solution. Carcasses and other debris were dissolved after 5–10 min, and resistant eggs (50% to 100% viable eggs) were collected and washed several times in an M-9 buffer [46]. The M-9 buffer was prepared with 1.5 g of KH_2PO_4, 3.0 g of Na_2HPO_4, 2.5 g of NaCl, 0.5 mL of 1 M $MgSO_4$, and sterile distilled water to a final volume of 500 mL [68]. Eggs were deposit in agar plates for 48 h, and young worms (larval stage 4 and 5) were harvested and resuspended in the M-9 buffer.

3.6. Preparation and Sensory Analyses of Dairy Prototypes

Ten mL of skimmed milk were mixed with purified PBP pigments at a final concentration of 1.2 mg PC/mL and 0.3 or 1.4 mg PE/mL. The skim milk that was used in this work was a commercial liquid product prepared by Colun (Cooperativa Agrícola y Lechera de la Unión Ltd., La Unión, Chile) with a content of 3.3% of total protein, 0.05% total fat, 4.7% carbohydrates, 32 mg% sodium, 115 mg% calcium and 90 mg% phosphorus. According with the manufactures, this skim milk was processed by UHT technology (Ultra High Temperature) at 138 °C for 4 s. The acceptability of PBP-containing dairy prototypes was evaluated by sensory preferences. The study used skimmed milk as a common matrix, and three dairy prototypes were designed: Prototype N° 1 (PC from the LLC-10 strain), Prototype N° 2 (PE from the CAQ-15 strain) and Prototype N° 3 (PE from the LLC-10 strain). A five-point hedonic scale test was used to measure appearance, smell, texture and flavor. The tests involved ten university students as impartial judges who did not have previous training in sensory-type analyses. Consumer acceptability was measured with five-point hedonic scale (1—dislike extremely; 2—dislike slightly; 3—neither like nor dislike; 4—like slightly; 5—like extremely). Data were subjected to variance analysis (ANOVA) with a significance level of $p < 0.05$.

3.7. Microbiological Analyses

Total coliforms, *Shigella* and *Salmonella* detection in the prototypes were performed in a BBL™ MacConkey II Agar (BD, Le Point Declaix, France), a selective and differential medium for the detection of coliforms and enteric pathogens. In addition, an XLD Agar (Xylose–Lysine–Deoxycholate Agar, BD, Le Point Declaix, France) was used to detect *Salmonella* and *Shigella*. Samples of each prototype were seeded on these agar plates and incubated at 37 °C for 24 h [69].

3.8. Statistical Analyses

Sensory test data were analyzed with an analysis of variance (ANOVA) with $\alpha = 0.05$. The PBP temperature stability data were gathered from two independent experiments, each in triplicate, and a non-parametric method was used in a stability test for comparisons of means values among groups ($\alpha = 0.05$). The Shapiro–Wilk normality test was done to determinate if the dataset is well-modeled by a normal distribution. Since the data did not show a normal distribution, a Kruskal–Wallis test was performed.

This research was approved by the Research Ethics Committee, CEIC, Universidad de Antofagasta (Document 044/2017).

4. Conclusions

PBPs purified from two Andean cyanobacteria from northern Chile showed chemical stability at a pH range from 5 to 8, and temperatures between 0 and 50 °C. The pigmented proteins from the LLC-10 strain had no toxic effects against *C. elegans*. The highest score at the sensory test was obtained by skim milk that was fortified with PE.

Colors are always attractive to children and induce consumption. The skim milk was selected to incorporate blue PC and red PE because of its low-fat content and its contribution to human health with vitamins, minerals and proteins, particularly to the overweight population.

We conclude that PBPs are natural proteins that can be used as colorants in the formulation of functional food products based on skim milk in the replacement of added artificial colorants. In addition, the native cyanobacterial PBPs have an appropriate level of antioxidant activity, and we propose their potential use as an innovative source of pigments in the pharmaceutical, cosmetic and food industries.

Finally, the UHT pasteurization (138 °C, 4 s) of Atacama cyanobacterial PBPs induced a minor denaturation of PE and PC; therefore, they can be added before the skim milk pasteurization process. However, milk usually is heated over 50 °C, and, in order to avoid the loss of added PBP coloration while maintaining bacteriological safety, we propose two alternatives for the use of cyanobacterial PBPs at high temperatures in dairy products before pasteurization. One is the application of molecular biology and genetic tools on PBP genes from mesophilic and thermophilic cyanobacteria in order to increase their stability for biotechnological processes and the safety of new functional foods. The second alternative is the use purified freeze-dried PBPs that are enclosed in sachets that can be added as powder to already pasteurized dairy food before their consumption, avoiding protein denaturation and the loss of antioxidant activity.

Author Contributions: Conceptualization, A.G.; methodology, A.G., I.N. and S.S.; investigation, A.G., F.S., V.G., R.G., C.V., J.N., C.T., I.N., J.C., S.S.; writing—original draft preparation A.G.; writing, reviewing and editing, A.G., I.N. and B.G.-S. All authors have read and agreed to the published version of the manuscript.

Funding: This research was funded by Semillero de Investigación, Universidad de Antofagasta (grant number: SI-5305); CONICYT-Chile (grant number: CeBiB FB-0001); Financiamiento Asistente de Investigación, VRIIP, Universidad de Antofagasta MINEDUC-UA (grant numbers: ANT1855 y ANT1856); Convenio Marco para Universidades Estatales, Ministerio de Educación Chile (grant number: NEXER ANT 1756).

Acknowledgments: Special thanks to Milton Urrutia, Universidad de Antofagasta, for his assistance with the statistics used in this work. We thank Catherine Lizama, Microbiology Laboratory, Universidad de Antofagasta her support in the microbiological analyses.

Conflicts of Interest: The authors declare no conflict of interest.

References

1. Marsac, T.D.N.; Houmard, J. Adaptation of cyanobacteria to environmental stimuli: New steps towards molecular mechanisms. *FEMS Microb.* **1993**, *104*, 119–189. [CrossRef]
2. Whitton, B.; Potts, M. Introduction of cyanobacteria. In *The Ecology of Cyanobacteria. Their Diversity in Time and Space*; Kluwer Academic Publishers: Dordrecht, the Netherlands, 2012; pp. 1–10.
3. Abed, R.M.; Dobretsov, S.; Sudesh, K. Applications of cyanobacteria in biotechnology. *J. Appl. Microbiol.* **2009**, *106*, 1–12. [CrossRef]
4. Vijayakumar, S.; Menakha, M. Pharmaceutical applications of cyanobacteria—A review. *J. Acute Med.* **2015**, *5*, 15–23. [CrossRef]
5. Bermejo, R.; Talavera, E.; Alvarez-Pez, J.; Orte, J. Chromatographic purification of biliproteins from *Spirulina platensis*. High-performance liquid chromatographic separation of their α and β subunits. *J. Chromatogr.* **1997**, *778*, 441–450. [CrossRef]
6. Saini, D.K.; Pabbi, S.; Shukla, P. Cyanobacterial pigments: Perspectives and biotechnological approaches. *Food Chem. Toxicol.* **2018**, *120*, 616–624. [CrossRef] [PubMed]
7. Samsonoff, W.A.; MacColl, R. Biliproteins and phycobilisomes from cyanobacteria and red algae at the extremes of habitat. *Arch. Microbiol.* **2001**, *176*, 400–405. [CrossRef] [PubMed]
8. Marín, J.; Pavón, N.; Llópiz, A.; Fernández, J.; Delgado, L.; Mendoza, Y. Pycocyanobilin promotes PC12 cell survival and modulates immune and inflammatory genes and oxidative stress markers in acute hypoperfusion in rats. *Toxicol. Appl. Pharmacol. Rev.* **2013**, *104*, 119–190.
9. Romay, C.H.; Armesto, J.; Remirez, D.; González, R.; Ledon, N.; García, I. Antioxidant and anti-inflammatory properties of C-phycocyanin from blue-g.reen algae. *Inflamm. Res.* **1998**, *47*, 36–41. [CrossRef] [PubMed]
10. González, R.; Rodríguez, S.; Romay, C.; Ancheta, O.; González, A.; Armesto, J.; Remirez, D.; Merino, N. Anti-inflammatory activity of phycocyanin extract in acetic acid-induced colitis in rats. *Pharmacol. Res.* **1999**, *39*, 55–59. [CrossRef]
11. Pleonsil, P.; Suwanwong, Y. An in vitro study of C-phycocyanin activity on protection of DNA and human erythrocyte membrane from oxidative damage. *J. Chem. Pharm. Res.* **2012**, *5*, 332–336.
12. Liua, Q.; Wanga, Y.; Caoa, M.; Yang, T.; Maoa, H.; Suna, L.; Liu, G. Anti-allergic activity of R-phycocyanin from *Porphyra haitanensis* in antigen-sensitized mice and mast cells. *Int. Immunopharmacol.* **2015**, *25*, 465–473. [CrossRef]
13. Liu, Y.; Xu, L. Inhibitory effect of phycocyanin from Spirulina platensis on growth of human leukemia K562 cells. *J. Appl. Phycol.* **2000**, *12*, 125–130. [CrossRef]
14. Sonani, R.; Rastogi, R.; Madamwar, D. Antioxidant potential of phycobiliproteins: Role in anti-aging research. *Biochem. Anal. Biochem.* **2015**, *4*, 2. [CrossRef]

15. Munir, N.; Sharif, S.; Manzoor, S. Algae: A potent antioxidant source. *Sky J. Microbiol. Res.* **2013**, *1*, 22–31.
16. Griffiths, J.C. Coloring foods and beverages. *Food Technol.* **2005**, *59*, 38–44.
17. Borzelleca, J.F.; Capen, C.C.; Hallagan, J.B. Lifetime toxicity/carcinogenicity study of fd & c red no. 3 (erythrosine) in rats. *Food Chem. Toxic.* **1987**, *10*, 723–733.
18. Mahmoud, N. Toxic effects of the synthetic food dye brilliant blue on liver, kidney and testes functions in rats. *J. Egypt. Soc. Toxicol.* **2006**, *34*, 77–84.
19. Mohammed, M.; Mohammed, Y.; Abo-EL-Sooud, K.; Eleiwa, M.M. Embryotoxic and teratogenic effects of tartrazine in rats. *Toxicol. Res.* **2019**, *35*, 75–81.
20. Shetty, M.J.S.; Geethalekshmi, P.; Mini, C. Natural Pigments as Potential Food Colourants: A Review. *Trends Biosci.* **2017**, *10*, 4057–4064.
21. Romay, C.H.; Remirez, D.; González, R. Actividad Antioxidante de la ficocianina frente a radicales peroxílicos y la peroxidación microsomal. *Rev. Cubana Investg. Biomed.* **2001**, *20*, 38–41.
22. Selvam, R.M.; Athinarayanan, G.; Nanthini, A.U.R.; Singh, A.J.A.; Kalirajan, K.; Selvakumar, P.M. Extraction of natural dyes from Curcuma longa, Trigonella foenum graecum and Nerium oleander, plants and their application in antimicrobial fabric. *Ind. Crops Prod.* **2015**, *70*, 84–90. [CrossRef]
23. Mirjalili, M.; Nazarpoor, K.; Karimi, L. Eco-friendly dyeing of wool using natural dye from weld as co-partner with synthetic dye. *J. Clean. Prod.* **2011**, *19*, 1045–1051. [CrossRef]
24. Frick, D. The coloration of food: Review of Progress in Coloration and Related Topics. *Color. Technol.* **2003**, *33*, 15–32. [CrossRef]
25. Dweck, A.C. Natural ingredients for coloring and styling. *Int. J. Cosmet. Sci.* **2002**, *24*, 287–302. [CrossRef] [PubMed]
26. Yusuf, M.; Shabbir, M.; Mohammad, F. Natural Colorants: Historical, Processing and Sustainable Prospects. *Nat. Prod. Bioprospect.* **2017**, *7*, 123–145. [CrossRef]
27. Chengaiah, B.; Rao, K.M.; Kumar, K.M.; Alagusundaram, M.; Chetty, C.M. Medicinal importance of natural dyes, a review. *Int. J. PharmTech Res.* **2010**, *2*, 144–154.
28. Andriamanantena, M.; Danthu, P.; Cardon, D.; Fawbush, C.D.; Raonizafinimanana, B.; Razafintsalama, V.E.; Rakotonandrasana, S.R.; Ethève, A.; Petit, T.; Caro, Y. Malagasy Dye Plant Species: A Promising Source of Novel Natural Colorants with Potential Applications—A Review. *Chem. Biodivers.* **2019**, *16*, e1900442. [CrossRef]
29. FDA. U.S. Food and Drug. Summary of Color Additives for Use in the United States in Foods, Drugs, Cosmetics and Medical Devices. Available online: https://www.fda.gov/industry/color-additive-inventories/summary-color-additives-use-united-states-foods-drugs-cosmetics-and-medical-devices (accessed on 14 January 2020).
30. Olsen, K.; Jespersen, L.; Strømdahl, L. Heat and light stability of three natural blue colorants for use in confectionery and beverages. *Eur. Food Res. Technol.* **2005**, *220*, 261–266.
31. Pandey, V.D.; Pandey, A.; Sharma, V. Biotechnological applications of cyanobacterial phycobiliproteins. *Int. J. Curr. Microbiol. App. Sci.* **2013**, *2*, 89–97.
32. Bermejo, R. Phycocianins. In *Cyanobacteria an Economic Perspective*; Sharma, N.K., Rai, A.K., Stal, L.J., Eds.; Willey Blackwell, John Willey and Sons Ltd.: Hoboken, NJ, USA, 2014; pp. 209–225.
33. Galetović, A.; Gómez-Silva, B. Inmovilización de cianobacterias filamentosas en perlas de alginato. *Rev. Cienc. Salud.* **1999**, *3*, 28–33.
34. Seibert, M.; Connolly, J.S. Fluorescence properties of C-phycocyanin isolated from a thermophilic cyanobacterium. *Photochem. Photobiol.* **1984**, *40*, 267–271. [CrossRef]
35. Couteau, C.; Baudry, S.; Roussakis, C.; Coiffard, L.J.M. Study of thermodegradation of phycocyanin from *Spirulina platensis*. *Sci. Aliments* **2004**, *24*, 415–421. [CrossRef]
36. Patel, A.; Pawar, R.; Mishra, S.; Sonawane, S.; Ghosh, P.K. Kinetic studies on thermal denaturation of C-phycocyanin. *Indian J. Biochem. Biophys.* **2004**, *41*, 254–257. [PubMed]
37. Antelo, F.S.; Costa, J.A.V.; Kalil, S.J. Thermal degradation kinetics of the phycocyanin from *Spirulina platensis*. *Biochem. Eng. J.* **2008**, *41*, 43–47. [CrossRef]
38. Zhao, J.; Brand, J.J. Specific bleaching of phycobiliproteins from cyanobacteria and red algae at high temperature in vivo. *Arch. Microbiol.* **1989**, *152*, 447–452. [CrossRef]

39. Tripathi, S.N.; Kapoor, S.; Shrivastava, A. Extraction and purification of an unusual phycoerythrin in a terrestrial desiccation tolerant cyanobacterium *Lyngbya arboricola*. *J. Appl. Phycol.* **2007**, *19*, 441–447. [CrossRef]
40. Kaur, S.; Khattar, J.I.S.; Singh, Y.; Singh, D.P.; Ahluwalia, A.S. Extraction, purification and characterisation of phycocyanin from *Anabaena fertilissima* PUPCCC 410.5: As a natural and food grade stable pigment. *Appl. Phycol.* **2019**, *31*, 1685–1696. [CrossRef]
41. Rastogi, R.P.; Sonani, R.R.; Madamwar, D. Physico-chemical factors affecting the in vitro stability of phycobiliproteins from *Phormidium rubidum* A09DM. *Bioresour. Technol.* **2015**, *190*, 219–226. [CrossRef]
42. Munawaroh, H.S.H.; Darojatun, K.; Gumilar, G.G.; Aisyah, S.; Wulandari, A.P. Characterization of phycocyanin from *Spirulina fusiformis* and its thermal stability. *J. Phys. Conf. Ser.* **2018**, *1013*, 012205. [CrossRef]
43. Pumas, C.; Peerapornpisal, Y.; Vacharapiyasophon, P.; Leelapornpisid, P.; Boonchum, W.; Ishii, M.; Khanongnuch, C. Purification and characterization of a thermostable phycoerythrin from hot spring cyanobacterium *Leptolyngbya* sp. KC45. *Int. J. Agric. Biol.* **2012**, *14*, 121–125.
44. Leu, J.; Lin, T.; Selvamani, M.J.P.; Chen, H.; Liang, J.; Pan, K. Characterization of a novel thermophilic cyanobacterial strain from taian hot springs in Taiwan for high CO2 mitigation and C-phycocyanin extraction. *Process Biochem.* **2013**, *48*, 41–48. [CrossRef]
45. Liang, Y.; Kaczmarek, M.B.; Kasprzak, A.K.; Tang, J.; Shah, M.M.R.; Jin, P.; Daroch, M. *Thermosynechococcaceae* as a source of thermostable C-phycocyanins: Properties and molecular insights. *Algal Res.* **2018**, *35*, 223–235. [CrossRef]
46. Rahman, D.Y.; Sarian, F.D.; van Wijk, A.; Martinez-Garcia, M.; van der Maarel, M.J.E.C. Thermostable phycocyanin from the red microalga *Cyanidioschyzon merolae*, a new natural blue food colorant. *J. Appl. Phycol.* **2017**, *29*, 1233–1239. [CrossRef] [PubMed]
47. Mishra, S.K.; Shrivastav, A.; Mishra, S. Effect of preservatives for food grade C-PC from *Spirulina platensis*. *Process Biochem.* **2008**, *43*, 339–345. [CrossRef]
48. Mishra, S.K.; Shrivastav, A.; Pancha, I.; Jain, D.; Mishra, S. Effect of preservatives for food grade C-phycoerythrin, isolated from marine cyanobacteria *Pseudanabaena* sp. *Int. J. Biol. Macromol.* **2010**, *47*, 597–602. [CrossRef] [PubMed]
49. Wu, H.; Wang, G.; Xiang, W.; Li, T.; He, H. Stability and antioxidant activity of food-grade phycocyanin isolated from *Spirulina platensis*. *Int. J. Food Prop.* **2016**, *19*, 2349–2362. [CrossRef]
50. Martelli, G.; Folli, C.; Visai, L.; Daglia, M.; Ferrari, D. Thermal stability improvement of blue colorant C-phycocyanin from *Spirulina platensis* for food industry applications. *Process Biochem.* **2014**, *49*, 154–159. [CrossRef]
51. Romay, C.H.; González, R.; Ledón, N.; Remirez, D.; Rimbau, V. C-Phycocyanin: A biliprotein with antioxidant, anti-Inflammatory and neuroprotective effects. *Curr. Protein Pept. Sci.* **2003**, *4*, 207–216. [CrossRef]
52. Patel, S.N.; Sonani, R.R.; Jakharia, K.; Bhastana, B.; Patel, H.M.; Chaubey, M.G.; Singh, N.K.; Madamwar, D. Antioxidant activity and associated structural attributes of *Halomicronema* phycoerythrin. *Int. J. Biol. Macromol.* **2018**, *111*, 359–369. [CrossRef]
53. Soni, B.; Trivedi, U.; Madamwar, D. A novel method of single step hydrophobic interaction chromatography for the purification of phycocyanin from *Phormidium fragile* and its characterization for antioxidant property. *Bioresour. Technol.* **2008**, *188*, 194. [CrossRef]
54. Sonani, R.R.; Singh, N.K.; Kumar, J.; Thakar, D.; Madamwar, D. Concurrent purification and antioxidant activity of phycobiliproteinsfrom *Lyngbya* sp. A09DM: An antioxidant and anti-aging potential of phycoerythrin in *Caenorhabditis elegans*. *Process Biochem.* **2017**, *49*, 1757–1766. [CrossRef]
55. Kuskoski, E.M.; Asuero, A.G.; Ana, M.; Troncoso, A.M.; Mancini-Filho, J.; Fett, R. Aplicación de diversos métodos químicos para determinar actividad antioxidante en pulpa de frutos. *Ciênc. Tecnol. Aliment. Camp.* **2005**, *25*, 726–732. [CrossRef]
56. Leung, M.C.K.; Williams, P.L.; Benedetto, A.; Au, C.; Helmcke, K.J.; Aschner, M.; Meyer, J.N. *Caenorhabditis elegans*: An emerging model in biomedical and environmental toxicology. *Toxicol. Sci.* **2008**, *106*, 5–28. [CrossRef] [PubMed]
57. Kaletta, T.; Hengartner, M.O. Finding function in novel targets: *C. elegans* as a model organism. *Nat. Rev. Drug Discov.* **2006**, *5*, 387–398. [CrossRef]

58. Ju, J.; Saul, N.; Kochan, C.; Putschew, A.; Pu, Y.; Yin, L.; Steinberg, C.E. Cyanobacterial xenobiotics as evaluated by a *Caenorhabditis elegans*. Neurotoxicity screening test. *Int. J. Environ. Res. Public Health* **2014**, *11*, 4589–4606. [CrossRef]
59. Gouveia, L.; Batista, A.; Miranda, A.; Empis, J.; Raymundo, A. *Chlorella vulgaris* biomass used as coloring source in traditional butter cookies. *Innov. Food Sci. Emerg. Technol.* **2007**, *8*, 433–436. [CrossRef]
60. Hossain, A.K.M.M.; Brennan, M.A.; Mason, S.L.; Guo, X.; Zeng, X.A.; Brennan, C.S. The Effect of astaxanthin-rich microalgae "*Haematococcus pluvialis*" and wholemeal flours incorporation in improving the physical and functional properties of cookies. *Foods* **2017**, *6*, 57. [CrossRef]
61. Caporgno, M.P.; Mathys, A. Trends in microalgae incorporation into innovative food products with potential health benefits. *Front. Nutr.* **2018**, *5*, 58. [CrossRef]
62. Allen, M.; Arnon, D. Studies on nitrogen-fixing blue-green algae. I. Growth and nitrogen fixation by *Anabaena cylindrica*. *Plant Physiol.* **1955**, *30*, 366–372. [CrossRef]
63. Ranjitha, K.; Kaushik, B.D. Purification of phycobiliproteins from *Nostoc muscorum*. *J. Sci. Ind. Res.* **2005**, *64*, 372–375.
64. Re, R.; Pellegrini, N.; Proteggente, A.; Pannala, A.; Yang, M.; Rice-Evans, C. Antioxidant activity applying an improved ABTS radical cation decolorization assay. *Free Radic. Biol. Med.* **1999**, *26*, 1231–1237. [CrossRef]
65. Benzie, I.F.F.; Strain, J.J. The Ferric Reducing Ability of Plasma (FRAP) as a measure of "Antioxidant Power": The FRAP Assay. *Anal. Biochem.* **1996**, *239*, 70–76. [CrossRef] [PubMed]
66. Lewis, J.A.; Fleming, J.T. Basic culture methods. In *Caenorhabditis Elegans: Modern Biological Analysis of an Organism*; Epstein, H.F., Shakes, D.C., Eds.; Academic Press: San Diego, CA, USA, 1995; pp. 4–27.
67. Brenner, S. The genetics of *Caenorhabditis elegans*. *Genetics* **1974**, *77*, 71–94. [PubMed]
68. Skantar, A.M.; Agama, K.; Meyer, S.L.F.; Carta, L.K.; Vinyard, B.T. Effects of geldanamycin on hatching and juvenile motility in *Caenorhabditis elegans* and *Heterodera glycines*. *J. Chem. Ecol.* **2005**, *31*, 2481–2491. [CrossRef]
69. BBL™ MacConkey II Agar (BD). Available online: http://legacy.bd.com/ds/technicalCenter/inserts/L007388(10).pdf (accessed on 15 January 2020).

© 2020 by the authors. Licensee MDPI, Basel, Switzerland. This article is an open access article distributed under the terms and conditions of the Creative Commons Attribution (CC BY) license (http://creativecommons.org/licenses/by/4.0/).

Article

A Microethnographic and Ethnobotanical Approach to Llayta Consumption among Andes Feeding Practices

Mailing Rivera [1], Alexandra Galetović [2], Romina Licuime [1] and Benito Gómez-Silva [2,*]

1. Departamento de Educación, Facultad de Educación, Universidad de Antofagasta, Antofagasta 124000, Chile; mailing.rivera@uantof.cl (M.R.); rlicuime1505@gmail.com (R.L.)
2. Departamento Biomédico, Facultad Ciencias de la Salud, and Centre for Biotechnology and Bioengineering, CeBiB, Universidad de Antofagasta; Antofagasta 1240000, Chile; alexandra.galetovic@uantof.cl
* Correspondence: benito.gomez@uantof.cl; Tel.: +56-9-98297844

Received: 29 October 2018; Accepted: 28 November 2018; Published: 9 December 2018

Abstract: Llayta is a dietary supplement that has been used by rural communities in Perú and northern Chile since pre-Columbian days. Llayta is the biomass of colonies of a *Nostoc* cyanobacterium grown in wetlands of the Andean highlands, harvested, sun-dried and sold as an ingredient for human consumption. The biomass has a substantial content of essential amino acids (58% of total amino acids) and polyunsaturated fatty acids (33% total fatty acids). This ancestral practice is being lost and the causes were investigated by an ethnographic approach to register the social representations of Llayta, to document how this Andean feeding practice is perceived and how much the community knows about Llayta. Only 37% of the participants (mostly adults) have had a direct experience with Llayta; other participants (mostly children) did not have any knowledge about it. These social responses reflect anthropological and cultural tensions associated with a lack of knowledge on Andean algae, sites where to find Llayta, where it is commercialized, how it is cooked and on its nutritional benefits. The loss of this ancestral feeding practice, mostly in northern Chile, is probably associated with cultural changes, migration of the rural communities, and very limited access to the available information. We propose that Llayta consumption can be revitalized by developing appropriate educational strategies and investigating potential new food derivatives based on the biomass from the isolated Llayta cyanobacterium.

Keywords: Andean microalgae consumption; Atacama; cyanobacteria; Llayta; microethnography; *Nostoc*

1. Introduction

The abundance and diversity of organisms in the Atacama Desert are severely limited by the high levels of desiccation and ultraviolet light [1,2]. In the Andes Mountains highlands, biodiversity is higher and plants have been used for centuries by local communities for feeding, foraging and ethnomedicine [3–8]. Microalgae and cyanobacteria are part of the Andes biodiversity but seldom acknowledged. Based on their nutritional and digestive benefits, microalgae and cyanobacteria (i.e., *Chlorella*, *Dunaliella*, *Arthrospira* and *Nostoc*) have been part of the human diet in South America, North America, Asia and Africa. Also, some species are natural resources for a variety of organic molecules with high interest to the biotechnological industry (proteins, amino acids, vitamins, polyunsaturated fatty acids, pigments) [3,4,9,10]. Edible members of the *Nostoc* genus are found in China where *Nostoc flagelliforme* has been consumed as a delicacy for centuries but its collection is prohibited today due to over-exploitation [11,12]. In South America, an indigenous foodstuff harvested in the Andes wetlands, known as Llayta, is the dry biomass of macrocolonies of a cyanobacterium from the genus *Nostoc* (Figure 1). Llayta consumption is a practice that can be traced back to pre-Columbian times and it has been recorded in documents from the 17th century [13,14] and, in a more recent botanical report [15].

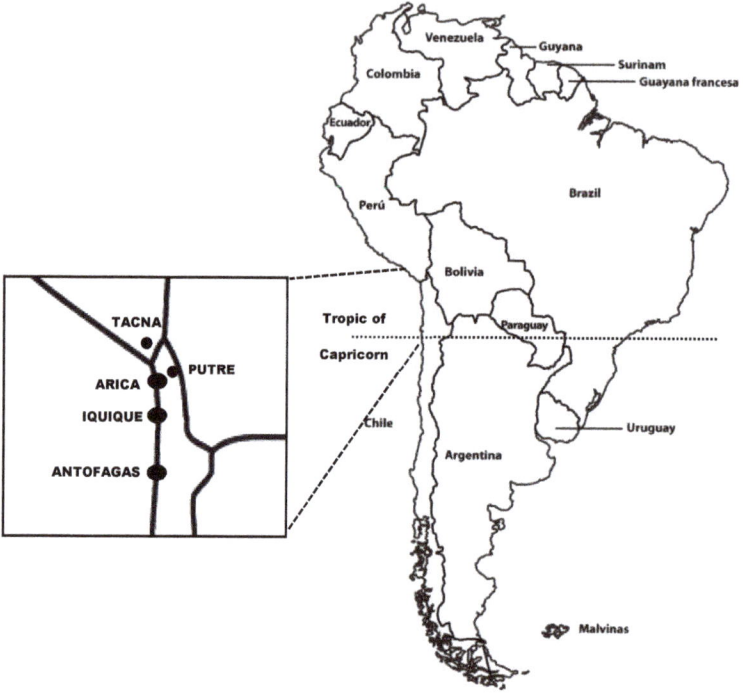

Figure 1. Locations, in southern Peru and northern Chile, where information on Llayta was acquired.

Thus, the genus *Nostoc* has been an old component of the human diet in South America, and it continues to be used today as a food additive in northern Chile (Arica y Parinacota and Tarapacá Regions) and in southern Peru (Tacna City) (Figure 2) [5,6,8,16–18]. However, and based on preliminary interviews, this ancient culinary legacy is disappearing and it is already unknown by the urban communities from other areas of the region; for example, at Antofagasta, the major coastal city in northern Chile, nearly 400 km south of Iquique (Figure 2).

Figure 2. A dry colony of Llayta obtained at a major food market in Arica, Chile. (Bar: 1 cm).

Our report provides the results of a microethnographic study conducted to learn how much people know about Llayta and their perception on this ancestral Andean ingredient, and is meant to be complementary to biochemical studies done on Llayta [4]. This work was based on the following considerations: (i) Llayta consumption is a feeding practice transmitted through generations in the rural Andean world of South America, without reports of adverse effects on human health; (ii) Llayta consumption is an old culinary legacy that is disappearing in regions of South America; (iii) Llayta is a nutritional ingredient containing essential amino acids (58% of total amino acids) and polyunsaturated fatty acids (33% total fatty acids); (iv) the prevalence of undernourishment in South America; (iv) Llayta consumption can be revitalized with appropriate educational strategies and new food derivatives can be developed from the biomass of the isolated cyanobacterium from Llayta [4,9,18,19].

We propose that this microethnographic approach will help us to explore explanations for an apparent decrease in Llayta consumption, and to provide arguments and suggestions for the revitalization of this feeding practice.

2. Materials and Methods

2.1. The Microethnographic Study

The microethnographic study on Llayta was designed to learn about the direct or indirect knowledge people have about Llayta by collecting social representations [20] from interviewing participants, including drawings prepared by children. The main expressions about Llayta were ethnographically registered and analyzed in order to explain the social worlds built by persons about their understanding of the surrounding natural, social and cultural environment [21].

2.2. Sample for the Microethnographic Study

Observations were carried out during the first half of 2014 in Tacna (Peru) and Putre (Chile) (Figure 2). Tacna is a Peruvian city located at the frontier between Peru and Chile with an active commercial exchange with Arica and Iquique, two coastal cities in Chile. Putre is a rural village at 2500 m above sea level in the Andes Range in northern Chile, close to sites where Llayta is harvested.

The participants were 19 active members of their community who were informed about the origin and purpose of the study. Their selection was based on their willingness to participate anonymously in the interviews.

The participants were 12 fourth-grade students (7 boys and 5 girls, 9–10 years old) and three middle-aged adults (three males and two females) from Putre, Chile. Participants from Tacna, Peru, were two vendors (one female and one male) working at the main food market of the city. Interviews and observations were conducted at sites normally used by the participants (street, market place, school, and hotel).

Oral interviews were conducted with all participants in order to learn their direct or indirect knowledge about Llayta (origin, physical description, uses, places where Llayta grows and is sold, its quality as food). In addition, children were also asked to draw representations of Llayta in order to evaluate how close their depictions were from the real subject.

3. Results

3.1. The Vocable Llayta: Alternative Names and Their Meaning

Llayta is the Aymara name that refers to colonies of a cyanobacterium that grows in the Andes highlands and is consumed by rural and urban communities in South America (Figures 1 and 2). Alternative names for Llayta can be found in several languages: Spanish, Quechua, Kunza and Mapudungun (Table 1). The variety of names for the vocable Llayta stresses the cultural and anthropological diversity of representations associated with this feeding practice.

Table 1. The Llayta vocable: alternative names, their ethnic origins and meaning.

Name	Language	Comments	Reference
Cushuro Llullucha Murmuta Crespito Yrurupa		The author did not indicate the language of origin of the words.	[12]
Chuncoro Murmunta	Aymara	"Vna yerua negra de comer, frutilla fe llama Chun-coro,o Murmunta" (Edible black herb called Chuncoro or murmuta)	[3]
Murmunta Chuncuru	Aymara	"(Prunus capulí Cav.) Cerezo. La infusión de las hojas de esta planta se usa como laxante. Sus frutos en Aymara se llaman". (The leaves infusion is used as a laxative. Their fruit are named in Aymara)	[22]
Cerezo	Aymara	"s. Cerezo. Plumas coloridas (10)" (Colored feathers)	[23]
Quchayuyo Murmunta	Aymara	"Bot. Cerezo. 2. Plumas coloridas, en Bolivia. Bot. Cochayuyo de agua dulce y del mar (p.e. alga comestible). Vte. QUCHAYUYU, MURMUNTA" (Colored feathers in Bolivia. Bot. Macroalga from freshwater and sea water) (edible alga))	[24]
Chungullu	Quechua	"una cianobacteria comestible, se encuentra en riachuelos y lagunitas del bofedal ubicado entre Isluga y Colchane, en las cercanías de la frontera entre Bolivia y la Región de Tarapacá en Chile". (An edible cyanobacterium, found in small rivers and lakes of wetlands located between Isluga and Colchane, near the border between Bolivia and the Tarapacá Region in Chile)	[7]
Murmunta, Chuncuro	Aymara Quechua	"..., probablemente derivado de su morfología, hábitat y uso. En Aymara: hierba de las ciénagas como granillos negros". (... derived probably from its morphology, habitat and use. In Aymara: black grain herb from wetlands)	[7]
Luche (lucha)	Mapuche	"..., es un símil de una alga roja, marina y comestible". (..., is like a marine edible red alga)	[7]
Yullucha	Aymara	"De llullu para referirse a formas vegetales que empiezan a desarrollarse o se pasman". (From llullu, to refer to plants starting or have stopped their growth).	[8]
Tchuckula Chucula	Kunza	"Nombre atacameño de la cianobacteria comestible Nostoc. Planta acuática que hay en la cordillera". (Name given by the Atacameños people to the edible cyanobacterium Nostoc. Aquatic plant found in the Andes Range).	[8]
Yoyo	Aymara Quechua	"Cianobacterias del género Nostoc" (cianobacteria from the genus Nostoc).	[9]
Chungulle, chungullo	Quechua	"Cianobacteria acuática de bofedales, procedentes de Chela. Se usa para la comida (caldo con papas chuño). Se indicó que "hay uno que se come, es especial. Se lava y se seca". (Aquatic cyanobacterium from Chela wetland. It is used as food, in soups with potatoes. It was mentioned that there is an edible one which is special. It is washed and dried).	[9]
Luche	Mapuche	"Símil de un alga roja comestible para designar a un alga verde-azulada de agua dulce, también comestible (Nostoc)". (Similar to an edible red alga, it is (Used to indicate an edible, freshwater blue-green alga, (Nostoc)).	[9]
Yoyo	Aymara	"Nombre de la cianobacteria Nostoc en Ollagüe, posiblemente aludiendo a su carácter comestible. Planta acuática comestible". (Name given in Ollagüe to the cyanobacterium Nostoc, possibly due to its edibility).	[9]

3.2. How Much People Know about Llayta

All participants were interviewed to assess the type and level of knowledge they have on Llayta. Table 2 provides extracts of the answers given by 10 participants (7 adults and 3 students). Only 8 of the 19 participants had some perception about Llayta; the others (58%) lack any knowledge about it.

The extracts in Table 2 corroborate that 7 participants have had personal experience of Llayta, i.e., direct knowledge. Only one teacher expressed indirect knowledge about Llayta since the information was second-hand (Table 2). Compared with adult participants, the oral expression of knowledge used by the fourth-grade students from Putre to refer to Llayta were few or absent. When asked to draw an image of Llayta, 11 out of 12 students were willing to participate and their drawings were far from a correct depiction of the colonies. As an exception, one student emphasized that his mother used to cook Llayta and her drawing was the closest image to it. Another student said: *"no, yo no"* (No, I do not (know Llayta)). A third student asked: *"¿Esa es la Llayta?* (Is this Llayta?), referring to a drawing made by another student. Table 2 is a compilation of the representations of Llayta.

Table 2. Transcripts of interviews conducted to adult and students participants about their knowledge on Llayta.

Informant	Type of Knowledge about Llayta
DIRECT KNOWLEDGE:	
Saleswoman at the food market, Tacna, Peru.	"Ahí huapé súcuros; de Súcuro se trae; de súcuro de ahí al fondo pues; ahí arriba de Puno. Otros caballeros traen y ahí compramos; para picante. Si, picante prepara rico ahí comen". (There, huapé súcuros; it is brought from Sucuro; down there; there, above Puno. Other people bring it and we buy it to make "picante"—A local dish; yes, a tasty picante is prepared up there).
Salesman at the food market, Tacna, Peru.	"Esta es nacional; ésta la traen de Camaná. Esta es de río muestra -y muestra Llayta-, es más rica y esa es de mar -muestra cochayuyo; a tres soles. Esto jefe chángalo, muélelo, jugo". (This is Peruvian; it is brought from Camaná. This from a river, it is better and this one is from the sea; it is worth three Peruvian new sols. Cut it and grind it for juice, boss).
Professional cook, Chile.	"Para el consumo lo comemos eh; la Llayta es un tipo de alga de mar, de agua dulce y de mar también hay. Lo traen y lo hacen secar, y seco lo venden en los negocios; para cocinarlo se la remoja. Los peruanos los comen la Llayta y el cochayuyo. De la altura, de Puno, de Juliaca, bofedales eso está en la altura, en agua dulce en los ríos crece por ahí, bueno acá en la frontera con Perú, tripartito, Visviri, ahí también crece". (We eat it for consumption, eh!; Llayta is a kind of marine alga; it is from freshwater and also marine. They bring it, dry it and sell it dried; they soak it before cooking. Peruvians eat Llayta and cochayuyo. From the highlands, from Puno, from Juliaca, from wetlands, which are in the highlands, in freshwater rivers, it grows around; right here, at the three parties' border, it also grows at Visviri).
Middle age woman, Putre, Chile.	"Llayta, arriba hay, arriba; Parinacota, ahí si hay Llayta; Llayta come..., el segundo come bonito, así que coce para ..., es como carne para ..., se prepara eso como carne, como picante cocino acá. Ahí no se pa' qué sea, en Parinacota hay río de esa, ahí florece". (Llayta is from up there; Llayta is at Parinacota; Llayta is eaten ..., nice as a second dish ..., it is cooked ..., it is like meat ..., it is prepared as meat ..., I cook it here as picante. I do not know what it is used for over there; in Parinacota there is a river where it grows).
Director, school at Putre, Chile.	"La Llayta es de por acá también. La Llayta se usa para el picante. La he comido no más, pero no sé qué me ha hecho; acá hacen mucho picante de pata con guata y Llayta. Hay en las lagunas, en Caqueña, en Tacna y aquí arriba Caquena". (Llayta is from here too. Llayta is used in picante. I have only eaten it; I do not know the effects on me; people right here prepares a lot of picante with meat and Llayta. There are some ponds, in Casqueña, in Tacna and up here in Caquena).
Teacher 1, at school in Putre, Chile.	"¡no! y los picantes de guatita, ahí le ponen la Llayta ... es un musgo que se trae de Caquena; es un musgo parecido al cochayuyo, es la misma que venden en el agro". (No! Llayta is added at the picante dishes ..., it is a moss brought from Caquena; it is a moss similar to cochayuyo marine macroalga, it is the same one that is sold at the food market).
Student 1, at school in Putre, Chile.	"Ah! yo sí, porque mi mamá cocina. En Caquena". (Mm! I do, since my mother cooks. At Caquena ...).
INDIRECT KNOWLEDGE:	
Teacher 2 at school in Putre, Chile.	"Eh... Yo no las he visto pero sí me han dicho que ahí en la laguna está la Llayta" pero de verla no. En Caqueña en Tacna y aquí arriba Caquena". (Eh ..., I have not seen it, but I have been told that Llayta is at the pond, but I have not seen it. At Caquena, in Tacna and up here in Caquena).
WITHOUT KNOWLEDGE:	
Student 2, at school in Putre, Chile.	"No, yo no". (No, I do not ... know Llayta).
Student 3, at school in Putre, Chile.	"¿esa es la Llayta?" (Is that Llayta?)

3.3. Fields of Representations for Llayta

Table 2 shows the extracts from the ethnographic registries. These are the descriptions and references that sustain the field of representation of Llayta for 7 adult participants, which can be organized in the following 3 semantic fields:

3.3.1. What is it?

"Es un musgo que se trae de Caquena"; "es un musgo parecido al cochayuyo, es la misma que venden en el agro". "Ésta es nacional"; "la Llayta es un tipo de alga de mar, de agua dulce y de mar también hay".

"It is a moss brought from Caquena"; "it is a moss similar to cochayuyo (seaweed), it is the same that is sold at the market", "This is national"; "Llayta is a kind of marine alga; from freshwater and also from the sea".

3.3.2. Where is it from?

"De súcuro se trae", "De súcuro de ahí al fondo pues"; "Ahí arriba de Puno"; "ésta es de río"; "en Parinacota hay río de esa, ahí florece"; "la Llayta es de por acá también"; "Hay en las lagunas, en Caqueña en Tacna y aquí arriba Caquena"; "de la altura, de Puno, de Juliaca, bofedales eso está en la altura, en agua dulce en los ríos crece por ahí, bueno acá en la frontera con Perú, tripartito, Visviri, ahí también crece"; "Llayta, arriba hay, arriba"; "Parinacota, ahí si hay Llayta"; "la Llayta es de por acá también"; "en Caquena"; "Eh ... Yo no las he visto pero sí me han dicho que ahí en la laguna está la Llayta, pero de verla no"; "en Caquena".

"It is brought from Sucuro"; "from Sucuro, back there"; "from Puno, up there"; "this is from a river"; "at Parinacota, there is a river where it blooms"; "Llayta is from here too"; "it is from ponds, at Caquena, in Tacna, and up here in Caquena"; "from highlands, at Puno, Juliaca, wetlands, this is at the highlands, it grows in freshwater rivers, well, here at the border with Peru, Visviri, where it also grows; "Llayta, it is up there, up high"; "Llayta is at Parinacota, for sure"; "Llayta is from here too"; "at Caquena"; "Eh ... , I have not seen it but I was told that over there at the pond, there is Llayta, but I have not seen it"; "at Caquena".

3.3.3. What is it for?

"Para picante"; "si picante prepara rico ahí comen"; "los peruanos los comen la Llayta y el cochayuyo"; "Llayta come... el segundo come bonito, así que coce para ... es como carne para ... se prepara eso"; "como carne, como picante cocino acá"; "la Llayta se usa para el picante"; "la he comido no más, pero no sé qué me ha hecho"; "¡no! y los picantes de guatita"; "ahí le ponen la Llayta ... "; "lo traen y lo hacen secar, y seco lo venden en los negocios para cocinarlo se la remoja"; "ah yo sí, porque mi mamá cocina".

"For making picante dish"; "yes, picante dish is good, they eat it"; "Peruvian people eat Llayta and cochayuyo"; "Llayta is for consumption ... , so you cook it ... , it is like meat ... ; like meat, like picante I cook it here"; "Llayta is used for picante"; "I have eaten it but I do not know how to prepare it"; "it is brought here, it is dried and it is sold dry to restaurants"; "it is soaked before cooking"; "yes I know it, my mother cooks it".

When asked for places where Llayta can be found, one informant from Tacna, Peru, used the vocable *"chuncuru"* (an Aymara synonym for Llayta) instead of using *"Sucuru"*, the right geographic site to where Llayta can be found. This mistake can be explained by the phonetic similarity between both words.

4. Discussion

4.1. Microethnographic Aspects

The ethnographic goal of this study was to document the anthropological and cultural tensions found in the social representations of the Llayta feeding practice and relate them to the knowledge and value given by the communities to Andean algae.

Ethnographic registries allow the collection of evidence and social representations from people on a particular subject, to discover what people think, believe, and know about their surroundings, and understand how people see it and fit it in their particular interpretations of realities [21,25,26]. The knowledge people may have on a particular natural situation is a good example of where social representations can be collected and interpreted from social and cultural perspectives [26]. Also, descriptions and references from participants are essential in the field of representation for an event of ethnographic interest [22].

In the study of the Llayta feeding practice, the ethnographic registry can be supported by anthropological, socio-cultural, and nutritional referents [20]. The first two provide information on the meaning(s) of the term Llayta, the identification of sites where Llayta grows naturally, and where it is commercialized and consumed.

Our results indicated that nearly 40% of the participants declared they knew Llayta and described it as a moss or an alga, without clarifying whether it grows in fresh water or seawater (Table 2). They also provide names for the sites of origin of Llayta: Parinacota, Putre, Caquena and Visviri in Chile, and Sucuro, Juliaca and Puno in Peru (Table 2, Figure 2). The participants knew that Llayta is used to prepare *"picante"*, a typical dish from rural areas in the Andes; however, they were unaware of its nutritional properties. All fourth-grade students from Putre did not know Llayta (11 out of 12 students; 58% of the participants of this study).

It is of considerable concern to confirm that a large proportion of the participants, all young people, did not know Llayta. This apparent loss may be explained by considering the impact of new technologies on rural life, cultural changes and migration from rural regions into urban centers, as described for the Aymara people in northern Chile [24,27].

The cultural and anthropological tensions observed in this study stemmed from a lack of knowledge on the following subjects: (a) conceptual limitations to explain what is an Andean alga or a microalga; (b) geographic locations where Llayta can be found and commercialized; (c) weak descriptions to explain how Llayta is cooked; and (d) the benefits obtained from its consumption. Also, this low or absence of knowledge on Llayta can be explained by how people perceive and relate to the world around them and, most stressfully, by a probable extinction of an Andean cultural practice. The massive ignorance about Llayta found in discussions with children in Putre is a clear example of it.

These anthropological tensions are indication of a paucity of information on Llayta. There is therefore an urgent need to educate people about all the cultural and nutritional knowledge accumulated on Llayta, so that it can be properly valued as a nutritional foodstuff.

4.2. Biochemical Characterization of Llayta

Llayta consumption is a traditional practice whose future revitalization can be supported by ethnographic information and evidence on the nutritional quality of Llayta [4].

The first biochemical evaluation of colonies of *Nostoc* cells from Peruvian wetlands were published by Aldave-Pajares [16,17]. Later, Gómez-Silva et al. [18] registered the proximal composition of Llayta colonies sold and consumed in Chilean territory, and also for the biomass from the isolated Llayta cyanobacterium. Both studies are in agreement on the total protein (30–35% w/w) and carbohydrate (50–60% w/w) content of the Andean *Nostoc* biomasses. More recently, it was informed that 60% of Llayta amino acids can be classified as indispensable; total lipids accounted for 2% of the biomass dry weight; and 32% of total fatty acids were polyunsaturated fatty acids, Vitamin E content was 4.3 mg% w/w, total polyphenols was 64 mg (as equivalent to gallic acid), with an antioxidant activity of 17.4 µmoles (as equivalent to Trolox), and total fiber content was 56% of dry weight [4]. Galetovic et al. [4] inferred that Llayta biomass is a nutritious dietary ingredient.

One reason for the need for safety assessments of foods and food ingredients based on microalgae and cyanobacteria biomasses to protect public health is the potential presence of cyanobacterial toxins active at low doses (e.g., microcystin and nodularin). In particular, *Arthrospira platensis* (Spirulina) is considered a safe food based on centuries of human consumption [23]. Comparatively, some species of the *Nostoc* genus have toxic members that synthesize microcystine-like cyanotoxin [28]. However, the genome of the colony-forming *Nostoc* strain isolated from the *Llayta* biomass did not show the presence of *mycE*, a key gene in the microcystine biosynthetic pathway, rendering it as a non-toxic *Nostoc* strain [4,28]. In addition, the absence of epidemiological records associated with Llayta consumption diminishes but does not remove the potential presence of toxic secondary metabolites in this cyanobacterial biomass [4].

5. Conclusions

Llayta consumption is a feeding practice with a know-how that has been transmitted through generations in the rural Andean world of South America, without untoward effects on human health. Nevertheless, this ancestral feeding legacy is being lost in young generations from northern Chile. New educational and anthropological strategies must be developed if there is interest in promoting the preservation and value of this traditional feeding practice and cultural legacy.

Llayta is a nutritious dietary ingredient for human consumption. This is supported by the absence of adverse epidemiological evidence, but also with interdisciplinary studies that complement ethnographic records with biochemical information.

Caution on the amount of Llayta consumed daily, based on its arsenic content, must be stressed. However, mass growth of the cyanobacterium isolated from Llayta under controlled conditions would provide arsenic-free biomass for the formulation of new food products. This biotechnological approach would revitalize the use of and add value to this ancestral food ingredient for the benefit of not only Andean communities, but also the whole population of South America and the world.

Author Contributions: Conceptualization, B.G.-S. and M.R.; Data curation, M.R. and R.L. Funding acquisition, B.G.-S. and A.G.; Investigation, A.G.; Methodology, M.R.; Supervision, B.G.-S. and A.G.; Validation, R.L.; Writing—original draft, B.G.-S. and M.R.; Writing—review and editing, B.G.-S.

Funding: This research was funded by Universidad de Antofagasta, Chile, grant number SI-5305, and CONICYT-CHILE, grant number CeBiB FB-0001.

Acknowledgments: The authors would like to thank Celedonio Maron Chura, Biblioteca de Antropología Andina (Instituto para el Estudio de la Cultura y Tecnología Andina, Arica, Chile) for his useful comments.

Conflicts of Interest: The authors declare no conflict of interest. The funders had no role in the design of the study; in the collection, analyses, or interpretation of data; in the writing of the manuscript; or in the decision to publish the results.

References

1. Gómez-Silva, B. On the Limits Imposed to Life by the Hyperarid Atacama Desert in Northern Chile. In *Astrobiology: Emergence, Search and Detection of Life*; Basiuk, V.A., Ed.; American Scientific Publishers: Los Angeles, CA, USA, 2010; pp. 199–213.
2. Gómez-Silva, B. Lithobiontic life: Atacama rocks are well and alive. *Anton Leeuwenhoek* **2018**, *111*, 1333–1345. [CrossRef] [PubMed]
3. Plaza, M.; Herrero, M.; Cifuentes, A.; Ibáñez, E. Innovative natural functional ingredients from microalgae. *J. Agric. Food Chem.* **2009**, *57*, 7159–7170. [CrossRef] [PubMed]
4. Galetovic, A.; Araya, J.E.; Gómez-Silva, B. Composición bioquímica y toxicidad de colonias comestibles de la cianobacteria andina Nostoc sp. Llayta. *Rev. Chil. Nutr.* **2017**, *44*, 360–370. [CrossRef]

5. Villagrán, C.; Castro, V.; Sánchez, G.; Hinojosa, F.; Latorre, C. La tradición altiplánica: Estudio etnobotánico en los Andes de Iquique, primera región, Chile. *Chungara* **1999**, *31*, 81–186.
6. Villagrán, C.; Castro, V. *Ciencia Indígena de Los Andes del Norte de Chile*; Editorial Universitaria: Santiago, Chile, 2003.
7. Villagrán, C.; Romo, M.; Castro, V. Etnobotánica del sur de los Andes de la primera región de Chile: Un enlace entre las culturas altiplánicas y las de quebradas altas del Loa Superior. *Chungara* **2003**, *35*, 73–124. [CrossRef]
8. Pardo, O.; Pizarro, J.L. *Chile: Plantas Alimentarias Prehispánicas*; Ediciones Parina: Arica, Chile, 2013; ISBN 978-956-9120-02-2.
9. Gantar, M.; Svircev, Z. Microalgae and cyanobacteria: Food for the thought. *J. Phycol.* **2008**, *44*, 260–268. [CrossRef] [PubMed]
10. Habib, M.; Ahsan, B.; Parvin, M.; Huntington, T.C.; Hasan, M.R. A Review on Culture, Production and Use of Spirulina as Food for Humans and Feeds for Domestic Animals and Fish. FAO Fisheries and Aquaculture. Circular No. 1034. 2008. Available online: http://www.fao.org/3/contents/b2d01d94-4707-54c1-9f65-01f699fc6d07/i0424e00.htm (accessed on 19 October 2018).
11. Gao, K. Chinese studies on the edible blue-green alga, Nostoc flagelliforme: A review. *J. Appl. Phycol.* **1998**, *10*, 37–49. [CrossRef]
12. Qiu, B.; Liu, J.; Liu, Z.; Liu, S. distribution and Ecology of the edible cyanobacterium Ge-Xian-Mi (Nostoc) in rice fields of Hefeng County in China. *J. Appl. Phycol.* **2002**, *14*, 423–429. [CrossRef]
13. Bertonio, L. Vocabulario de la Lengua Aymara. Colección Biblioteca Nacional de Chile: Impreso en la Compañía de Jesús, Perú. Electronic Document. 1612. Available online: http://www.memoriachilena.cl/602/w3-article-8656.html (accessed on 19 October 2018).
14. Mannheim, B. Poetic form in Guaman Poma's Wariqsa Arawi. *Amerindian* **1986**, *11*, 41–67.
15. Lagerheim, M.G. La Yuyucha. La Nuova Notarisia 3:1376-1377. 1892. Available online: https://archive.org/stream/lanotaristacomme7293levi/lanotarisiacomme7293levi_djvu.txt (accessed on 29 November 2018).
16. Aldave-Pajares, A. Cushuro, algas azul-verde utilizadas como alimento en la región altoandina peruana. *Soc. Bot. Libertad* **1969**, *1*, 9–22.
17. Aldave-Pajares, A. Algas andino peruanas como recurso hidrobiológico alimentario. *Bol. Lima* **1985**, *7*, 66–72.
18. Gómez-Silva, B.; Mendizabal, C.; Tapia, I.; Olivares, H. Microalgas del norte de Chile. IV. Composición química de Nostoc commune Llaita [Microalgae from northern Chile: Chemical composition of Nostoc commune Llaita]. *Rev. Invest. Sci. Tecnol.* **1994**, *43*, 19–25.
19. FAO. Food Security & Nutrition around the World. 2018. Available online: http://www.fao.org/state-of-food-security-nutrition/en/ (accessed on 29 October 2018).
20. Pereira de Sá, C. The Central Nucleus Approach to Social Representations. Electronic Document. 1996. Available online: http://www.lse.ac.uk/methodology/pdf/QualPapers/CELSO-Core-periphery.pdf (accessed on 29 November 2018).
21. Lam, M.R.; Gómez, W.C.; Vega, M.B.; Peña, R.A. Representaciones del monumento natural La Portada de Antofagasta. *Zona Próxima* **2012**, *17*, 58–75.
22. Restrepo, B. Investigación en educación. In *Especialización en Teoría, Métodos y Técnicas de Investigación Social*; ARFO Editores e Impresores Ltda.: Bogotá, Colombia, 2002.
23. Banchs, M.A. Concepto de representaciones sociales: Análisis comparativo. *Rev. Costarric. Psicol.* **1986**, *5*, 27–40.
24. Jodelet, D. La representación social: Fenómenos, concepto y teoría. In *Psicología Social II. Pensamiento y Vida Social*; Moscovici, S., Ed.; Paidós: Barcelona, Spain, 1984.
25. Niccolai, A.; Bagagli, E.; Biondi, N.; Rodolfi, L.; Cinci, L.; Luceri, C.; Tredici, M.R. In vitro toxicity of microalgal and cyanobacterial strains of interest as food source. *J. Appl. Phycol.* **2017**, *29*, 199–209. [CrossRef]
26. González, H.; Gundermann, H.; Rojas, R. *Diagnóstico y Estrategia de Desarrollo Campesino en la I Región de Tarapacá*; Serie Documentos de Trabajo; Taller de Estudios Andinos; Corporación Norte Grande, Ed.; Impresos Publicitarios: Arica, Chile, 1991; 246p.

27. Carrasco-Gutiérrez, A.M.; González-Cortez, H. Movilidad poblacional y procesos de articulación rural urbano entre los Aymara del norte de Chile. *Si Somos Americanos* **2014**, *14*, 217–231. [CrossRef]
28. Rantala, A.; Rajaniemi-Wacklin, P.; Lyra, C.; Lepistö, L.; Rintala, J.; Mankiewick-Boczek, J.; Sivonen, K. Detection of microcystin-producing cyanobacteria in Finnish lakes with genus-specific microcystin synthetase gene E (mcyE) PCR and associations with environmental factors. *Appl. Environ. Microbiol.* **2006**, *72*, 6101–6110. [CrossRef] [PubMed]

© 2018 by the authors. Licensee MDPI, Basel, Switzerland. This article is an open access article distributed under the terms and conditions of the Creative Commons Attribution (CC BY) license (http://creativecommons.org/licenses/by/4.0/).

MDPI
St. Alban-Anlage 66
4052 Basel
Switzerland
Tel. +41 61 683 77 34
Fax +41 61 302 89 18
www.mdpi.com

Foods Editorial Office
E-mail: foods@mdpi.com
www.mdpi.com/journal/foods